Leonardo's Choice

Carol Gigliotti
Editor

Leonardo's Choice

Genetic Technologies and Animals

Editor
Carol Gigliotti
Emily Carr University of Art + Design
1399 Johnston Street
Vancouver BC V6H 3R9
Granville Island
Canada
gigliott@ecuad.ca

ISBN 978-90-481-2478-7 (hardcover) e-ISBN 978-90-481-2479-4
ISBN 978-94-007-0523-4 (softcover)
DOI 10.1007/978-90-481-2479-4
Springer Dordrecht Heidelberg London New York

Library of Congress Control Number: 2009927334

Printed on acid-free paper

Springer is part of Springer Science+Business Media (www.springer.com)

Preface

Defending animals' rights is often a contentious activity even among friends and colleagues. This volume emerged from just such a situation and it is to the credit of my good friend and colleague, Victoria Vesna of the University of California LA (UCLA), that it first took shape. Having read my essay "Leonardo's Choice: the ethics of artists working with genetic technologies" and witnessing negative reactions to the idea of critiquing work from an "animal standpoint," she invited me to guest edit a special issue on this topic for the journal *AI & Society*. Out of this issue grew the impetus for the present volume. I wish to thank her for her foresight and support, as well as the Editor-in-Chief of the journal *AI & Society*, Karamjit S. Gill. I owe everyone with whom I have worked at Springer, both for the journal issue and the book, a debt of gratitude: Beverly Ford, Executive Editor of Computer Science, Fritz Schmhul, now in Life Sciences, and my editors in Bioethics/ Philosophy, Jolanda Voogd and Marion Wagenaar. At all times, their professionalism and patience has been a gift.

An enormous debt of gratitude is owed to each and every author of the essays included here. Their fearlessness in speaking out has been a constant source of inspiration and their eloquence a joy to read. Without their generous participation the publication of this book could not have taken place. Each has contributed enormously to this book and to the growing area of animal studies in positive ways affecting our changing relationship with animals and the entire non-human world.

My thanks are due to the organizations and individuals who have allowed their photographs to be reprinted here: ITV/Carlton TV, All Creatures.org, Brian Gunn/IAAPEA, Scala/Art Resource, NY, Rick Warren, The National Museum of Scotland, Tina Mahoney, English Lakes.com, Jerry Ohlinger's, The Australian Museums and The Museums of Cape Town.

A number of friends, colleagues and loved ones have given comments, support and critiques over the time this book has come to fruition. Among them are Calder G. Lorenz, Sharon Romero, Karolle Wall, M. Simon Levin, Char Davies, Leslie Bishko, Mercedes Lawry, Chris Garvin, Niranjan Rajah, Amritha Fernandes+Bakshi, Sanjit Fernandes+Bakshi, Arthur and Marilouise Kroker, Julie Andreyev, Steve Gibson, Tom Regan, Alyce Miller, Susan McHugh, Annie Potts, Philip Armstrong, Leeza Fawcett, Jody Castricano, Ralph Acampora, Cary Wolfe, Lisa Brown, Lisa Kremmerer, Richard Kahn, Steven Best, and David Woods.

I would also like to thank several organizations that offered support during the development of this book through gifts of time or funding: the Reverie Foundation, Sitka Center for the Arts, and Emily Carr University of Art and Design.

This book is dedicated to the memories of Nik Williams, Laurie Long, Dan O'Connor and, of course, Radicchio.

Vancouver, BC, Canada
Carol Gigliotti

Contents

Part III

Contributors

Steve Baker Professor Emeritus of Art History, University of Central Lancashire, Preston, Lancashire, UK, sbaker1@uclan.ac.uk

Steven Best Department of Philosophy, University of Texas, El Paso, TX, USA, best@utep.edu

Lynda Birke Anthrozoology Unit, University of Chester, Chester, UK, l.birke@chester.ac.uk

Taimie L. Bryant University of California, Los Angeles, School of Law, Los Angeles, CA, USA, bryant@law.ucla.edu

Beth Carruthers Independent Scholar Galiano Island, BC, Canada, beth@bethcarruthers.com

David Delafenêtre Simon Fraser University, Burnaby, BC, Canada, caodaserradeaires@caodaserradeaires.ca

Carol Freeman School of Geography and Environmental Studies, University of Tasmania, Hobart, TAS, Australia, carol.freeman@utas.edu.au

Carol Gigliotti Dynamic Media and Critical & Cultural Studies, Emily Carr University, Vancouver, BC, Canada, gigliott@ecuad.ca

Vincent J. Guihan Carleton University, Ottawa, ONT, Canada, vjguihan@connect.carleton.ca

Caroline Seck Langill Liberal Arts, Ontario College of Art and Design, Toronto, ONT, Canada, clangill@faculty.ocad.ca

Susan McHugh University of New England, Biddeford, ME, USA, smchugh@une.edu

Kelty Miyoshi McKinnon School of Architecture and Landscape Architecture, University of British Columbia, Vancouver, BC, Canada, keltymc@gmail.com

Traci Warkentin Department of Geography, Hunter College, The City University of New York, USA, twarkentin@hunter.cuny.edu

Introduction

Carol Gigliotti

Abstract *Leonardo's Choice: Genetic Technologies and Animals* is an edited interdisciplinary collection of 12 essays and 1 dialogue focusing on the use of animals in biotechnology and the profoundly disastrous effects of that use for all inhabitants of this planet. As editor of this collection, my essay "Leonardo's Choice: The Ethics of Artists Working with Genetic Technologies" grew out of an increasing concern with the risks of genetic technologies for animals in scientific research and the genre of art practice involving genetic technologies and the non-human. While some of the work in this art genre aims to question the corporate uses of genetic technologies, much of the work is based on an acceptance of the inevitability of these technologies. I wanted to investigate if using the methodologies of a science still positing human beings as the centre and rationale of all endeavours, and animals as mere resources, would serve only to reinforce that anthropocentric view in the arts and a corresponding commitment to this view in broad cultural perceptions. I began with the belief that whether the object of genetic modification or transference is plant, animal, or tissue, one needs to question, confront and act on the ethical impact that instance of commodification and colonization will have on the future of a naturally occurring biodiversity and on the individual lives of non-humans involved.

Keywords Animals · Genetic technologies · Ethics · Art · Animal studies

The title of this book, *Leonardo's Choice*, and the title of the essay from which it came, refers to Leonardo's view of himself as both a scientist and an artist. Like Leonardo, who had compassion for animals and yet used them for his art, contemporary scientists and artists are faced with a choice: to view their creative human drive as limitless or to acknowledge real and possible consequences of their use of living beings in these "creative" pursuits. The latter choice would entail a new and more responsible understanding and practice of the organic creativity of which all beings, including humans, are a part.

C. Gigliotti (✉)
Dynamic Media and Cultural & Critical Studies, Emily Carr University, Vancouver, BC, Canada

This collection focuses on the profound effect, for both humans and non-humans, of using animals in genetic technologies. Unlike the majority of discussions on biotechnology, whether endorsing or critical, this volume, as a whole, views seriously the disastrous impact of these technologies on animals themselves. Amidst the wealth of human intelligence and imagination invested in the development of technologies, the natural world and non-human beings have been regulated to proprietary roles, even though our technological innovations could not exist without them. Our long-standing pre-occupation with technological outlooks and technological solutions have obscured the reality and agency of the more-than-human world, or what is left of it. If we have appeared, up to now, sanguine about the erosion of our "real" home while we have been busy in our "virtual" one, the results of this disconnection from our physical legacies are beginning to unravel that complacency.

The consensus among biologists is that we now are moving towards the sixth great species extinction, the first to be caused largely by the activities of a single species—us (Levin & Leakey, 1995). The effects of global climate change, one of this extinction's major causes (Mayhew et al., 2008), are now grasped both through media and first-hand knowledge. In addition, these effects have encouraged a mounting awareness of the negative consequences of our long history of a self-serving anthropocentrism, enacted, mediated and created through technological means. Coinciding, but not coincidental, with this sporadic and often rationalized and sublimated comprehension is a growing concern for biotechnologies' impacts on the food we eat, our health care and our environment, both natural and built. These worries have encouraged the recent growth of bioethics committees and institutes ranging from those who seem to serve either as apologists for an inevitably biotechnological future or those few who, against great odds, question this same inevitability. For some, the increasingly invasive uses of animals in genetic technologies have supplied a warning sign to back up and survey our handiwork. What kind of future would include a legless pig or a featherless chicken, we may ask? And to our dismay, we learn both already exist in varying forms.

In 2000, the image of a green fluorescent rabbit appeared in various media across the globe. Newspapers, magazines, television and the Web told the story of artist Eduardo Kac who had "commissioned" the transgenic process of taking green fluorescent protein from a little Pacific jelly fish, *Aequorea victoria*, and inserting it into the zygote of a rabbit, an art piece he called "GFP Bunny". Many of the reports mentioned the fact that the documented process of making transgenic animals using mice had been going on since 1981, and since 1985 using other animals, such as pigs, sheep, rabbits and fish. The public reaction to this well-publicized example, however, was overwhelmingly one of shock and discomfort. The idea of an artist taking control of a transgenic process was fertile ground for a sudden public realization: biotechnological activity involving animals was not just the stuff of science fiction, but was actively being accomplished in such a way that a non-scientist could direct the process. The viability of the genetic modification of animals noisily entered the public consciousness.

The Context of Animal Rights

If this example was a shock to the general public, it was not surprising to a number of those involved in animal rights and animal advocacy. One of the major intellectual leaders of the animal rights movement, philosopher Tom Regan (2001) recognized the significance of these developments when he said,

> Few areas of applied philosophy have witnessed more dramatic growth in the recent past than has bioethics, moreover, given the pace of advances in the life sciences, from developments in preventive medicine to the cloning of sheep and mice, few areas of ethical concern are likely to grow more dramatically in the foreseeable future. . .Whatever the future holds, one thing is certain: other-than-human animals will be used in the name of advancing scientific knowledge, both basic and applied. What is less certain is whether in doing so, those who use them will act wisely and well. (p. 1)

Discussions and perspectives about the morality of using animals for scientific purposes are informed by a long history of humankind's attitudes towards the other animals, one divergent in its origins and in its normative views. It can be argued that if ancient theories based on the notion of harmony both among humans and between humans and non-humans had prevailed, our wholesale acceptance of the inferior status of animals, and thus our assumptions that they exist for our use, would be an atypical perspective.

Italian philosopher Paola Cavalieri forcefully argues just this point by re-examining three critical moments in the history of our dominion over animals. For Cavalieri (2006), the first moment is the

> . . .struggle within the Classical Greek world between the idea of an original bond among all conscious beings and a contrasting global plan of rationalization of human and nonhuman exploitation. (p. 54)

Cavalieri (2006) and other authors[1] view this struggle, based as it was on the construction of initial political and economic justifications of protecting the order of the *polis*, as an important turning point in the triumph of the exploitation of animals. Commitments to both kinship with and justice for animals in the early Greek thought of Pythagoras were distorted as a consequence of this turning point. While critics have overlooked or trivialized Pythagoras' vegetarianism and his teaching of such, this way of life was due to his integrated worldview of the notion of harmony. He saw friendship to both humans and non-humans as a crucial contributing virtue in this worldview.

An alternative view of the human–animal relationship existed in the diverse history of ethical vegetarianism dating at least as far back as Pythagoras (c. 580 BCE–500 BCE) in the West and in Hinduism (c. 6500 BCE), Jainism (c. 7 BCE), Taoism (c. 6 BCE) and Buddhism (c. 6 BCE) in the East (Lucas, 2005). Thinkers in these religious practices spoke out against two of the most visible forms of animal suffering during these ancient times—meat eating and religious sacrifice. Harmony with nature and respect and compassion for *all* life forms were tenets of these geographically separated but spiritually connected movements. Recognizing the importance of these religious movements to the development of the major ethical and philosophical ideas shaping human thought can only give one pause in imagining a present quite different from the global market and technological culture we now inhabit, based as it is on the deaths of approximately 55 billion land animals alone killed annually for food worldwide.[2]

The second pivotal moment Cavalieri and others have described in the path to our present-day uses of animals is the scientific revolution of the seventeenth century led by Descartes and his followers. In his search for a stable foundation on which to base the "truth" of scientific discoveries, Descartes insisted one could rely only on one's ability to doubt and therefore think. The human body—a mere machine—was a vessel in which the human mind, the self, was enclosed. Only humans possessed this self that Descartes considered to be the soul. Animals, uttering horribly wrenching cries of pain and fear as they were being eviscerated while alive, had no soul, so were not really experiencing pain. Experimentation on them now was not only possible, but also considered to be necessary as a component of methods for doing a new and modern science based on rational and testable evidence. While this view and use of animals was contested by contemporaries of Descartes as well as those who wrote during the Enlightenment and beyond,[3] Descartes' insistence on the necessity of these methods for "true" knowledge opened the door to the wide use of animal experimentation. The number of animals used globally today in various experiments is approximately 180 million each year, an underestimation due to the lack of reporting in many countries and the non-reporting of birds, rats and mice in the United States.[4]

Cavalieri sees the last few decades' rapid process of industrialization and mechanization of farming practices as the third moment in the move to control and dominate other animals. Developing at an alarming rate, however, is an even more persistent and overwhelming trajectory that may, in fact, enforce irreparably the status of animals as inferior and existing solely for our use. I see this as a fourth and, perhaps, catastrophic moment in the centuries-long shift from our understanding of our communion and solidarity with the non-human, ensouled world to a world in which we see ourselves as the creators of all life.

This moment is our moment: the advent and growth of biotechnologies. A great deal of discourse and practice about the creative possibilities of these technologies is influenced by commitments to capital-fueled ideas of progress. Unfortunately, even the push towards more "sustainable" and presumably "nature friendly" ideas about creativity has produced examples such as salmon sperm, seen as bio-waste from the fishing industry, being used to make nanotechnological "green" LED displays.[5] While this may be touted as a creative collaboration with nature, it is in reality pure exploitation of our fellow beings.

The long road from the once flourishing acceptance of animals as allied beings in the sixth century BC with Pythagoras, into the fifth and fourth centuries and even the third century BC with thinkers such as Porphyry, has degraded into the currently growing wholesale acceptance of technologies in which animals are seen as mere objects of use. These changes are the result of shifts to more expedient worldviews at key historical moments in reaction to changing political and economic realities and choices, not unlike our realities and choices today.

Conterminous with today's programs of globalization and biopolitics with their mass negative effects on humans[6] and, consequently even more negative effects on animals, there has been a slow and steady building of arguments and support for possibilities, as Cavaleiri concludes "... to defend the idea that animals' lives have value...and to consider it wrong to kill them" (p. 66). The upheavals involved in the success of these arguments from philosophers such as Regan, Cavalieri and Best, among others, have been felt in the growth of animal rights organizations across the world, as well as protests and direct action against uses of animals in animal experimentation, factory farming, fur farming, hunting and entertainment, among other uses. Voices in the arts, humanities and sciences have begun making overtly visible the importance of our relationship with non-human others in our own construction of a human identity. But, as Carey Wolfe (2003a) points out, "...the US public has long since gotten the point that is just beginning to dawn on our critical practice" (p. 1), at least in terms of more inclusive attitudes towards animal cognition and consciousness. It is also true and needs to be articulated clearly that the current goals of Western science and technology, bound up as they are with entrenched ideas of animals and nature existing solely for our use, are antithetical to these challenges and are still driving the development of "transformative" biotechnologies.

The Context of Genetic Technologies

This collection's central questions revolve around how Western ideas and practices of creative freedom are disassociated from the impacts they have on the non-human world. This disassociation has contributed to shifting an organic understanding of nature to a mechanistic model in which the image of the non-human world is one of an (mere) inert, soulless machine[7] and in which the agency of animals is obscured. Contemporary ramifications of this shift include an emerging emphasis on the technological *replacement* of the naturally occurring world, a *technodevolution* that devalues a naturally occurring biodiverse earth and the uniquely occurring characteristics of its multi-species inhabitants. The reductive nature of much technological thought devolves and flattens the creative organic biodiversity of the natural world.

The term "biotechnology" was coined in 1919 by the Hungarian engineer Karl Ereky, and supporters of its growth often cite the use of yeast to make bread and beer, as well as dog breeding, as earlier examples of the long history of what is now biotechnology (Bud, 1993). The advent of modern biotechnology in the 1970s, however, is commonly considered to be the discovery of the first restriction enzyme, an enzyme that cuts specific sequences of double-stranded DNA, leading to the development of recombinant DNA technology (Nobel Prize Foundation, 1978). The first practical use of this work was the manipulation of *Escherichia coli* bacteria to produce human insulin for diabetics (Villa-Komaroff et al., 1978).

Modern biotechnologies, including genetic transfer and artificial cloning, based on similar discoveries throughout the end of the twentieth century and into the twenty-first, contrary to the insistence of some genetic technology proponents, are radically different from traditional plant and animal breeding. Unlike traditional breeding, genetic technology methods disrupt the sequence of the genetic code of the host, disturbing the functioning of neighbouring genes. Even more important, however, is the ability in genetic engineering to transfer genes across species barriers. The practical outcomes of this are unanticipated side-affects for the recipient organism as well as special risks that come with the use of viral genes and vectors in genetic engineering. The instability of this genetic material and its propensity to recombine with infecting viruses may give rise to new viruses that may become potentially dangerous pathogens for plants, animals and human beings (Antoniou et al., 1997).

The scientific community is still divided about the effects of genetically modified "products", of both animals and plants, even though the overarching received view is that genetic technologies specifically, and biotechnologies in general, are supported by the majority of the scientific community. A study by the Cornell University science faculty, however, found that while almost half of the scientists polled had reservations and criticism about genetically modified food and crops, these scientists were less comfortable expressing their views with colleagues than those scientists with pro-genetically modified food views (Kuehn, 2004). In fact, there have been a number of documented cases where university scientists whose research has turned up negative consequences of transgenic technologies have been stifled at every turn by both large biotech companies and their own university administration afraid of losing funding from those same companies. (Charman, 2001; Dowie, 2004). Novelist Michael Crichton's take on the entwinement of genetic research and commerce, to which Carol Freeman refers in her essay in this volume, is a pithy comment on the state of the art: "Crichton calls the commercialization of molecular biology 'the most stunning ethical event in the history of science'" (as cited in Freeman, 2009). Judith Roof (Roof, 2007) clarifies a probable determinant for the ease with which genetic technologies and corporate goals have merged when she calls DNA "the perfect commodity" (p. 198). In its tiny, neat package of information of instruction and operation, so easily transportable, and ability to last, she adds, it is "the perfect version of an imaginary entity that in itself embodies a shift in our ideas of history, identity, commodities, and commodity systems" (p. 198).

One of the most influential books describing the assumptions (other than monetary gain) driving the concepts in scientific enquiry, particularly those involving the gene, are Evelyn Fox Keller's (Keller, 1995, 2000) *Refiguring Life* and *The Century of the Gene*. Keller's purpose in both books is to demolish the widely held but simplistic concept of the gene as the smallest unit constituting a "program" for making an organism. She emphasizes how assumed gender metaphors in everyday language, combined with computer terminology, influenced by military, cybernetic and reductionism assumptions, have played a powerful role in the development of genetic sciences. Many scientists in molecular biology, as Keller explains in *The Century of the Gene*, have now realized this as an unhelpful causal model for evolutionary complexity. Consequently, the field of evolutionary developmental biology, formed largely in the 1990s, is a synthesis of findings from molecular developmental biology and evolutionary biology. "Evo devo", as it is sometimes called, considers the diversity of the organismal form in an evolutionary context and emphasizes the *linked* process and context of evolutionary development for animals and humans (Carroll, 2005).

Notwithstanding influences of this growing field on biological thought, the incorporation of information theory and cybernetics, both building blocks of contemporary informatic thought and practice, still holds sway in molecular biology and genetics. This combination of informatics and genetic technologies has encouraged a new concept of biological materiality, one particularly suited to transportable commodification. "Despite this newfound materiality made possible via informatics, there is also a strong emphasis on the valuation of genes or DNA. . .. A database, can exist, in effect, without computers of any kind" (Thacker, 2005, p. 101).

The BAC, or "bacterial artificial chromosome", is a kind of wet biological "library". A free-floating circular loop of DNA called a "plasmid" found in bacteria such as *E. coli* and spliced with a gene sequence from a human or animal is used to investigate samples of any desired gene. A section heading, in an essay written by three researchers at Yale and Mt. Sinai Schools of Medicine, is entitled "Farm animals: an unexploited gold mine for biotech". This article is found in Oxford University's *Nucleic Acids Research Journal* online and includes a short paragraph on the "Ethical reservations of farm animal genomic study". It exudes enthusiasm, however, for utilizing animals for many uses:

> Farm animal genomic studies continue to attract audiences excited by the multitude of applications. The meat industry can now use cow and chicken genomic data to confirm the quality of meat products.... In the healthcare arena, farm animal genomic work will aid in enterprises such as xenotransplantation (the transfer of animal tissues or organs into humans). (Fadiel et al., 2005)

The rhetoric found in this quote is instructive in understanding crucial aspects in what Thacker calls "informatic essentialism" in the context of biotechnologies. Instead of the dematerialization of the body written of so eloquently in much posthumanist discourse, genetic technologies in combination with database technologies are used to redefine biological materiality (Thacker, 2003, p. 89). Farm animals, already redefined as such by centuries of use in human food and labour, are now approached by the life sciences and medical practices as data warehouses of information. As information, animals are now able to be reconfigured, recoded and most importantly redesigned for, as the above quote makes clear, commercial enterprises: food, health, military, even "eco-friendly" or "sustainable" undertakings.

While researchers in comparative ethology, the study of animals in the field, are contributing to comprehension of the cognitive and emotional lives of other beings, much of the work in genetic technologies is reinforcing an understanding of animals as suited to act as a material language, a symbolic technique, without concern for their intrinsic value as beings with whom we share this planet. Animals have been conscripted into these technologies to further an agenda of controlling the creation of all life through the manipulation of various manifestations of code. *In today's biotechnologies, animals have become code.*

The history of animals being used in genetic technologies begins in 1980, with the creation of the first "transgenic animal, a mouse—in such a way that the gene would be expressed in the mouse and in its future offspring" (Ihlman, 1996). At the time of this writing, the most recent transgenic "advance" was reported in a *ScienceDaily* headline: "Scientists have developed the first genetically altered monkey model that replicates some symptoms observed in patients with Huntington's disease" (NIH/National Center for Research Resources, 2008).

In July 2005, the results of a committee made up of invited stem cell scientists, primatologists, philosophers and lawyers were published in *Science Magazine*. The group was brought together at John Hopkins University to deliberate on the potential effects of grafting human stem cells into the brains of non-human primates, the first instance of which was in 2001. Organized two years after the initial experiment, the 22-member panel took two more years to agree "...to disagree about whether non-human primates should be used for invasive biomedical procedures at all, and to focus instead on whether experiments with stem cells and the brain posed any new, unique ethical dilemmas." Mark Greene, Ph.D., and a member of the panel, said,

> Many of us expected that, once we'd pooled our expertise, we'd be able to say why human cells would not produce significant changes in non-human brains. But the cell biologists and neurologists couldn't specify limits on what implanted human cells might do, and the primatologists explained that gaps in our knowledge of normal non-human primate abilities make it difficult to detect changes. And there's no philosophical consensus on the moral significance of changes in abilities if we could detect them. (Greene, 2005)

The organizers of the panel may be commended for initiating this discussion. The fact, however, that 22 so-called experts could not specify what kind of effect implanting human cells might have on non-human primates does not speak particularly well for arguments for the blanket acceptance of this or similar techniques of genetic technologies. This group, instead, concluded,

> ...cognitive and emotional changes are least likely to occur when such work is conducted on healthy adult members of species *distantly* related to humans, such as macaques, rather than early in the brain development of our closest biological relatives, the chimpanzees and other great apes. [p. 386, my emphasis]

Sidestepping the larger and unresolved "old" issue of using animals in research for any purpose, the committee produced in its quest for "new ethical dilemmas" a cowardly, but ironic and revealing tautology. As a group they decided that in lieu of any real knowledge of whether grafting human stem cells into the brains of "higher" primates would cause them to become more like humans than they already obviously are, researchers should graft those cells into the brains of "lesser" monkeys, which again, in lieu of any real knowledge, may or may not have the effect of making them more like humans. Philosopher Mary Midgley's famous and succinct quote about the ethics and efficacy of animal experimentation would have been helpful if only the committee had been open to reasoning along with her about animals in general, "...if they are sufficiently like us to be really comparable, they may be too like us to be used freely as experimental subjects" (Midgley, 2003, p. 147).

Tom Regan (as cited in Svoboda, 2008), speaking about the untold numbers of animals upon whose death and suffering one successful transgenic experiment is based, insists,

> The animals used for these purposes are in fundamental ways like us—their behavior tells us they're like us, evolutionary theory tells us they're like us...What we have with transgenic research is another incentive for reducing animals to something whose purpose for being in the world is to serve human interests. And that's fundamentally flawed (para 14).

In response to the objection that using animals in research is worth it because it saves lives, a recent report from the Medical Research Modernization Committee states the opposite case very clearly:

> The value of animal experimentation has been grossly exaggerated by those with a vested economic interest in its preservation. Because animal experimentation focuses on artificially created pathology, involves confounding variables, and is undermined by differences in human and nonhuman anatomy, physiology, and pathology, it is an inherently unsound method to investigate human disease processes. The billions of dollars invested annually in animal research would be put to much more efficient, effective, and humane use if redirected to clinical and epidemiological research and public health programs. (Anderegg et al., 2002, p. 18)

Leonardo's Choices

Leonardo's Choice: Genetic Technologies and Animals is an edited interdisciplinary collection of 12 essays and 1 dialogue focusing on the use of animals in biotechnology and the profoundly disastrous effects of this use for both animals and us. As editor of this collection, my essay "Leonardo's Choice: The Ethics of Artists Working with Genetic Technologies" grew out of an increasing concern, not only about the risks of genetic technologies in general, but also with a growing genre of art practice involving genetic technologies and the non-human. While some of the work in this art genre aims to question the corporate uses of genetic technologies, I wanted to investigate if using the methodologies of a science that still posits human beings as the centre and rationale of all endeavour, and nature and the non-human as mere resources, would only serve to reinforce that anthropocentric view in the arts and corresponding cultural arenas. I began with the belief that whether the object of genetic modification or transference is plant, animal or tissue, one needs to question and confront the ethical impact that instance of commodification and colonization will have on the future of a naturally occurring biodiversity and on the individual lives of non-humans involved.

In this way, the collection makes a useful contribution to a growing discussion in both academic and public forums concerning ethics and animals. Seven of the essays were published in 2006 with an introduction and photos of animals in laboratory settings in a special issue of the Springer journal *AI and Society*.[8] As guest editor, I invited contributors from the disciplines of philosophy, cultural, art and literary theory and history and theory of science, as well as environmental studies, to respond to the topics in my essay. The authors replied with unique perspectives on the broad and multiple layers of meanings and values called into question by these themes. The volume at hand continues to be structured and integrated around the central theme of the use of animals in biotechnologies, but adds perspectives from law, landscape architecture, history, geography and cultural studies. Included authors span three continents and four countries. Since the publication of the journal issue, the growth of biotech and genetic technologies has been formidable, but the questions and issues forthcoming from the use of animals in these areas have only grown more urgent.

The included essays contribute significantly to a growing scholarship surrounding "the question of the animal" as well as counter discussions hoping to disqualify the general way that rather abstract phrase is posed, vapourizing the actual specificity of animal's lives. Emanating from philosophical, cultural and activist discourses, this question is currently being debated in post-humanist theoretical circles as well as post-colonial ones. While a number of authors refer to, and sometimes add to, ethical and ontological views towards animals in analytic philosophy, others concentrate on perspectives and methodologies of the Continental tradition. It is hoped the collection will also contribute a critical animal studies perspective[9] to the flourishing area of human–animal studies in the humanities and the widespread discussion of culture, technology and nature. The volume's authors speak to an audience eager for more sophisticated investigations of the complex relationships between humans and animals and what these relationships might offer to disciplines whose most basic assumptions continue to concern the centrality of the human.

Audiences for this collection include, but are not limited to

- philosophers, lawyers, artists, activists and scholars and their students from many disciplines wishing to extend the idea of justice and intrinsic worth to the non-human;
- theorists and activists who perceive biotechnologies' invasion of the self-organizing and generational capacities of the natural world as yet another bid for control by corporate-led globalization;
- cultural theorists and students of critical and cultural theory interested in human–animal relationships as rich areas of investigation for shifting concepts of identity and otherness.

Other edited books in this field inviting comparison include Cary Wolfe's (2003b) *Zoontologies: The Question of the Animal*; H. Peter Steves' (1999) *Animal Others: On Ethics, Ontology, and Animal Life*; and The Animals Studies Group's (2006) *Killing Animals*. All three are important milestones in this nascent area of thought.

Leonardo's Choice: Genetic Technologies and Animals differs from these collections in its focus on this most contemporary use of animals and possibly the most irreparable: biotechnology. Along with Cary Wolfe (2003a), it disregards "the humanist habit of making even the *possibility* of subjectivity conterminous with the species barrier" (p. 1). Its significance, however, lies in its urgency in critiquing the continuing blindness towards animal subjectivity involved in the use of animals in genetic technologies, as well as the control or erasure of that subjectivity through those uses. The topic of genetic technologies, as one of the most pressing challenges to a growing concern about our relationship with the natural world, is thrown into high relief in this volume through perspectives, by and large, hoping to refute the inevitability of a biotechnological future and the rationales behind it.

This volume places animals at the centre of such discussions, refusing to dismiss the effects of these technologies on their lives and agency. This stance opens at least three related and useful paths through the jumble of conflicted assumptions and contradictions about the rationales for biotechnologically driven applications of animals. Concentrating on the central issue of the use of animals in genetic technologies elicits ethical and political viewpoints about the necessity of public involvement in any decision-making process related to biotechnologies. It also prioritizes the consideration of animals in attitudes questioning the assumed *inescapability* of these technologies. As Steven Best points out in his powerful opening essay, "Genetic science, animal exploitation and the challenge for democracy", the unpredictable variables in biotechnological experimentation using inherently uncertain techniques combined with the instrumental use of animals cause great suffering. These unpredictabilities deeply challenge

> ...existing definitions of life and death, demand a rethinking of fundamental notions of ethics and moral value, and pose unique challenges for democracy. (Best, 2009)

The second useful path is the consideration of the paradoxical quality of the human–animal relationship and how it is utilized and for what purpose. A pressing question in my understanding of how to write about artists working with genetic technologies, for instance, concerned what role not only uncovering but also confronting ethical choices in this arena played in artists' thinking and practice. The intent of many of the essays included is not only to investigate and acknowledge the complexity of the topic but also to confront and act on the ethical choices involved. Some authors use these paradoxes as places of creative investigation in which to question our use of animals as only objects for our use, while other authors see these juxtapositions as indications of the fascination with the erasure of boundaries prevalent in today's post-humanist thinking. Still others distinguish this fascination with boundary breaching as locations where animals are made to pay for our resistance to acknowledging their intrinsic worth.

The third concept vexing these discussions is that most valued trait of the human species, creativity. How should one look at these ideas in art or science? Does curiosity, freedom of expression or invention always take precedence, or is the wider focus to see the ethical implications of these practices first and then to adjust what our goals for art or science are? Creative freedom, one of the most highly valued aspects of the human species as a social form, what Susan McHugh calls "the central cultural work of ordering species in the distinction of human species being",[10] is also a major player in maintaining dominance over non-human animals. Scientists and artists consider creative freedom an important ingredient in the development of transgenic technologies, but ironically, dominance over animals based on ideas of human centred creativity may be hard to maintain as the genetic makeup of animals is moved closer to humans.

Concern over the future, often the location where a great deal of human creativity is focused, is a common thread that runs through these essays as well as the minds of most of us in these difficult times. The essays in Part I of the collection offer differing and enlarged perspectives on the juxtapositions of animals, humans and genetic technologies and how these perspectives might shift the future towards a more ethical relationship with animals and involvement with biotechnologies. Two of the essays, those of Steven Best and Vincent Guihan, emerge from decidedly animal rights and animal liberation viewpoints, while Beth Carruthers' essay ponders an alternative stance hoping to sidestep or diffuse the clash of human and animal needs. All three cite the shared bodily being of humans and other animals as central to providing a way forward.

Philosopher Steven Best blames the current devastating impact of industrial biotechnology for animals, the natural world and shifts in how human beings visualize a future, squarely on an anthropocentric co-construction of science and technology fuelled by capitalist and corporate imperatives. He challenges the notion that a single disciplinary approach both to understanding and questioning the values, methodologies and impacts of genetic technologies will prove helpful. Insisting the future is not inevitable, but still ambiguous and open to political will and struggle, he argues instead for a "supradisciplinary" approach incorporating ethical and political values developed through an educated and participatory democracy coupled with a new sensitivity for nature.

Vincent Guihan, doctoral candidate in the Cultural Mediations program at Carleton University, builds upon Foucault's ideas of "bio-power" and "man-as-species" to reassess Darwin's influence on how we have arrived at the present moment in our relationship with animals. Guihan describes this moment as holding within it two poles of "cultural" understanding of animals: the reduction of animals (as well as humans and all of nature) into mere products for use *and* contemporary animal rights theory. He sees this latter as the "reverse discourse" of the former trend. Outlining genetic technologies' lineage in the eugenics movement of improvement of human and non-human animals, he clarifies retrograde qualities inherent in eugenic's emphasis on biology as destiny and its prioritizing of the perfection of the "human species being" ahead of all naturally occurring differences and specificities. Against this, the rights of animals not to be used, to be able to operate outside the power framework of human control, to be able to demonstrate agency, fulfill needs and meet wants are the driving goals of both Best's and Guihan's essays.

Independent scholar Beth Carruthers' essay considers the flaws in what she calls the foundational ontology of Western ethics in a search of a "shared ontology" between humans and the entire natural world. Drawing on Val Plumwood's arguments against what both she and Plumwood see as problems with rights theory, she surmises that only through accepting the unbearable intimacy of *knowing* we both feed on life and are food to it, can we begin to come to terms with our embodied relationship with the entire natural world, including animals.

Part II includes four essays and a dialogue focusing on the most visible, politically ambiguous, and debated use of animals in genetic technologies today, that of the use of live animals, animal tissue and cells in bioart—a practice in which the medium is living matter and the works of art are produced with biotechnological tools.

My own essay, "Leonardo's Choice: The Ethics of Artists Working with Genetic Technologies", closely questions the notion of the radical, and hence, assumed progressive nature of biotechnological practices either for science or for the purposes of an "art form". These questions are asked in light of growing calls by those inside and outside academe for a greater understanding of the intrinsically valuable biodiversity of nature and the impact of these art and science practices on the lives of all the animals involved. Comparing the artists' somewhat abstract rhetoric about their work with the actuality of the life of laboratory animals, artists' forays into the manipulation of life-forms with genetic technological practices are critiqued within the contexts of linked ethical, political, social and economic values driving the development of these technologies.

After reading my essay, cultural theorist Steve Baker suggested engaging in a dialogue which he entitled "We have always been transgenic" after a phrase of mine from the "Leonardo's Choice" essay and what he felt might be a pivotal meeting point in our thought. Our hope in engaging and publishing this dialogue was that we might be able to explore both our common interest in contemporary artists' engagement with questions of ethics and animal life and the significant differences in our own approaches to those questions. Our further hope is that the readers will find this helpful and stimulating for their own use.

Artist and writer Caroline Seck Langill's essay "Negotiating the Hybrid: Art, Theory and Genetic Technologies" addresses the issue of artist's forays into work both critical of and involved with genetic technologies from a historical perspective. Tracing contributing scientific and cultural sources from the seventeenth century on, Langill guides us into the present where contemporary artists and cultural theorists grapple with paradoxical abstractions of the freedoms of hybridity and plurality at the expense of the material reality of the natural world.

Biologist and animal behaviourist Lynda Birke's contribution, "Meddling with Medusa: On Genetic Manipulation, Art and Animals", challenges the notion that making transgenic organisms is radical for any purpose, whether it be for an "art-form" or for the purposes of developmental biology, due to nature's own complexity. Birke's related theme concerns the public unease with these activities. She maintains this unease is based not on ignorance, but on a concern over what meanings these reductionist manipulations might have for the future.

UCLA law professor Taimie Bryant's carefully considered essay outlines the complex and ambiguous relationship between the US legal system and the political and social will to protect animals and nature. The issue of whether "bioart" falls into the category of science or art, while viewed by many critics and supporters in the arts as a marginal issue in contemporary aesthetic thought, becomes a substantive question in any legal action involving the harm done to animals in these projects. Since, as she explains, scientific endeavours receive preferential treatment under the law, artistic collaborations involving scientists or scientific laboratories undertaken for non-scientific reasons, in many cases, have been protected from such intrusion. As Byrant points out, the landmark 1980 US Supreme court decision of Diamond vs. Chakrabarty validating the patentability of genetically altered beings as an inevitable outcome of the "scientific mind" emphasizes the uselessness of the law itself to generate the will to "protect" nature. In fact, as it now stands, US law protects those involved in exploiting animals.

Part III includes investigations of the making of species identity through close readings of novelists' visions of a genetically controlled future, as well as case studies of our current attitudes, both critical and accepting, towards forays into uses of genetic technologies.

Literary theorist Susan McHugh's essay included here, "The Call of the Other 0.1%: Genetic Aesthetics and the New Moreaus", investigates the multiple film versions of the H.G. Wells classic, *The Island of Dr. Moreau*, for clues as to how species has become a primary form of identity previously through genetic breeding and more recently through genetic aesthetics.

Environmentalist Traci Warkentin's essay investigates the concepts of the natural and the artificial, contamination and purity, integrity and fragmentation through a close reading of Margaret Atwood's recent dystopian novel, *Oryx and Crake*. Focusing on Atwood's speculative look at what the future might hold for animals used in current xenotransplantation experiments, particularly pigs, and current trends in factory farming of animals bred for consumption, such as chickens, Warkentin questions the implications of these developing biotechnologies for the future of our embodied sensibilities so necessary for ethical thought and action.

Historian David Delefêntre's essay provides a historical case study of the activist program in Australasia to ban cosmetic surgery—particularly tail docking and ear cropping—in dogs. While not involving genetic technologies, the issues of "natural breeding", whether for cosmetic purposes or to breed a dog with traits geared towards human desires, emerge in this discussion as well. The Australian success in bringing the ban into law offers ideas for generating a shift in public opinion towards using animals in genetic technologies or more generally for human uses. Delefêntre sees this as a move towards a more global shift in non-speciest attitudes towards animals.

Landscape architect Kelty Miyoshi McKinnon places the "distancing abstraction of contemporary genetic manipulation" within the context of a Bateseon ecological understanding of the long history of sheep, humans and the land. This placement allows unique views of both the contemporary methods of "pharming"—the use of genetic engineering to insert genes into plants or animals to produce pharmaceuticals—and human redemption, via the promise of cloning, from the guilt associated with causing the current species extinction.

The promise of redemption is also discussed in Carol Freeman's "Ending Extinction". Similar to ethnic cleansing, the mass extinction occurring today is, like the widespread use of genetic technologies, at the fullest reach of human power, control and domination of animals. While projects involving genetic technologies attempting to revive extinct species such as the Quagga or the Thylacine may initially seem to be possible solutions to current species disaster, Freeman questions the more covert but deep-seated motives of those involved in these projects.

The photographs placed throughout the book are predominantly of my own choosing except for the photos chosen by the authors Carol Freeman, Susan McHugh and Kelty Miyoshi McKinnon for their own essays. The remaining images are from many different sources, some of which are uncredited. My decision to use photos in the original journal version and in this book was based on a desire to ground discussions about the role of animals used in these technologies in the realities of life for them in experimental situations. I chose photos emphasizing the individuality of the animal shown and indications for possibilities for agency and flourishing that were being either controlled or destroyed by their unwilling insertion into the experimental arena. A majority of the photos are not a documentation of animals being used in genetic technology research. This is due to the fact that gaining copyright for a number of photos of genetically modified animals was in most cases denied to me. The controversial nature of their inclusion was cited as a reason. Still, I felt including available photos of animals in experimental situations would attest to the brutality under which laboratory animals live and die.

The roles creativity might play in scenarios of the future loom large since creativity is the human ability on which we have most relied on until now to meet our needs. It is my hope, as editor of this collection, to spark new concepts, combined with more nuanced understanding of animals' right to life and to agency, about sources of creativity we share with animals. It is also my hope that these ideas lead us to very different conclusions about how we might share a future with animals than those now operating in the sciences and arts of genetic biotechnologies. As Henry Beston (1928) so eloquently put it,

> For the animal shall not be measured by man. In a world older and more complete than ours, they move finished and complete, gifted with extensions of the senses we have lost or never attained, living by voices we shall never hear. They are not brethren; they are not underlings; they are other nations, caught with ourselves in the net of life and time, fellow prisoners of the splendour and travail of the earth. (20)

Notes

1. See Stuart (2006), Phelps (2007), Lucas (2005), and Ryder (2000).
2. "The number of land-based animals killed for food in 2005 world-wide was approximately 55 billion, according to the U.N. Food and Agriculture Organization. This conservative figure does not account for non-slaughter deaths and under-reporting by developing nations. Again, the many billions of fishes and other aquatic animals killed for food are not reported at all" (as cited in Farm Animal Reform Movement, 2004).
3. See Ryder (2000), particularly Chapters 4 and 5.
4. "An estimated 180 million animals are used in experiments every year across the globe. Not all countries keep accurate records of their animal use, and some official figures are likely to be underestimates. In the USA, for example, 80% of animals used (birds, rats and mice) are

not included in official figures at all. Across Europe an estimated 13 million animals are used each year, with the UK (nearly 3 million animals) consistently the largest user of laboratory animals. In many cases (including the UK) there are other significant omissions in official statistics. For example, in the UK animals who are bred for research, but subsequently not used, will be killed as 'surplus' but not appear in the statistics. Also excluded are animals killed purely for biological products such as blood, or those involved in longer term experiments after the initial first year (any subsequent years of suffering simply disappear from the statistics)" (Dr. Harden Trust, 2008). Also see Knight (2008).
5. An example of this would be a recent announcement of a new "green nanotechnology" in Nanotechnology Today (2007). A researcher at University of Cincinnati together with the Air Force Research Laboratory has developed a new approach to making green electronics, salmon sperm. As the researcher points out: "The driving force, of course, is cost..." and "Salmon sperm is considered a waste product of the fishing industry. It's thrown away by the ton". This researcher thinks that other animal or plant sources might be equally as useful, given the waste of the US agricultural industry.
6. See Foucault's (1990) ideas on biopolitics in *History of Sexuality, Volume 1* and see also Esposito (2008) *Bios: Biopolitics and Philosophy*.
7. Some of the most important studies in this area are Carolyn Merchant's (1980) *The Death of Nature: Women, Nature and the Scientific Revolution* 1980 and her *Reinventing Eden: The Fate of Nature in Western Culture* (Merchant, 2003).
8. See also (Gigliotti, 2006) "Introduction: Genetic Technologies and Animals". *AI and Society* 20 (2006): 3–5. Retrieved on February 25, 2007 from http://www.springerlink.com/content/1435-5655/
9. See "What is Critical Animal Studies?" on the Institute for Critical Animal Studies (2008) website. http://www.criticalanimalstudies.org/?p=6.
10. See Susan McHugh, "The Call of the Other 0.1%: Genetic Aesthetics and the New Moreaus" in this volume.

References

Anderegg, C., Cohen, M. J., et al. (2002). *A critical look at animal experimentation*. Cleveland, OH: Medical Research Modernization Committee.

Animal Studies Group. (Ed.). (2006). *Killing animals*. Urbana and Chicago: University of Illinois.

Antoniou, M., Cummins, J., Fagan, J., Ho, M. H. & Midtvedt, T. (1997). The difference between traditional breeding methods and genetic engineering – consequences for safety and labeling regulation policy. Retrieved April 15, 2008, from http://www.psrast.org/diffbrd.htm

Best, S. (2009). Genetic science, animal exploitation and the challenge for democracy. In C. Gigliotti (Ed.), *Leonardo's choice: genetic technologies and animals*. Dorchedt: Springer.

Beston, H. (1928). *The outermost house*. New York: Ballentine Books.

Bud, R. (1993). *The uses of life: A history of biotechnology*. Cambridge: Cambridge University Press.

Carroll, S. (2005). *Endless forms most beautiful: The new science of Evo Devo and the making of the animal kingdom*. New York: Norton.

Cavalieri, P. (2006). The animal debate: A reexamination. In P. Singer (Ed.), *In defense of animals: the second wave* (pp. 54–68). Malden, Massachusettes: Blackwell.

Charman, K. (2001). Spinning science into gold. *Sierra Mag*., July/Aug.

Dowie, M. (Sunday, January 11, 2004). Biotech critics at risk: Economics calls the shots in the debate [Electronic Version]. *SF Chronicle*. Retrieved Feb. 21, 2008, from http://www.mindfully.org/GE/2004/Biotech-Risk-Debate11jan04.htm.

Dr. Harden Trust. (2008). Dr. Harden Trust: faqs [Electronic Version]. Retrieved May 27, 2008 from http://www.drhadwentrust.org/faqs.

Esposito, R. (2008). *Bios: Biopolitics and philosophy* (T. Campbell, Trns.). Minneapolis and London: University of Minnesota.

Fadiel, A., Anidi, I. & Eichenbaum, K. D. (2005). Farm animal genomics and informatics: An update. *Nucleic Acids Research, 33*(19), 6308–6318.

Farm Animal Reform Movement. (2004). Animal death statistics report 2004 [Electronic Version]. Retrieved January 12, 2007 from http://www.farmusa.org/literature.htm.

Foucault, M. (1990). *History of Sexuality. Volume 1: An Introduction*. New York: Vintage.

Freeman, C. (2009). Ending extinction: The quagga, the thylacine, and the "smart human" In C. Gigliotti (Ed.), *Leonardo's choice: Genetic technologies and animals*. Dorchedt: Springer.

Gigliotti, C. (2006). "Introduction: Genetic technologies and animals." *AI and Society, 20*(1), 3–5.

Greene, Mark, et al. (2005). "Ethics: Moral issues of human-non-human primate neural grafting." *Science, 309*(5733), 385–386.

Ihlman, D. L. (1996). The transgenic mouse. *Pathbreakers*, 1982, Retrieved May 15, 2008 from http://www.washington.edu/research/pathbreakers/1982b.html

Keller, E. F. (1995). *Refiguring life: Metaphors of twentieth-century biology*. Cambridge, Mass: Harvard University Press.

Keller, E. F. (2000). *The century of the gene*. Cambridge, Mass: Harvard University Press.

Knight, A. (2008). Animal experiments: Critical assessments and reviews. Retrieved May 27, 2008, from http://www.animalexperiments.info/studies/studies.htm

Kuehn, R. R. (2004). Suppression of environmental science. *American Journal of Law & Medicine, 30*, 333.

Levin, R. & Leakey, R. (1995). *The sixth extinction: Patterns of life and the future of humankind*. New York: Doubleday.

Lucas, S. (2005). A defense of the feminist-vegetarian connection. *Hypatia, 20*(1), 160.

Mayhew, P., Jenkins, G. B. & Benton, T. G. (2008). A long-term association between global temperature and biodiversity, origination and extinction in the fossil record. *Proceedings of the Royal Society B-Biological Sciences, 275*(1630), 47–53.

Merchant, C. (1980). *The death of nature: Women, nature and the scientific revolution*. San Francisco: Harper Collins.

Merchant, C. (2003). *Reinventing Eden: The fate of nature in western culture*. London and New York: Routledge.

Midgley, M. (2003). *Myths we live by*. London: Routledge.

NIH/National Center for Research Resources (May 19, 2008). First transgenic monkey model of Huntington's Disease developed. *ScienceDaily*. Retrieved June 29, 2008 from http://www.sciencedaily.com/releases/2008/05/080518152643.htm

Nobel Prize Foundation. (1978). The Nobel Prize in physiology or medicine. Retrieved 2008-06-07, from http://nobelprize.org/nobel_prizes/medicine/laureates/1978/

Phelps, N. (2007). *The longest struggle: Animal advocacy from Pythagoras to PETA*. New York: Lantern Books.

Regan, T. (2001). *Defending animal rights*. Urbana and Chicago: University of Illinois Press.

Roof, J. (2007). *The poetics of DNA*. Minneapolis and London: University of Minnesota.

Ryder, R. D. (2000). *Animal revolution: Changing attitudes towards speciesm*. Oxford, New York: Berg.

Steves, H. P. (Ed.). (1999). *Animal others: On ethics, ontology, and animal Life*. Albany: State University of New York.

Stuart, T. (2006). *The bloodless revolution: A cultural history of vegetarianism from 1600 to modern times*. New York and London: W. W. Norton and Company.

Svoboda, E. (June 11, 2008). Old McDonald had a pharm [Electronic Version]. *salon.com*. Retrieved June 12, 2008 from http://www.salon.com/news/feature/2008/06/11/transgenic_goats/index.html.

Thacker, E. (2003). Data made flesh: Biotechnology and the discourse of the posthuman. *Cultural Critique, 53*(Winter), 72–97.

Thacker, E. (2005). *The Global Genome: Biotechnology, politics and culture*. Cambridge, Mass: MIT Press.

Villa-Komaroff, L., Efstratiadis, A., Broome, S., Lomedico, P., Tizard, R., Naber, S. P., et al. (1978). A bacterial clone synthesizing proinsulin. *Proceedings of the National Academy of Sciences* U.S.A., *75*(8), 3727–3731.

Wolfe, C. (2003a). *Animal rites: American culture, the discourse of species, and posthumanist theory*. Chicago: University of Chicago.

Wolfe, C. (Ed.). (2003b). *Zoontologies: The question of the animal*. Minneapolis and London: University of Minnesota.

Part I

Genetic Science, Animal Exploitation, and the Challenge for Democracy

Steven Best

Anyone who thinks that things will move slowly is being very naïve—Lee Silver, Molecular Biologist.

Abstract As the debates over cloning and stem cell research indicate, issues raised by biotechnology combine research into the genetic sciences, perspectives and contexts articulated by the social sciences, and the ethical and anthropological concerns of philosophy. Consequently, I argue that intervening in the debates over biotechnology requires supra-disciplinary critical philosophy and social theory to illuminate the problems and their stakes. In addition, debates over cloning and stem cell research raise exceptionally important challenges to bioethics and to a democratic politics of communication.

Keywords Cloning · Ethics · Animal rights · Democracy · Biotechnology

As we move into a new millennium fraught with terror and danger, a global postmodern condition is unfolding in the midst of rapid evolutionary and social changes co-constructed by science, technology, and the restructuring of global capital. We are quickly morphing into a new biological and social existence that is evermore mediated and shaped by computers, mass media, and biotechnology, all driven by the logic of capital and a powerful emergent technoscience. In this global context, science is no longer merely an interpretation of the natural and social worlds; rather it has become an active force in changing them and the very nature of life. In an era where life can be created and redesigned in a petridish, and genetic codes can be edited like digital text, the distinction between "natural" and "artificial" has become greatly complex. The new techniques of manipulation call into question existing definitions of life and death, demand rethinking of fundamental notions of ethics and moral values, and pose unique challenges for democracy.

S. Best (✉)
Department of Philosophy, University of Texas, El Paso, TX, USA
e-mail: best@utep.edu

C. Gigliotti (ed.), *Leonardo's Choice*, DOI 10.1007/978-90-481-2479-4_1,

As technoscience develops by leaps and bounds, and as genetics rapidly advances, the science–industrial complex has come to a point where it is creating new transgenic species and is rushing toward a posthuman culture that unfolds in the increasingly intimate merging of technology and biology. The posthuman involves both new conceptions of the "human" in an age of information and communication, and new modes of existence, as human flesh merges with steel, circuitry, and genes from other animal species. Exploiting more animals than ever before, technoscience intensifies research and experimentation into human cloning. This process is accelerated because genetic engineering and cloning are developed for commercial purposes, anticipating enormous profits on the horizon for the biotech industry. Consequently, all natural reality—from microorganisms and plants to animals and human beings—is subject to genetic reconstruction in a commodified "Second Genesis."

At present, the issues of cloning and biotechnology are being heatedly debated in the halls of science, in political circles, among religious communities, throughout academia, and, more broadly, in the media and public spheres. Not surprisingly, the discourses on biotechnology are polarized. Defenders of biotechnology extol its potential to increase food production and quality, to cure diseases, to endow us with "improved" human traits, and to prolong human life. Its critics claim that genetic engineering of food will produce Frankenfoods, which pollute the food supply with potentially harmful products that could devastate the environment, biodiversity, and human life itself. They also argue that animal and human cloning will breed monstrosities; that dangerous new eugenics is on the horizon; and that the manipulation of embryonic stem cells violates the principle of respect for life and destroys a bona fide "human being."

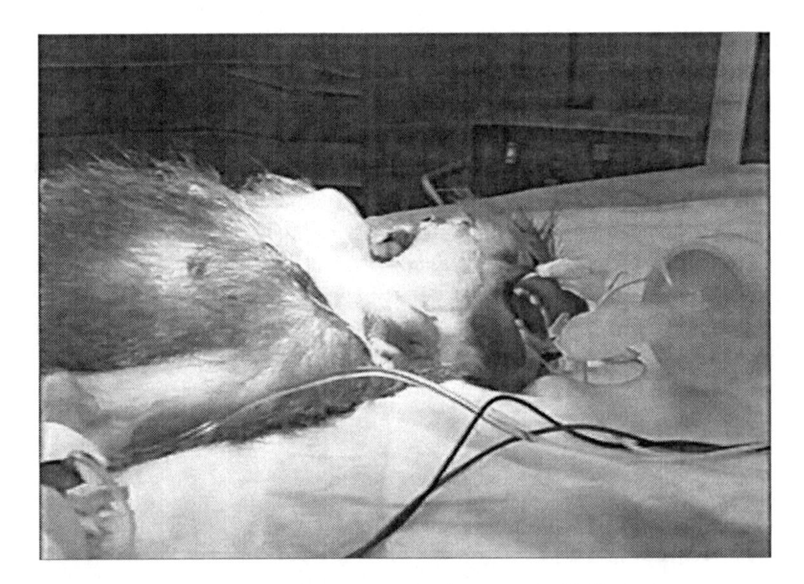

Fig. 1 Baboon used for xenotransplantation (credit: Courtesy of ITV/Carlton)

Interestingly, the same dichotomies that have polarized information-technology discourses into one-sided technophobic and technophilic positions are reproduced in debates over biotechnology. Just as I believe that critical theories of technology are needed to produce more dialectical perspectives that distinguish between positive and negative aspects and effects of information technology (Best and Kellner, 2001), so too would I claim that similar approaches are required to articulate the potentially beneficial and perhaps destructive aspects of biotechnology. Indeed, current debates over cloning and stem cell research suggest powerful contradictions and ambiguities in these phenomena that render one-sided positions superficial and dangerous. Parallels and similar complexities in communication and biotechnology are not surprising given that information technology provides the infrastructure to biotechnology that has been constituted by computer-mediated technologies involved in the Human Genome Project, and, conversely, genetic science is being used to push the power and speed of computers through phenomena such as "gene chips."

As the debates over cloning and stem cell research indicate, issues raised by biotechnology combine research into the genetic sciences, perspectives and contexts articulated by the social sciences, and the ethical and anthropological concerns of philosophy. Consequently, I argue that intervening in the debates over biotechnology require supra-disciplinary critical philosophy and social theory to illuminate the problems and their stakes. In addition, debates over cloning and stem cell research raise exceptionally important challenges to bioethics and a democratic politics of communication. Biotechnology is thus a critical flashpoint for ethics and democratic theory and practice. Contemporary biotechnology underscores the need for more widespread knowledge of important scientific issues; participatory debate over science, technology, values and our very concept of human life; and regulation of new developments in the biosciences, which have high economic, political, and social consequences.

New genetic technologies like stem cell research contain positive potential for medical advances that should not be blocked by problematic conservative positions. Nonetheless, I believe that the entire realm of biotechnology is fraught with dangers and problems that require careful study and democratic debate. The emerging genomic sciences should thus be undertaken by scientists with a keen sense of responsibility and accountability, and be subject to intense public scrutiny and open discussion. Finally, in the light of the dangers and potentially deadly consequences of biotechnology, I maintain that the positive potential of biotechnology can be realized only in the new context of cultivating new sensibilities toward nature, engaging in ethical and political debate, and participating in political struggles over technoscience and its effects.

Brave New Barnyard: The Advent of Animal Cloning

> The idea is to arrive at the ideal animal and repeatedly copy it exactly as it is—Dr. Mark Hardy

From its entrenched standpoint of unqualified human superiority, science typically first targets objects of nature and animals with its analytic gaze and instruments. The current momentous turn toward cloning is largely undertaken by way of animals, although some scientists have already directly focused on cloning human beings. While genetic engineering creates new "transgenic" species by inserting the gene from one species into another, cloning replicates cells to produce identical copies of the host organism by injecting its DNA into an enucleated egg. In a potent combination, genetic engineering and cloning technologies are used together in order, first, to custom design a transgenic animal to suit the needs of science and industry (the distinction is irrevocably blurred) and, second, to mass reproduce the hybrid creation endlessly for profitable peddling in medical and agricultural markets.

Cloning is a return to asexual reproduction and bypasses the caprice of the genetic lottery and random shuffling of genes. It dispenses with the need to inject a gene into thousands of newly fertilized eggs to get a successful result. Rather, much as the printing press replaced the scribe, cloning allows mass reproduction of a devised type, and thus opens genetic engineering to vast commercial possibilities. Life science companies are poised to make billions of dollars in profits, as numerous organizations, universities, and corporations move toward cloning animals and human stem cells, and patenting the methods and results of their research.

To date, science has engineered myriads of transgenic animals and has cloned animals such as sheep, calves, buffalo, goats, bulls, pigs, mice, and cats. Although still far from precise, cloning nevertheless has become routine. What is radically new and startling is not cloning itself, since from 1952 scientists have replicated organisms from embryonic cells. Rather, the new techniques of cloning, or "nuclear somatic transfer," from adult mammal body cells constitute a novel form of animal reproduction. These methods accomplish what scientists long considered impossible—reverting adult (specialized) cells to their original (nonspecialized) embryonic state where they can be reprogrammed to form a new organism. In effect, this startling process creates the identical twin of the adult that provided the original donor cell. This technique was used first to create Dolly, the first mammal cloned from a cell from an adult animal, and subsequently all of her varied offspring.

Dolly and Her Progeny

Traditionally, scientists considered cloning beyond the reach of human ingenuity. But when Ian Wilmut and his associates from the Roslin Institute near Edinburgh, Scotland, announced their earth-shattering discovery in March 1997, the "impossible" appeared in the form of a sheep named Dolly, and "natural law" had been broken. Dolly's donor cells came from a 6-year-old Finn Dorset ewe. Wilmut starved mammary cells in low-nutrient tissue culture where they became quiescent and subject to reprogramming. He then removed the nucleus containing genetic material from an unfertilized egg cell of second sheep, a Scottish Blackface, and, in a nice Frankenstein touch, fused the two cells with a spark of electricity. After 277 failed attempts, the resulting embryo was then implanted into a third sheep, a surrogate mother who gave birth to Dolly in July 1996.

Many critics said Dolly was either not a real clone or was just a fluke. Yet, less than 2 years after Dolly's emergence, scientists had cloned numerous species, including mice, pigs, cows, and goats, and had even made clones of clones of clones, producing genetic simulacra in mass batches as Huxley (1989) envisioned happening to human beings in his 1931 novel *Brave New World*. The commercial possibilities of cloning animals were dramatic and obvious for all to behold. The race was on to patent novel cloning technologies and the transgenic offspring they would engender.

Animals are being designed and bred as living drug and organ factories, as their bodies are disrupted, refashioned, and mutilated to benefit meat and dairy industries. Genetic engineering is employed in biomedical research by infecting animals with diseases that become a part of their genetic make up and are transmitted to their offspring, as in the case of researchers trying to replicate the effects of cystic fibrosis in sheep. Most infamously, Harvard University, with funding from Du Pont, patented a mouse—OncoMouse—that has human cancer genes built into its genome and are expressed in its offspring (Haraway, 1997).

In the booming industry of "pharming" (pharmaceutical farming), animals are genetically modified to secrete therapeutic proteins and medicines in their milk. The first major breakthrough came in January 1998, when Genzyme Transgenics created transgenic cattle named George and Charlie. The result of splicing human genes and bovine cells, they were cloned to make milk that contains human proteins such as the blood-clotting factor needed by hemophiliacs. Co-creator James Robl said, "I look at this as being a major step toward the commercialization of this [cloning] technology."[1]

In early January 2002, the biotech company PPL announced that they had cloned a litter of pigs, which could aid in human organ transplants. On the eve of the publication of an article by another company, Immerge Bio Therapeutics claimed that they had achieved a similar breakthrough.[2] The new process involved creation of the first "knockout" pigs, in which a single gene in pig DNA is deleted to eliminate protein that is present in pigs, which is usually violently rejected by the human immune system. This meant that a big step could be made in the merging of humans and animals, and creating animals as harvest-machines for human organs. In May 2009, Japanese scientists announced the crossing of a threshold with the successful creation of the first transgenic primates, marmoset mokeys, that can mimic human diseases for better than mice, rats, or pigs (Madrigal, 2009).

Strolling through the brave new barnyard, one can find incredible beings that appear normal, but in fact are genetic satyrs and chimera. Cows generate lactoferrin, human protein useful for treating infections. Goats manufacture antithrombin III, a human protein that can prevent blood clotting, and serum albumin, which regulates the transfer of fluids in the body. Sheep produce alpha antitrypsin, a drug used to treat cystic fibrosis. Pigs secrete phytase, a bacterial protein that enables them to emit less of the pollutant phosphorous in their manure, and chickens make lysozyme, an antibiotic, in their eggs to keep their own infections down.

"BioSteel" presents an example of the bizarre wonders of genetic technology that points to the erasure of boundaries between animate and inanimate matter, as well as among different species. In producing this substance, scientists have implanted a spider gene into goats, so that their milk produces superstrong material that can be used for bulletproof vests, medical supplies, and aerospace and engineering projects. In order to produce vast quantities of BioSteel, Nexia Biotechnologies intends to house thousands of goats in 15 weapons-storage buildings, confining them in small holding pens.[3]

As we see, animals are genetically engineered and cloned to produce stocks of organs for human transplants. Given the severe shortage of human organs, thousands of patients every year languish and die before they can receive a healthy kidneys, livers, or hearts. Rather than encouraging preventative medicine and finding ways to encourage more organ donations, medical science has turned to xenotransplantation, and has begun breeding herds of animals (with pigs as a favored medium) to be used as organ sources for human transplantation.

Clearly, this is a very hazardous enterprise due to the possibility of animal viruses causing new plagues and diseases in the human population (a danger which exists also in pharmaceutical milk). For many scientists, however, the main concern is that the human body rejects animal organs as foreign and destroys them within minutes. Researchers seek to overcome this problem by genetically modifying the donor organ so that they knock out markers in pig cells and add genes that make their protein surfaces identical to those in humans. Geneticists envision cloning entire herds of altered pigs and other transgenic animals so that an inexhaustible warehouse of organs and tissues would be available for human use. In the process of conducting experiments such as transplanting pig hearts modified with a human gene into the bodies of monkeys, companies such as Imutran have caused horrific suffering, with no evident value to be gained given the crucial differences among species and introducing the danger of new diseases into human populations.[4]

As if billions of animals were not already exploited enough in laboratories, factory farms, and slaughterhouses, genetic engineering and cloning exacerbate the killing and pain with new institutions of confinement and bodily invasion that demand billions more captive bodies. Whereas genetic and cloning technologies in the cases described at least have the potential to benefit human beings, they have also been appropriated by the meat and dairy industries for purposes of increased profit through the exploitation of animals and biotechnology. It's the nightmarish materialization of the H.G. Wells scenario where, in his prophetic 1904 novel *The Food of the Gods*, scientists invent a substance that prompts every living being that consumes it to grow to gargantuan proportions (Best and Kellner, 2001). Having located the genes responsible for regulating growth and metabolism, university and corporate researchers immediately exploited this knowledge for profit. Thus, for the glories of carcinogenic carnivorous consumption, corporations such as MetaMorphix and Cape Aquaculture Technologies have created giant pigs, sheep, cattle, lobsters, and fish that grow faster and larger than the limits set by evolution.

Amidst the surreality of Wellsian gigantism, cattle and dairy industries are engineering and cloning designer animals that are larger, leaner, and faster-growing value producers. With synthetic chemicals and DNA alteration, pharmers can produce pigs that mature twice as fast and provide at least twice the normal amount of sows per litter as they eat 25% less feed, and cows that produce at least 40% more milk. Since 1997, at least one country, Japan, has sold cloned beef to its citizens.[5] But there is strong reason to believe that U.S. consumers—already a nation of guinea pigs in their consumption of genetically modified foods—have eaten cloned meat and dairy products. For years, corporations have cloned farmed animals with the express purpose of someday introducing them to the market, and insiders claim many already have been consumed.[6] The National Institute of Science and Technology has provided two companies, Origen Therapeutics of California, and Embrex of North Carolina, with almost $5 million to fund research into factory farming billions of cloned chickens for consumption.[7] With the Food and Drug Administration pondering whether to regulate cloned meat and dairy products, it is a good bet they are many steps behind an industry determined to increase their profits through biotechnology. The future to come seems to be one of cloned humans eating cloned animals.

While anomalies such as self-shearing sheep and broiler chickens with fewer feathers have already been assembled, some macabre visionaries foresee engineering pigs and chickens with flesh that is tender or can be easily microwaved, and chickens that are wingless so they won't need bigger cages. The next step would be to just create and replicate animal's torsos—sheer organ sacks—and dispense with superfluous heads and limbs. In fact, scientists have already created headless embryos of mice and frogs in grotesque manifestations of the kinds of life they can now construct at will.

Clearly, there is nothing genetic engineers will not do to alter or clone an animal. Transgenic "artist" Eduardo Kac, for instance, commissioned scientists at the National Institute of Agronomic Research in France to create Alba, a rabbit that carries fluorescent protein from jellyfish and thus glows in the dark. This experiment enabled Mr. Kac to demonstrate his supremely erudite postmodern thesis that "genetic engineering [is] in a social context in which the relationship between the private and public spheres are negotiated."[8]

Although millions of healthy animals are euthanized every year in U.S. animal "shelters," corporations are working to clone domestic animals, either to bring them back from the dead, or to prevent them from "dying" (such as in the Missyplicity Project, initiated by the wealthy "owners" of a dog who they wanted to keep alive indefinitely).[9] Despite alternatives to coping with allergy problems and the dangers with cloning animals, Transgenic Pets LLC is working to create transgenic cats that are allergen-free.[10] In 2002, the biotechnology company Genetic Savings and Clone showcased the world's first cloned cat, named CC, for Carbon Copy. Pandering to animal guardians' misconceived desires to immortalize their cat, for a price of $50,000 each, Genetic Savings since has cloned additional cats, and hopes to cash in on what could be a booming market in feline simulacra at great risk for health problems and premature aging.

Transgenic Travesties

The agricultural use of genetics and cloning has produced horrible monstrosities. Transgenic animals often are born deformed and suffer from fatal bleeding disorders, arthritis, tumors, stomach ailments, kidney disease, diabetes, inability to nurse and reproduce, behavioral and metabolic disturbances, high mortality rates, and large offspring syndrome. In order to genetically engineer animals for maximal weight and profit, a Maryland team of scientists created the infamous "Beltway pig" afflicted with arthritis, deformities, and respiratory disease. Cows engineered with bovine growth hormone (rBGH) have mastitis, hoof and leg maladies, reproductive problems, numerous abnormalities, and die prematurely. Giant supermice endure tumors, damage to internal organs, and shorter life spans. Numerous animals born from cloning are missing internal organs such as hearts and kidneys. A Maine lab specialized in breeding sick and abnormal mice that go by names such as Fathead, Fidget, Hairless, Dumpy, and Greasy. Similarly, experiments in the genetic engineering of salmon have led to rapid growth and various aberrations and deformities, with some growing up to ten times their normal body weight (Fox, 1999). Cloned cows are ten times more likely to be unhealthy as their natural counterparts. After 3 years of efforts to clone monkeys, Dr. Tanja Dominko fled in horror from her well-funded Oregon laboratory. Telling cautionary tales of the "gallery of horrors" she experienced, Dominko said that 300 attempts at cloning monkeys produced nothing but freakishly abnormal embryos that contained cells either without chromosomes or with up to nine nuclei.[11]

For Dominko, a "successful" clone like Dolly is the exception, not the rule. But even Dolly became inexplicably overweight and arthritic, and may have been prematurely aging. In February 2003, suffering from progressive lung disease, poor Dolly was euthanized by her "creators," bringing to a premature end the first experiment with adult animal cloning and raising questions concerning its ethics.

A report from newscientist.com argues that genes are disrupted when cultured in a lab, and this explains why so many cloned animals die or are grossly abnormal. On this account, it is not the cloning or IVF process that is at cause, but the culturing of the stem cells in petri dishes, creating major difficulties in cloning since so far there is no way around cloning through cultured cells in laboratory conditions.[12]

A team of U.S. scientists at the MIT Whitehead Institute examined 38 cloned mice and learned that even clones which look healthy suffer genetic maladies, as mice cloned from embryonic stem cells had abnormalities in the placenta, kidneys, heart, and liver. Scientists feared that the defective gene functioning in clones could wreak havoc with organs and trigger foul-ups in the brain later in life and that embryonic stem cells are highly unstable.[13] "There are almost no normal clones," study author and MIT biology professor Rudolf Jaenisch, explained. Jaenisch claims that

Fig. 2 Housing for small mammals from birth to death (credit: © Brian Gunn/IAAPEA)

only 1–5% of all cloned animals survive, and even those that survive to birth often have severe abnormalities and die prematurely.[14]

As I argue below, these risks make human cloning a deeply problematic undertaking. Pro-cloning researchers claim that the "glitches" in animal cloning eventually can be worked out. In January 2001, for example, researchers at Texas A&M University and the Roslin Institute claimed to have discovered a gene that causes abnormally large cloned fetuses, a discovery they believe will allow them to predict and prevent this type of mutation. It is conceivable science someday will work out the kinks, but for many critics this assumes that science can master what arguably are inherent uncertainties and unpredictable variables in the expression of genes in a developing organism. One study showed that some mouse clones seem to develop normally until an age the equivalent of 30 years for human beings; then there is spurt of growth and they suddenly become obese.[15] Mark Westhusin, cloning expert at Texas A&M, points out that the problem is not that of genetic mutation, but of "genetic expression," such that genes are inherently unstable and unpredictable in their functioning. Another report indicates that a few misplaced carbon atoms can lead to cloning failures.[16] Thus, as suggested by chaos theory, small errors in the cloning process could lead to huge disasters, and the prevention of all such "small errors" seems to presume something close to omniscience.

Yet, while most scientists are opposed to cloning human beings (rather than stem cells) and decry it as "unacceptable," few condemn the suffering caused to animals or position animal cloning research itself as morally wrong or problematic, and many scientists aggressively defend animal cloning. Quite callously and arbitrarily, for example, Jaenisch proclaims, "You can dispose of these animals, but tell me—what do you do with abnormal humans?"[17] The attitude that animals are disposable resources or commodities rather than subjects of a life with inherent value and rights is a good indication of the problems inherent in the mechanistic science that still prevails in twenty-first century technoscience and a symptom of callousness toward human life that worries conservatives.

Despite the claims of its champions, the genetic engineering of animals is a radical departure from natural evolution and traditional forms of breeding. Genetic engineering involves manipulation of genes rather than whole organisms. Moreover, scientists engineer change at unprecedented rates, and can create novel beings across species boundaries that previously were unbridgeable. Ours is a world where cloned calves and sheep carry human genes, human embryo cells are merged with enucleated cows' eggs, monkeys and rabbits are bred with jellyfish DNA, a surrogate horse gives birth to a zebra, a dairy cow spawns an endangered gaur, and tiger cubs emerge from the womb of an ordinary housecat.

The ability to clone a desired genetic type brings the animal kingdom into entirely new avenues of exploitation and commercialization. From the new scientific perspective, animals are framed as genetic information that can be edited, transposed, and copied endlessly. Pharming and xenotransplantation build on the system of factory farming that dates from the postwar period and is based on the confinement and intensive management of animals within enclosed buildings that are prison-houses of suffering.

The proclivity of the science–industrial complex to instrumentalize animals as nothing but resources for human use and profit intensifies in an era in which genetic engineering and cloning are perceived as a source of immense profit and power. Still confined for maximal control, animals are no longer seen as whole species, but rather as fragments of genetic information to be manipulated for any purpose.

Weighty ethical and ecological concerns in the new modes of animal appropriation are largely ignored, as animals are still framed in the seventeenth century Cartesian worldview that sees them as nonsentient machines. As Rifkin (1998, 35) puts it, "Reducing the animal kingdom to customized, mass-produced replications of specific genotypes is the final articulation of the mechanistic, industrial frame of mind. A world where all life is transformed into engineering standards and made to conform to market values is a dystopian nightmare, and needs to be opposed by every caring and compassionate human being who believes in the intrinsic value of life."[18]

Patenting of genetically modified animals has become a huge industry for multinational corporations and chemical companies. PPL Therapeutics, Genzyme Transgenics, Advanced Cell Technology, and other enterprises are issuing broad patent claims on methods of cloning nonhuman animals. PPL Therapeutics, the company that "invented" Dolly, has applied for the patents and agricultural rights to the production of all genetically altered mammals that could secrete therapeutic proteins in their milk. Nexia Biotechnologies obtained exclusive rights to all results from spider silk research. Patent number 4,736,866 was granted to Du Pont for Oncomouse, which the Patent Office described as a new "composition of matter." Infigen holds the U.S. patent for activating human egg division through any means (mechanical, chemical, or otherwise) in the cloning process.

One might argue that genetics does not augur solely negative developments for animals. Given the reality of dramatic species extinction and loss of biodiversity, scientists are collecting the sperm and eggs of endangered species like the giant panda in order to preserve them in a "frozen zoo," such as exists in the San Diego

Zoo. It is indeed exciting to ponder the possibilities of a Jurassic Park scenario of reconstructing extinct species (for example, scientists recently have uncovered the well-preserved remains of a Tasmanian tiger and a woolly mammoth). In October 2001, European scientists cloned a seemingly healthy mouflon lamb, member of an endangered species of sheep. In April 2003, ACT produced the first successful interspecies clone when a dairy cow gave birth to a pair of bantengs, a species of wild cattle, cloned from an animal that died over 20 years ago. One of the pair, however, was thereafter euthanized because it was born twice the normal size and was suffering. Currently, working with preserved tissue samples, ACT is working to bring back from extinction the last bucardo mountain goat, which was killed by a falling tree in January 2000.[19]

But critics dismiss these efforts as a misguided search for a technofix that distracts focus from the real problem of preserving habitat and biodiversity. Even if animals could be cloned, there is no way to replicate habitats lost forever to chainsaws, bulldozers, and invading human armies. Moreover, the behaviors of cloned animals would unavoidably be altered and they would end up in zoos or exploitative entertainment settings, where they would exist as spectacle and simulacra.

Animals raised through interspecies cloning such as the gaur produced by ACT will not have the same disposition as if raised by their own species and so, for other reasons will not be less than "real." Additionally, there is the likelihood that genetic engineering and cloning would aggravate biodiversity loss to the extent it creates monolithic super breeds that could crowd out other species or be easily wiped out by disease. There is also great potential for ecological disaster when new beings enter an environment, and genetically modified organisms are especially unpredictable in their behavior and effects.

Still, cloning may prove a valuable tool in preserving what can be salvaged from the current extinction crisis. Moreover, advances in genetics also may bypass and obviate pharming and xenotransplantation through the use of stem cell technologies that clone human cells, tissues, or perhaps even entire organs and limbs from human embryos or an individual's own cells. Successful stem cell technologies could eliminate at once the problem of immune rejection and the need for animals. There is also the intriguing possibility of developing medicines and vaccines in plants, rather than animals, thus producing a safer source of pharmaceuticals and nutraceuticals and sparing animals' tremendous suffering. None of these promises, however, brighten the dark cloud cloning casts over the animal kingdom, or dispel the dangers of the dramatic alteration of agriculture and human life.

Deferring the Brave New World: Challenges for Ethics and Democracy

> Human history becomes more and more a race between education and catastrophe—H.G. Wells

By summer of 2001, technical and esoteric debate over stem cells, confined within the scientific community during the past years, had moved to the headlines to become the forefront of the ongoing science wars—battles over the cultural, ethical, political, and economic implications of science. The scientific debate over stem cell research in large part is a disguised culture war, and conservatives, liberals, and radicals have all jumped into the fray. Coming from a perspective of critical theory and radical democratic politics, I reject conservative theologies and argue against conflations of religion and the state. Likewise, I question neoliberal acceptance of corporate capitalism and underscore the implications of the privatization of research and the monopolization of knowledge and patents by huge biotech corporations. In addition, I urge a deeper level of public participation in science debates than do conservatives or liberals and believe that the public can be adequately educated to have meaningful and intelligent input into technical issues such as cloning and stem cell research that have tremendous social and ethical implications.

As I have shown, numerous issues are at stake in the debate over cloning, having to do not only with science, but also with religion, politics, economics, democracy, ethics, the meaning and nature of human beings, and all life forms as they undergo a process of genetic reconstruction. Thus, my goal throughout this paper has been to question the validity of the cloning project, particularly within the context of a global capitalist economy and its profit imperative, a modernist paradigm of reductionism, and a Western sensibility organized around the deluded attempt to dominate nature. Until science is recontextualized within a new holistic paradigm informed by respect for living processes, by democratic decision making, and by a new ethic toward nonhuman animals and nature, it will remain a dangerous and regressive force. Moreover, politicians beholden to corporate interests have no grasp of the momentous issues involved, requiring that those interested in democratic politics and progressive social change must educate and involve themselves in the science and politics of biotechnology.

We have already entered a new stage of the postmodern adventure in which animal cloning is highly advanced and human cloning is on the horizon, if not now underway. Perhaps human clones are already emerging, with failures being discarded, as were the reportedly hundreds of botched attempts to create Louise Brown, the first test-tube baby, in 1978. At this stage, human cloning is indefensible in the light of the possibility of monstrosities, dangers to the mother, burdens to society, failure to reach consensus on the viability and desirability of the project, and the lack of compelling reasons to warrant this fateful move. The case is much different, however, for therapeutic cloning, which is incredibly promising and offers new hope for curing numerous debilitating diseases. But even stem cell research, and the cloning of human embryos is problematic, in part because it is the logical first step toward reproductive cloning and the mass production of desired types, which unavoidably brings about new (genetic) hierarchies and modes of discrimination.

We thus need to discuss the numerous issues involved in the shift to a posthuman, postbiological existence where the boundaries between bodies and technologies begin to erode as we morph toward a cyborg nature. Our technologies are no longer extensions of our bodies, as Marshall McLuhan stated, but rather are intimately merging with our bodies, as we implode with other species through the genetic crossings of transgenic species. In an era of rapid flux, our genotypes, phenotypes, and identities are all mutating. Under the pressure of new philosophies and technological change, the humanist mode of understanding the self as a centered, rational subject has transformed into new paradigms of communication and intersubjectivity (Hayles 1999) and information and cybernetics (Habermas, 1979, 1984, 1987).

Despite these shifts, it is imperative that elements of the modern enlightenment tradition be retained, as it is simultaneously radicalized. Now more than ever, as science embarks on the incredible project of manipulating atoms and genes through nanotechnology, genetic engineering, and cloning, its awesome powers must be measured and tempered through ethical, ecological, and democratic norms in a process of public debate and participation. The walls between "experts" and "lay people" must be broken down along with the elitist norms that form their foundation. Scientists need to engage the public to discuss the complexities of matters such as cloning and stem cell research, while citizens public intellectuals and activists need to become educated in biotechnology in order to debate biotechnology issues in the media or other public arenas.

Scientists should recognize that their endeavors embody specific biases and value choices, subject them to critical scrutiny, and seek more humane, life-enhancing, and democratic values to guide their work. Respect for nature and life, preserving the natural environment, and serving human needs over corporate profits should be primary values embedded in science.

This approach is quite unlike how science so far has conducted itself in many areas. Most blatantly, perhaps, scientists, hand in hand with corporations, have rushed the genetic manipulation of agriculture, animals, and the world's food supply while ignoring important environmental, health, and ethical concerns. Immense power brings enormous responsibility, and it is time for scientists to awaken to this fact and make public accountability integral to their ethos and research. A schizoid modern science that rigidly splits facts from values must give way to a postmodern metascience which grounds the production of knowledge in a social context of dialogue and communication with citizens. The shift from cold and detached "neutrality" to participatory understanding of life that deconstructs the modern subject/object dichotomy derails realist claims to unmediated access to the world, as it opens the door to an empathetic and ecological understanding of nature (Keller, 1983; Birke, and Hubbard, 1995).

In addition, scientists need to take up the issues of democratic accountability and ethical responsibility. As Bill Joy argued in a much-discussed *Wired* article in July 2000, uncontrolled genetic technology, artificial intelligence, and nanotechnology could create catastrophic disasters, as well as utopian benefits. Joy's article set off a firestorm of controversy, especially his call for government regulation of new technology and "relinquishment" of development of potentially dangerous new technologies, as he claimed biologists called for in the early days of genetic engineering when the consequences of the technology were not yet clear.[20] Arguing that scientists must assume responsibility for their productions, Joy warned that humans should be careful about the technologies they develop, as the changes they usher in may have unforeseen consequences. Joy noted that robotics was producing increasingly intelligent machines that might generate creative robots which could be superior to humans, produce copies of themselves, and assume control of the design and future of humans. Likewise, genetic engineering could create new species, some perhaps dangerous to humans and nature, while nanotechnology might build horrific "engines of destruction" as well as the "engines of creation" envisioned by Eric Drexler (1987).

Science and technology, however, not only require responsibility and accountability on the part of scientists, but also regulation by government and democratic debate and participation by the public. The public needs to agree on rules and regulations for cloning and stem cell research, and there should be laws, guidelines, and regulatory agencies open to public input and scrutiny. To be rational and informed, citizens must be educated about the complexities of genetic engineering and cloning, a process that can unfold through vehicles such as public forums, teach-ins, and creative uses of the broadcast media and Internet. The Internet is a treasure-trove of information, ranging from informative sites such as the Council for Responsible Genetics (http://www.gene-watch.org) and The Institute of Science in Society (http://www.i-sis.org.uk), to lists servers and blogs.

But to publicize and politicize biotechnology issues, social movements will have to take up issues like the cloning and stem cell debates into their public pedagogies and struggles. Movements like the antinuclear coalitions and organized struggles against genetically modified foods have had major successes in educating the public, promoting debate, and influencing legislation and public opinion. It will not do, however, to simply let the market decide what technologies will or will not be allowed, nor should bans be accepted on technologies that can benefit human life. Instead, citizens and those involved in social movements should engage issues of biotechnology and aid in public education and debate.

An intellectual revolution is needed to remedy the deficiencies in the education of both scientists and citizens, such that each can have, in Habermas' framework, "communicative competency" informed by sound value thinking, skills in reasoning, and democratic sensibilities. A Deweyean reconstruction of education would have scientists take more humanities and philosophy courses and engage ethical and political issues involved in the development and implementation of science and technology, and would have students in other fields take more science and technology courses to become literate in some of the major material and social forces of the epoch. C.P. Snow's classic analysis of the "two cultures" problem provides a challenge for democratic reconstruction of education to overcome in an increasing scientific and technological age that requires more and better knowledge of science and humanities.

Critical and self-reflexive scrutiny of scientific means, ends, and procedures should be a crucial part of the enterprise. "Critical," in Haraway's analysis, signifies "evaluative, public, multiactor, multiagenda, oriented to equality and heterogeneous well-being" (Haraway, 1997, 95). Indeed, there should be debates concerning precisely what values are incorporated into specific scientific projects and whether these serve legitimate ends and goals. In the case of mapping the human genome, for instance, enormous amounts of money and energy were spent, but almost no resources went toward educating the public about the ethical implications of having a genome map. The Human Genome Project spent only 3–5% of its $3 billion budget on legal, ethical, and social issues, and Celera spent even less.[21]

A democratic biopolitics and reconstruction of education would involve the emergence of new perspectives, understandings, sensibilities, values, and paradigms that put in question the assumptions, methods, values, and interpretations of modern sciences, calling for a reconstruction of science and ultimately of capitalist society itself.[22] At the same time, as science and technology co-construct each other, and both co-evolve in conjunction with capitalist growth, profit, and power imperatives, science is reconstructing—not always for the better—the natural and social worlds, as well as our very identities and bodies. There is considerable ambiguity and tension in how science will play out given the different trajectories it can take. Unlike both the salvationist promises of the technoscientific ideology and the apocalyptic dystopias of some of its critics, I see the future of science and technology to be entirely ambiguous, contested, and open. For now, the only certainty is that the juggernaut of the genetic revolution is rapidly advancing and that in the name of medical progress, animals are being victimized and exploited in new ways, while the replication and redesign of human beings is looming.

The human species is thus at a terribly difficult and complex crossroads. Whatever steps we take, it is imperative that we do not leave the decisions to the scientists, anymore than we would to the theologians (or corporate-hired bioethicists for that matter), for their judgment and objectivity is less than perfect, especially for the majority who are employed by biotechnology corporations and have a vested interest in the hastening and patenting of the brave new world of biotechnology.[23] The issues involving genetics are so important that scientific, political, and moral debate must take place squarely within the public sphere. The fate of human beings, animals, and nature hangs in the balance; thus it is imperative that the public become informed on the latest developments in biotechnology and that lively and substantive democratic debate take place concerning the crucial issues raised by the new technosciences.

Notes

1. Cited in Carey Goldberg, and Gina Kolata, "Scientists Announce Births of Cows Cloned in New Way". *The New York Times*. January 21, 1998: A 14. Companies are now preparing to sell milk from cloned cows; see Jennifer Mitol, "Got cloned milk?" http://www.abcnews.com/, July 16, 2001. For the story of Dolly and animal cloning, see Kolata (1998).
2. See Sheryl Gay Stolberg, "Breakthrough in Pig Cloning Could Aide Organ Transplants," (*The New York Times*, January 4, 2001). In July 2002, the Australian government announced draft guidelines that would regulate transplanting animal organs into humans and anticipated research with pig organs translated into humans within 2 years; see Benjamin Haslem, "Animal-to-human transplants get nod," *The Australian*, July 8, 2002: A1.

3. See http://abcnews.go.com/sections/DailyNews/biotechgoats.000618.html.
4. See Heather Moore, "The Modern-Day Island of Dr, Moreau," *Alternet*, http://www. alternet.org/ story.html?StoryID=11703, October 12, 2001. For a vivid description of the horrors of animal experimentation, see Singer (1975); for an acute diagnosis of the unscientific nature of vivisection, see Greek and Greek (2000).
5. See "In Test, Japanese Have No Beef With Cloned Beef," *The Washington Post,* http://www.washingtonpost.com/wpsrv/inatl/daily/sept99/japan10.htm. According to one report, it is more accurate to refer to this beef as being produced by "embryo twinning," and not the kind of cloning process that produced Dolly; see "Cloned' Beef Scare Lacks Meat," *Wired,* http://www.wired. com/news/technology/0,1282,19146,00.html. As just one indicator of the corporate will to clone animals for mass consumption, the National Institute of Science and Technology has donated $4.7 million to two industries to fund research into cloning chickens for food. See "Cloned chickens on the menu," *New Scientist.com,* August 15, 2001.
6. See Heather Moore, "The Modern-Day Island of Dr, Moreau," op. cit., and Sharon Schmickle, "It's what's for dinner: milk and meat from clones," *The Star Tribune,* http://www.startribune. com/stories/462/868271.html, December 2, 2001.
7. "Clonefarm: Billions of identical chickens could soon be rolling off production lines," *NewScientist.com* http://www.newscientist.com/hottopics/cloning/cloning.jsp?id= 23040300, August 18, 2001.
8. Cited in Heather Moore, "The Modern Day Island of Dr. Moreau," op. cit.
9. The Missyplicity Project boasts strong code of bioethics; see http://www.missyplicity.com/.
10. See http://www.transgenicpets.com/.
11. "In Cloning, Failure Far Exceeds Success," Gina Kolata, *The New York Times,* http:// www.nytimes. com/2001/12/11/science/11CLON.html.
12. See "Clones contain hidden DNA damage," http://www.newscientist.com/news/ news.jsp?id=ns9999982; see also the study published in *Science* (July 6, 2001), which discusses why so many clone pregnancies fail and why some cloned animals suffer strange maladies in their hearts, joints, and immune system.
13. "Clone Study Casts Doubt in Stem Cells: Variations in Mice Raise Human Research Issues," *The Washington Post,* http://www.washingtonpost.com/ac2/wp-dyn/A23967-2001 Jul5?language= printer, July 6, 2001.
14. See "Scientists Warn of Dangers of Human Cloning," *ABC News,* http://www.abcnews. com/. See also the commentaries in Gareth Cook, "Scientists say cloning may lead to long-term ills," The *Boston Globe*, July 6, 2001; Steve Connor, "Human cloning will never be safe," *Independent*, July 6, 2001; Carolyn Abraham, "Clone creatures carry genetic glitches," July 6, 2001; Connor cites Dolly-cloner Ian Wilmut who noted: "It surely adds yet more evidence that there should be moratorium against copying people. How can anybody take the risk of cloning a baby when its outcome is so unpredictable?"
15. See "Report Says Scientists See Cloning Problems", *ABC News,* http://abcnews.go.com/ wire.US/reuters200103525_573.html.
16. The Westhusin quote is at http://abcnews.go.com/cloningflaw010705.htm; the "misplaced carbons" quote is in Philip Cohen, "Clone Killer," http://www. newscientist.com/news.
17. "Human Clone Moves Sparks Global Outrage," *Sydney Morning Herald,* http://www.smh.com.au/, March 11, 2001.
18. Given this attitude, it is no surprise that in September, 2001, Texas A&M University, the same institution working on cloning cats and dogs, showed off newly cloned pigs, who joined the bulls and goat already cloned by the school, as part of the "world's first cloned animal fair."
19. See "Back from the Brink: Cloning Endangered Species," Pamela Weintraub, http://news.bmn.com/hmsbeagle/109/notes/ feature2, August 31, 2001. "Gene Find No Small Fetus," *Wired,* http://www.wired.com/news/print/ 0,1294,41513,00.html
20. See the collection of responses to Joy's article in *Wired* 8.07 (July 2000). Agreeing with Joy that there need to be firm guidelines regulating nanotechnology, the Foresight Institute

has written a set of guidelines for its development that take into account problems such as commercialization, unjust distribution of benefits, and potential dangers to the environment. See http://www.foresight.org/guidelines/current.html. I encourage such critical dialog on both the benefits and dangers of new technologies and hope to contribute to these debates.

21. See http://www.wired.com/news/0,1294,36886,00.html.
22. On "new science" and "new sensibilities," see Herbert Marcuse, *One-Dimensional Man* (Beacon Press: Boston, 1964) and *An Essay on Liberation* (Beacon Press, Boston 1969).
23. For a sharp critique of how bioethicists are bought off and co-opted by corporations in their bid for legitimacy, see "Bioethicists Fall Under Familiar Scrutiny," *The New York Times*, http://www.nytimes.com/2001/08/02/health/genetics/02BIOE.html.

References

Best, S. & Kellner, D. (2001). *The postmodern adventure: Science, technology, and cultural studies at the third millennium.* New York: Guilford Press.
Birke, L. & Hubbard, R. (1995). *Reinventing biology: Respect for life and the creation of knowledge.* Bloomington: Indiana University Press.
Drexler, E. (1987). *Engines of creation: The coming era of nanotechnology*. New York: Anchor Books.
Fox, M.W. (1999). *Beyond evolution: The genetically altered future of plants, animals, the earth and humans.* New York: The Lyons Press.
Greek, R. & Greek, J.S. (2000). *Sacred cows and golden geese: The human cost of experiments on animals.* New York: Continuum.
Habermas, J. (1979). *Communication and the evolution of society.* Boston, MA: Beacon Press.
Habermas, J. (1984). *Theory of communicative action, Vol. 1.* Boston, MA: Beacon Press.
Habermas, J. (1987). *Theory of communicative action, Vol. 2.* Boston, MA: Beacon Press.
Haraway, D. (1997). *Modest Witness@Second Millenium. FemaleMan Meets OncoMouse: Feminism and Technoscience*. New York: Routledge.
Hayles, K. (1999) *How we became posthuman: Virtual bodies in cybernetics, literature, and informatics*. Chicago: University of Chicago Press.
Huxley, A. (1989). *Brave new world.* New York: Perennial Library.
Kass, L. (1998). The wisdom of repugnance. In G. Pence (Ed.), *Flesh of my flesh: The ethics of human cloning*. (pp. 13–37). Lanham, MD: Rowman and Littlefield Publishers.
Keller, E.F. (1983). *A feeling for the organism: The life and work of Barbara McClintock.* New York: WH.Freeman and Co.
Kolata, G. (1998). *Dolly: The road to dolly and the path ahead.* New York: Morrow.
Madrigal, A. (2009, March 27). Glowing Monkeys Make More Glowing Monkeys the Old-Fashioned Way. *Wired Magazine*. Retrieved June 8, 2009 from http://www.wired.com/wiredscience/2009/05/glowing-monkeys-make-more-glowing-monkeys-the-old-fashioned-way/
Marcuse, H. (1964). *One-dimensional man.* Boston, MA: Beacon Press.
Marcuse, H. (1969). *An essay on liberation.* Boston, MA: Beacon Press.
Rifkin, J. (1998). *The biotech century: Harnessing the gene and remaking the world*. New York: Tarcher/Putnam.
Singer, P. (1975). *Animal liberation.* New York: Avon Books.

Darwin's Progeny: Eugenics, Genetics and Animal Rights

Vincent J. Guihan

Animals, who we have made our slaves, we do not like to regard as our equals (Darwin, 1837–8, Notebook B, para 232)

Abstract *On the Origin of Species* reflects and furthers a shift towards the systemic organization of scientific knowledge about human beings, as well as other animal species that give rise to an enormous body of juridical and biological knowledge. I argue that eugenics itself, to a great degree, emerges as a prospect of human improvement after the successful "improvement" of non-human animals through controlled breeding over the Victorian period, and that this discourse, although it may be nominally repudiated with respect to human beings, goes largely unchallenged with respect to thinking about non-human animals well into the latter half of the twentieth century. Finally, I argue that the animal advocacy movement emerges at the end of the twentieth century and early twenty-first to constitute what can properly be termed a Darwinian response to eugenics that embodies an attempt to incorporate animals within the broader Enlightenment project of rights.

Keywords Eugenics · Genetics · Darwinism · Discourse of species · Biopower · Animal rights

If, as Foucault (1977) describes the torture of poor Damiens the regicide, his flesh "torn from his breasts, arms, thighs and calves with red hot pincers" (p. 3), in the opening to *Discipline and Punish*, I were to open this paper with a description of the torture that a factory farmed animal endures from birth to table, or with a description of one of Galen's many public vivisections of live, unanaesthetized pigs, monkeys and dogs in second century Rome, this would be an unpopular essay. And yet, *Discipline and Punish*, one of Foucault's great works, is required reading for anyone who works in cultural studies. Foucault draws out the torture of Damiens in an effort to show a historical transition away from torture to more subtle, refined

---

V.J. Guihan (✉)
Carleton University, Ottawa, ONT, Canada
e-mail: vjguihan@connect.carleton.ca

C. Gigliotti (ed.), *Leonardo's Choice*, DOI 10.1007/978-90-481-2479-4_2,

and invisible techniques for disciplining human beings. Historically, a similar shift has happened for animals, but mostly in reverse, moving from husbandry to factory farms—although, as with the disciplining of human animals, our dominance over non-humans has been increasingly made invisible. "Since Darwin," as Richard Ryder (1970, para 4) argues, "scientists have agreed that there is no 'magical' essential difference between human and other animals, biologically speaking. Why, then, do we make an almost total distinction morally?"

Although animal ethicists frequently go back to Descartes or Aristotle, to understand the role that animals play in Western thought, their influence on how we consider animals in relation to ourselves pales in comparison with Darwin's influence on the popular imagination.[1] In Victorian England, *On the Origin of Species* "was a bestseller. The publisher John Murray ran off 1,250 copies and took orders for 1,500 even before the publication day, including 500 for a circulating library. A month later, he produced another 3,000 copies" (Radford, 2008, para 10). As Keith Francis (2006) puts it, "in the public view, however, the theory of evolution became more and more the truth. To most nonscientists, the evidence to support what they supposed were Darwin's theories became overwhelming" (p. 73). Today, in a broad field of discourse that intersects hard and social sciences, philosophy and political theory, as well as cultural representation, the Animal as a figure constitutes and is constituted by bodies of knowledge that can only be described as prolific; and Darwinism is largely responsible.

On the Origin of Species reflects and furthers a shift towards the systemic organization of scientific knowledge about human beings, as well as other animal species that give rise to an enormous body of juridical and biological knowledge. At the time of *Origin*, Darwin (1859) remarks that "the laws governing inheritance are largely unknown" (p. 36), but already, the question of genetics and its relationship to evolution were becoming important in Western thought. Wallace, Lamark, Herbert Spencer and other contemporaries of Darwin were addressing similar questions. The question is not whether Darwin himself (or Descartes, or any other individual figure) sets off this process. The question is what historical and economic conditions encourage Darwin (and before him, Lamark, Descartes, Carolinus Linnaeus and others) to pose questions about the relationship between human and non-human animals and what made Darwin's ideas excite a century and a half of intensive scientific study. Although *Origin* is a defining work, what Darwin (1872) writes in *The Expression of the Emotions in Man and Animals* exemplifies what was really at stake with what was controversial about his thought:

> No doubt as long as man and all other animals are viewed as independent creations, an effectual stop is put to our natural desire to investigate as far as possible the causes of Expression. By this doctrine, anything and everything can be equally well explained; and it has proved as pernicious with respect to Expression as to every other branch of natural history. [...] He who admits on general grounds that the structure and habits of all animals have been gradually evolved, will look at the whole subject of Expression in a new and interesting light. (p. 12)

What was controversial was Darwin's suggestion that humans and non-human animals were not created separately but that human beings evolved from other species (usually referred to as macro-evolution). Both *Origins* and *Expression* reaffirmed a close relationship between human beings and non-human animals. For natural historians familiar with Aristotelian views, this was not unusual. However, it provided a new way of thinking for public audiences that were more accustomed to a creationist view that emphasized the ontological dissimilarities between humans and non-human animals as well as their separate creation.

Darwin's work coincides with and furthers a shift in the conditions of possibility. Animals go from being elements of nature to be catalogued during the sixteenth to the eighteenth century primarily, to productive labourers and commodities to be studied and improved. This shift has a number of consequences for contemporary understandings of human and animal difference, as well as for the attending ethics of today. Further, I argue that eugenics itself, to a great degree, emerges as a prospect of human improvement after the successful "improvement" of non-human animals through controlled breeding over the Victorian period, and that this discourse, although it may be nominally repudiated with respect to human beings, goes largely unchallenged with respect to thinking about non-human animals well into the latter half of the twentieth century. Finally, I argue that the animal advocacy movement emerges at the end of the twentieth century and early twenty-first to constitute what can properly be termed a Darwinian response to eugenics that embodies an attempt to incorporate animals within the broader Enlightenment project of rights.

My chapter is a genealogy of these discursive shifts over the course of the long twentieth century; it is an effort to better understand the contemporary conditions of possibility with respect to a reverse discourse on behalf of animals. For Foucault, discourse is what "transmits and produces power; it reinforces it, but also undermines and exposes it, renders it fragile" (*History,* 1978, p. 101). A reverse discourse is one that relies on a body of knowledge that has been produced by a certain set of ruling relations in order to argue against those ruling relations. As I argue later in this chapter, the same body of knowledge produced by an understanding of genetics that was largely the result of Darwin's work provided a "scientific" basis for eugenics but also laid the ground-work for contemporary animal rights theory. That is, the knowledge that human beings and animals were not so distinct genetically provided justification for the view that human and non-human animals should be bred to improve the species. However, along with contemporary genetic technologies and the knowledge that they have produced, that same body of knowledge has provided a basis for a discourse that calls instead for the rights of all sentient beings.

Darwin or Descartes?

Animal ethicists have often emphasized Descartes, or humanism, or a lack of contact with animals as prime movers behind the increasingly unethical nature of the relationship between human and non-human animals. However, these attempts at defining origins or definitive causes are problematic for a number of reasons. This is not to say Descartes had no influence. He had a remarkable influence on scientific discourse. But in terms of popular discourse, Descartes' views on animals were not shared popularly at the time of his writing and the religious elements of Descartes' understanding of animals much more greatly preoccupied his contemporaries (Harrison, 1992, p. 219). Of course, if it were popular wisdom that animals were machines, there would be no need for Descartes to problematize anthropomorphic approaches to animals. His need to justify experimentation on animals by putting forward the argument that animals were machines (i.e. that animals are not sentient like human beings) signals with relative clarity at least some basic, historically continuous moral concerns for animals, as well as a general belief in their sentience and biological similarity. And, in fact, Malenbranche, Voltaire and Thomas More vociferously attacked Descartes for his views on animals (Harrison, 1992, p. 219). Further, following the Second World War, a number of cultural critics, philosophers, and scientists only marginally concerned with animals have rejected the Cartesian view, the animal rights and animal welfare movements have rejected this view and virtually every children's film and work of children's literature involving an animal has rejected this view. Most popular Western discourses take it for granted that animals are not machines, and yet, this is precisely how science and agribusiness treats them and only a very small number of people see this as morally problematic; it is the desire to profit that motivates these industries, while Cartesian thought provides them with ample justification to do so.

Nevertheless, in a certain sense, what many animal ethicists have argued to be the historical drivers behind a lack of moral concern about animal use is problematic. By the time Descartes was arguing that animals are machines, the Great Chain of Being was already under question and animal use, even live vivisection, was already a

Fig. 1 Baboon waiting to be used in xenotransplantation research (credit: Courtesy of ITV/Carlton)

part of Western culture. Historically speaking, the Cartesian view that animals were automata was not without its critics. Hume, for example, believed rather strongly (based on the empiricism that led Aristotle to very different conclusions) that animals could speak and reason in their own limited ways.[2] Rousseau argued that they should have protection under natural law,[3] and Bentham argued that animals should have the law's formal protection.[4] Similarly, it is problematic to point to the rise of Humanism. Animals were held as property in the West long before Humanism, and it is historically unclear how much influence Humanist values have had on the last 150 odd years of Western political history. The First and Second World War suggest rather strongly that one of Humanism's basic tenets, that we have a moral duty to other humans because they are human, has never received serious acceptance. Similarly, it is problematic to argue that we lack moral concern because we no longer have a relationship with animals. Many of us have relationships with what Deleuze and Guattari refer to as "individual animals, family pets, sentimental, Oedipal animals each with its own pretty history, 'my' cat, 'my' dog" (1987, p. 241). The daily experience of many Westerners is a personal and direct contact with other animals in the form of their companion animals. These arguments also do not easily account for urban animals (squirrels, raccoons, ground hogs and the like). Although urban animals are frequently considered to be pests, not many "pet owners" would agree with Descartes that their particular companions are machines. This is not to suggest that Descartes, Humanism and our gradual distancing from animals (except under the most controlled circumstances) have not been influential in shaping Western attitudes toward non-human beings. It is only to situate them within a broader discursive field and examine them within a post-Darwinian context that has witnessed the rise of intense factory farming and increased animal testing, in spite of an aggressive discourse of husbandry, and later animal welfare, that has been historically continuous since the origins of Western thought. There is something larger at work than Descartes' writings, the Enlightenment or Humanism, the dissociation of human beings from wild animals. Over the long twentieth century, the view that animals are machines for our use—in practice if not in popular discourse—has become more prominent than ever. Darwin and Darwinism provide a focal point to understand this shift.

Genetics and Eugenics: The Rise of Biopower and Man-as-a-Species

Beginning with its publication in 1859, Darwin's *Origin* provoked a great deal of debate over the end of the nineteenth and twentieth centuries. In the popular press, Darwin was vilified and heroized. His work destabilized what was long taken to be understood: that human subjectivity, what it means to be human, was completely distinguishable from what it meant to be an animal. Darwin's insistence on the evolutionary and genetic relatedness of humans to other animals effectively denaturalized a process of domination in which the Animal as a discursive figure props up Man as a discursive figure while regulating both. In that sense, anthropocentrism regulated both the human and the non-human, in much the same way that patriarchy, empire and capital and other regimes of power have regulated and authorized other subject positions. Darwin's work provides a historical focal point that begins a prolific scientific and not-so-scientific outpouring that endeavoured to better understand the differences between human and non-human animals from both an anatomical and ethological perspective.

It is in this Darwinian moment, with the rise of biopower, just following the revolutions of 1848, and not in the classical period, not in the Renaissance with Bacon and Descartes, and not in the Enlightenment with Kant and the Cartesians, that a discourse of species with political weight truly begins to emerge. In *Society Must be Defended*, Foucault (1997) argues that "in the seventeenth and eighteenth centuries we saw the emergence of techniques of power that were essentially centered on the body, on the individual body [. . .]. Attempts were made to increase their productive force through exercise, drill, and so on" (p. 242). Furthermore, Foucault (*Society*, 1997) argues, over the nineteenth century,

> unlike discipline, which is addressed to bodies, the new nondisciplinary power is applied not to man-as-body, but to the living man, to man as living being; ultimately, if you like, to man-as-species [. . .] So far a first seizure of power over the body in an individualizing mode, we have a second seizure of power that is not individualizing but, if you like, massifying, that is directed not at man-as-body, but at man-as-species. After the anatomo-politics of the human body established in the course of the eighteenth century, we have, at the end of that century, the emergence of something that is no longer an anatomo-politics of the human body, but what I would call a "biopolitics" of the human race. (pp. 242–3)

Certainly, the organizing of man-as-a-species has its consequences on the organization of other species-as-species. This process is what gives rise to the knowledge that establishes both ethology and behaviourism as bodies of knowledge separable from one another and from zoology and psychology. The attempts to make human animals more productive also find their corollaries in attempts to make other animals species more productive, both through advances in farming, controlled breeding and with the process by which individual species were identified and catalogued. Foucault argues further that "Biopolitics' domain is, finally[. . .] control over relations between the human race or human beings insofar as they are a species, insofar as they are living beings, and their environment" (pp. 244–5), a concern that one can note in both the exploitation of the environment by the forces of capital, and those environmentalists who seek to create a sustainable relationship between human beings and their ecosystem—both represent anthropocentric strategies of management and control. Neither management strategy contests, in fact, both fundamentally assume, the right or the duty of human beings to structure the environment as it suits them, in some cases, down to the genetic level. The application of biopower to animals may not have been as rapid historically because rapidity was less necessary or because animals, even as producers, have remained entirely commodities. The application has certainly been more brutal, perhaps because anything else has been less necessary. After all, non-human animals cannot strike, cannot organize and cannot resist as a mass group, and human beings, redirected by a discourse of human supremacy, have failed to notice just how much biopolitics has organized us as one species among many. We have not become masters; we have become mere overseers.

In *A Thousand Plateaus*, Deleuze and Guatarri (1987) describe a similar historical shift. Before the nineteenth century, they argue, "Nature is conceived as an enormous mimesis in the form of a chain of beings perpetually imitating one another, progressively, and tending toward the divine higher term they all imitate by graduate resemblance" (p. 235). With the nineteenth century, however, cultural understandings of the Animal as a discursive figure change. With the nineteenth century and afterwards, they argue that

> We must distinguish three kinds of animals. First, individuate animals, family pets, sentimental, Oedipal animals each with its own pretty history, "my" cat, "my" dog. These animals invite us to regress into a narcissistic contemplation. [...] And then there is a second kind: animals with characteristics or attributes: genus, classification or State animals; animals as they are treated in the great divine myths in such a way as to extract from them series or structures, archetypes or models [...] Finally, there are more demonic animals, pack or affect animals that form a multiplicity, a becoming, a population, a tale ... Or once again, cannot any animal be treated in all three ways? (1987, p. 241)

As with Foucault, Deleuze and Guatarri sense a shift in how animals are used figurally in cultural work to understand human subjectivity and to support its various ideological apparatuses: the family, the state and the civilization. Although the shift from cataloguing animals to a focus on the regulation of their birth, death and productivity is not entirely historically contiguous, it is this shift that takes us from the naturalism of early zoologists and explorers to the factory farms of the post-Second World War period. Animals (including human beings) were certainly commodities (in the sense that they could be exchanged based on a monetary value, not just bartered for another good) from the sixteenth century onward.

It is not so surprising then that the prospect that human and non-human animals shared a common evolution in combination with a view that non-human animals existed for human use coincided with and furthered a highly opportunistic scramble to mobilize these new bodies of knowledge for political or economic benefit. These scientific advances in combination with political will and economic demand transformed non-human animals into both commodities and commodities producers. In 1862, Pasteur developed pasteurization, which facilitated the commodification of cow's milk and then later anthrax immunizations to further commodify cattle in 1871. His work either developed or led to the development of many other vaccines for rabies, chicken cholera and other common animal diseases. Mendel's genetic experiments were published in 1865. Friedrich Miescher discovered DNA in 1871. It is in this context, of the production and advancement of scientific knowledge, with clear economic benefits, and benefits to general human health and welfare that Louis Galton first advocated "eugenics" in 1889. The widespread European and American eugenics programs over the first half of the twentieth century and "ethnic cleansing" over the second half represent a relatively continuous, if evolving, discourse of eugenics, predicated on a view of Darwin's work that understands biology as a destiny.

The role of genetic knowledge and genetic technologies in this process has been considerable. Specific scientific advances and techniques, such as pasteurization, the development of antibiotics, an increasing understanding of breeding and vivisection, have all helped to industrialize and commodify agriculture and doubled world-wide agricultural production between 1820 and 1920. None of this would have been possible without a basic knowledge of both genetics and the shared genetic make-up of human beings and animals, even if only in implicit forms (such as a knowledge of breeding, a knowledge that vaccines work in both human and non-human animals, and so on.) If there is a villain in contemporary history, it is Pasteur. Without his work with respect to pasteurization and the development of antibiotics for animals, the development of contemporary agribusinesses and factory farms would be difficult to imagine today. But as with Darwin, without Pasteur himself, economic drivers would have simply driven someone else to do similar work. Eugenics itself, to a great degree, emerges, not as the work of any given individual, but as a prospect of human improvement after the successful "improvement" of non-human animals through controlled breeding over the Victorian period.

Sir Francis Galton, who coined the term, was far from being a jack-booted proto-Nazi. Darwin's half-cousin, Galton was a respected scientist, publishing multiple articles on eugenics in the *Journal of American Sociology, Nature* and *Popular Science Monthly.* It is worth noting here that, in a paper published in the *Journal of American Sociology*, Galton (1904) clearly links eugenics with the improvement of human and non-human animals:

> A fable will best explain what is meant [by the term eugenics]. Let the scene be the zoological gardens in the quiet hours of the night, and suppose that, as in old fables, the animals are able to converse, and that some very wise creature who had easy access to all the cages, say a philosophic sparrow or rat, was engaged in collecting the opinions of all sorts of animals with a view of elaborating a system of absolute morality. [...] Though no agreement could be reached as to absolute morality, the essentials of eugenics may be easily defined. All creatures would agree that it was better to be healthy than sick, vigorous than weak, well-fitted than ill-fitted for their part in life; in short, that it was better to be good rather than bad specimens of their kind, whatever that kind might be. So with men. (p. 1–2)

Clearly, Galton does not elide the comparison between human beings and animals in his work, but rather, makes copious use of animals as metaphors; indeed, considers human beings just as he would any other animal. His rhetorical technique "all creatures would agree" bears more than a striking resemblance to Aristotle's frequent rhetorical preface "all men would agree". The similarity between Galton's view and Nazism is not surprising, but its resonance with the view of later behaviourists suggests both the historical continuity of attempts to erase species difference as well as the liminality of eugenics—so long as it is applied to non-humans.

It would be tempting to mark the start of our broad, modern scientific preoccupation with animals with Darwin, but that would only be partly true. Darwin's importance lies in the fact that his work represents the first formal refusal of a moral hierarchy as the natural corollary of a stable phylogeny. His refutation of Aristotle prepends a wealth of scientific inquiry and experimentation upon non-human animals in both the natural and social sciences, some good, some bad but all ugly. With Darwin, the nature of animals and the nature of the relationship between human and non-human animals all become a matter of both cultural and scientific discourse with a number of intersections in popular discourse. Pushed into the public sphere, what it means to be human and what it means to be an animal comes under increasing scrutiny with Darwin; it is this increased scrutiny that dereifies and unravels the understood nature of the relationship between human beings and other animals.

To a great degree, the study of non-human animals and the production of knowledge about their species difference have outstripped the study and knowledge of human beings over the last century and a half. Almost certainly, this has a great deal to do with the prospect that human beings represent only one animal species in a world in which scientists have categorized over 1.8 million other animal species. It is reasonable to assume one of the causes of this knowledge production is the further economic exploitation of domesticated species. Still another driver would be the underlying assumption of comparative psychology and biology over the last 150 years: that understanding the behaviour (and the underlying psychology and biology) of other species illuminates human behaviour (and its underlying psychology and biology). The rise of eugenics biology, phrenology and other discourses over the nineteenth century produces the "scientific" knowledge(s) that justify the racism, sexism and homophobia of the nineteenth century. Other discursive strands coincide with these forceful relations to form a "discourse of species", in which human beings are marginalized *as though they were animals* in an effort to exploit them economically. Post-WWII, both the humanities and the sciences have begun to reject the idea of a subjectivity that springs from human anatomy, but have been slow—resistant perhaps, to accept both a similar possibility for non-human animals and/or the prospect that in some cases, anatomy may still play an important role in the overdetermination of the subject.

With the *scala naturae* no longer defining absolutely the relationship between human beings and animals, a number of emerging theories throw open the door to renegotiation—mostly to the detriment of both human and non-human animals. Across the nineteenth and twentieth century, knowledge about animals (including human beings) was produced, categorized and organized, and turned to the advantage of various competing force relations. The question is not, "what does it mean to be a human being or an animal?" but rather what does *species* mean and how does it function? Following Aristotle's *Great Chain of Being*, comparative psychology was the dominant model for understanding animal differences. Broadly understood, comparative psychology is the study of animal psychology, which includes behaviourism, ethology and other methodological approaches. In that usage, comparative psychology emerges with Darwin in the nineteenth century and continues with the work of George Romanes (among others). It splits into subfields of inquiry in the early part of the twentieth century, including behavioural ecology ethology. Over the latter half of the twentieth century these subfields have produced their own subfields including behavioural ecology, cognitive ethology, zoosemiotics and others.

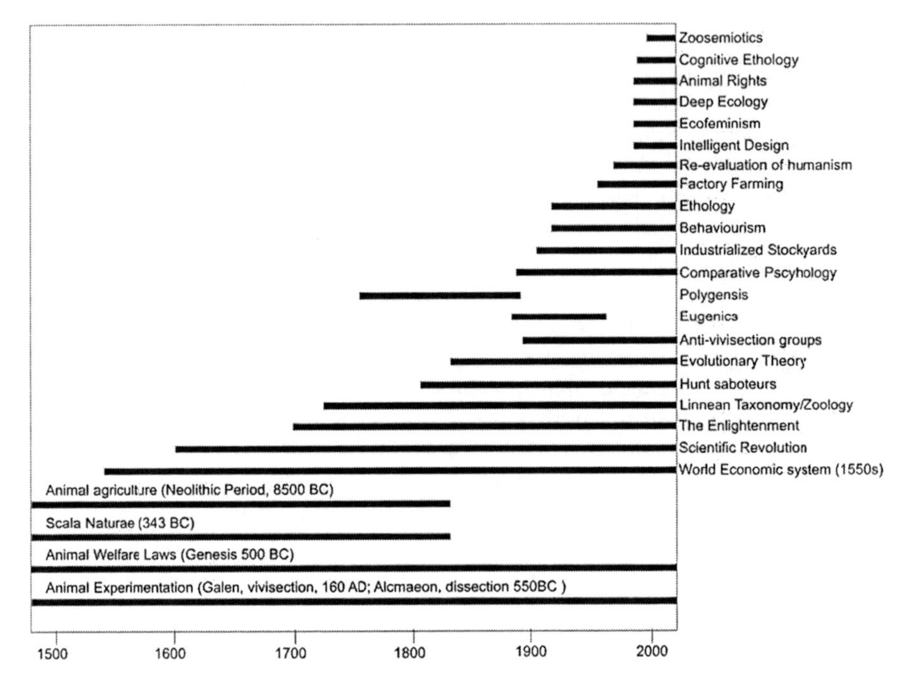

Fig. 2 The rise of a discourse of species (credit: © Copyright 2008 Vincent J. Guihan)

Aside from its copious body of knowledge, comparative psychology has produced a number of iconic memes with respect to popular understanding of animal differences. Lab rats running through mazes, dogs salivating at the sound of a bell and chimpanzees using sign language are all the result of comparative psychology. As early as 1894, C. Lloyd Morgan (a British zoologist and early psychologist) wrote that "in no case may we interpret an action as the outcome of a higher psychical faculty, if it can be interpreted as the outcome of one which stands lower on the psychological scale" (p. 53). Morgan was concerned by what he took to be anthropomorphism in science, or, more aptly, attempts to understand what happens in animal consciousness by human understandings of consciousness. Ethology and behaviourism emerge as critical bodies of knowledge that critically, if only nominally, address anthropomorphism in comparative psychology.

In most respects, behaviourism begins with this assumption, attempting to understand all human and non-human animal behaviour in terms of its lowest common psychological denominators. Across the early twentieth century, Ivan Pavlov, Skinner and J. B. Watson emerge as the principle definers of behaviourism. Pavlov, of course, is famous for his experiments on dogs. He is the first to describe what becomes known as "classical conditioning", pairing feedings with the ringing of a bell. In contrast to Pavlov, Skinner's thought is more broadly based and philosophical. Skinner's thought is structuralist and modernist on one hand, but clearly opposed to either a romantic or a theological humanism on the other. In *Science and*

Human Behavior, Skinner (1953) argues that "science is more than the mere description of events as they occur. It is an attempt to *discover order*, to show that certain events stand in lawful relations to other events" (p. 6, my emphasis). Watson (1913) is even more direct; in "*Psychology as the Behaviorist Views It*", he argues that "the behaviorist, in his efforts to get a unitary scheme of animal response, recognizes no dividing line between man and brute" (p. 158). Although behaviourism reaches an apex of influence in North America in the 1970s and 1980s, its roots go back across the twentieth century and are likely to continue well into the twenty-first.

As a differentiated field of study, ethology emerges in the 1920s, pioneered by Konrad Lorenz and Nikolaas Trinbergen. William Morton Wheeler introduces the term in 1902. However, many consider Darwin and his book *The Expression of Emotions in Man and Animals* to be one of the first modern examples of ethology. Like behaviourism, ethology concerns itself with biology and anatomy as primary drivers of animal behaviour. Unlike behaviourism, however, ethology has no broad philosophical opinions on the human condition and it includes the prospect of animal cognition and local ecology within the framework of those behaviours. Both differences are important. Ethologists are also typically less concerned about anthropomorphism per se, and more concerned about understanding individual and species differences (Lorenz, 1952). Following WWII ethology splintered into several subfields. The most notable is cognitive ethology, which works from the position that if human beings have cognitive capabilities, it is only reasonable to assume that other animal species do as well. The work in cognitive ethology is to understand both human and animal behaviour, similarities and differences, within a continuum of cognitive capability that mitigates patterned behaviours. Where behaviourism attempts a universalizing, totalizing understanding of behaviour as stimulus in, response out, ethology approaches behaviour from a perspective of difference. On the other hand, ethology entails some anthropocentrism with its exclusion, by and large, of human beings as animals fit for study. Regardless of the motivations behind this scientific study, the knowledge produced has been re-marshalled to suit various ideological views. With caveats about origins in mind, the rise of capitalism beginning in Renaissance Europe begins a slow reorganization of all life, human and environmental, into property-based systems (Wallerstein, 1974). With the rise of neo-colonialism in the late nineteenth century, this reorganization becomes global and sufficiently hegemonic. One of the notable consequences of this reorganization is the assimilation of virtually all beings into the property system, either as property-holders (the slim minority) or as property. Property-holders have rights. Property does not. Although which groups of human beings fall into which category has varied substantially over the last 150 years, this has not been the case of other animal species, which have always been property.

The concomitant rise of eugenics discourses over the nineteenth century produce the "scientific" knowledge(s) that justify the racism, sexism and homophobia of the nineteenth century, oppressing people *as though they were animals*. Other discursive strands coincide with these force relations to form a "discourse of species", in which human beings are marginalized *as though they were animals* in an effort to exploit them economically. Once this politicization is complete, there is little left to distinguish rigidly between literary, philosophical or scientific bodies of knowledge about species. This historical tendency culminates in the enormity of Nazism, a regime that makes the mobilization of "species" as a discursive construct quite plain. As Charles Patterson (2002) argues, in *Eternal Treblinka,* the "industrialized slaughter of cattle, pigs, sheep and other animals paved the way, at least indirectly, for the Final Solution" (p. 109). The American slaughterhouse and the Nazi gas chamber share a common technological framework and a common interest in eugenics. That archetypal American, Henry Ford, whose system of mass production inspired Nazi death camps was himself influenced by Chicago's stockyards.[5] It is not a surprise that Heidegger (1954) claims that agriculture has become a "mechanized food industry, in essence the same as the manufacture of corpses in the gas chambers and death camps, the same thing as the blockades and reduction of countries to famine, the same thing as the manufacture of hydrogen bombs" (p. 15).

This search for a specific, localized point of origin is ill-fated. If the need for commodities and commodity producers has been constant in the West since the mid-1500s, then the need for human and animal slavery should be constant. The end of institutionalized, legalized human slavery as a Western phenomenon did not coincide with an end to the economic needs of the West. Taken as a whole, the end of literal chattel slavery represents a modest shift from unwaged to waged labour to better suit the accumulation of capital. In contrast, the use of animals as commodities and commodities producers still suits the accumulation of capital quite well; if anything, demand has increased with the rise of consumer society in the post-WWII period. The consequence of this property status for most animals is that "the bodies and functions of animals have been completely appropriated by capital, and, subsequently, put to use in a single way only, subordinating the total animal being to this single productivity activity" (Torres, 2007, p. 40). "Animals", according to Torres, "are not only commodities and property themselves, they also produce commodities, and in a sense, serve as either the "raw" inputs or the product labor power of business" (Torres, 2007, p. 58). It is only expected then, but not inevitable, that factory farming and animal testing would increase during this period. In a broad sense, economic factors drive the production of knowledge, whether Descartes', Bacon's, Liebnitz', Agassiz', Lamark's and finally Darwin's inquiry and those who followed Darwin. But the bodies of knowledge that science has produced about animals, and the corollary understanding of animals as sentient beings like ourselves, has also compelled us to reconsider the ethical nature of the relationship. If it is wrong for one human being to harm another because we are all sentient, and animal species share the same mechanics of sentience (as biologists suggest), evince roughly similar self-awareness (as ethologists suggest), then how can we justify the exploitation of non-human animals as a matter of economic expediency, luxury and personal pleasure?

What is the contemporary nature of the discourse of species, then? The discourse of species reflects, largely, the interplay of science and philosophy—how animals actually are, and how they function as an instrument to better understand human beings—finding their intersection in what can only be broadly described as "cultural" understandings of animals. There are two poles. One pole is constituted by

Fig. 3 Piglets born and raised for xenotransplantation organ use (credit: Courtesy of ITV/Carlton)

a set of discursive tendencies that bears great similarity to patriarchy, empire, capital and other Western regimes of domination. This tendency organizes all human beings, animals and nature into property for the use of a ruling class. Unlike the domination involved with gender, class and race, however, the constituents of this ruling class are always human beings. There are no non-human apologists in reality, but it is worth noting that anthropomorphic representations of animals are often used as apologists for the regime. The other pole reflects a number of discursive tendencies that bear similarity to feminism, anti-imperialism and socialism; these include ecofeminism, radical ecology, the green movement(s) as well as the animal rights and animal welfare movements. Very much like the reverse discourses that emerge as a response to patriarchy, empire and capital, the reverse discourses that emerge as a response to the anthropocentrism of biopower are fragmented, proposing differing critiques, differing tactics for resistance and differing solutions.[6] To over-generalize, these tendencies attempt to reorganize animals (including human beings) within a framework that is either free of domination altogether, or in which that domination is regulated but accepted.

Animal Rights as a Reverse Discourse

The term "animal" in our daily language has almost ceased to have its literal usage. When one hears the phrase "treated like an animal", one does not envision a literal chicken in a battery cage, her beak burned off, only one square foot of room for the entirety of her already abbreviated life, crammed in with several of her fellows.

When one hears the phrase "acted like an animal", one does not think of cows or horses or sheep grazing peacefully, of the dove that returned Noah's olive branch, or of the elephants that mourn their dead. In popular usage the word "animal" has been emptied of all but the most metaphorical meanings. And yet never have we had a better understanding of the factual relationship between humans and non-human animals. Completed in 2003, the Human Genome Project maps the entire set of genes in human DNA, allowing biologists to confirm that at "the genomic level, for example, chimpanzees and humans share 96 percent of their DNA—a figure that rises to 99 percent for genes that actually encode proteins" (Bradshaw & Sapolsky, 2006, para 10). The similarity between humans and non-humans, what was a seemingly legitimate controversy in Darwin's days is now a matter of arguing with hard science. And yet, although non-human animals remain the subject of cheerful cultural work (books, television, theatre, movies, even music), we show less interest in the suffering and exploitation of actual animals. Why do we sentimentalize them, on one hand, and subject them to the most appalling exploitation on the other hand? This cognitive dissonance, long treated as though it were invisible, is now coming under increasing cultural scrutiny.

This is not to say that pro-animal discourses have not been common in the West. Hume presents one example; and Jeremy Bentham, another. Further, the British SPCA was founded in 1824, coinciding roughly with the formation of the Bands of Mercy (groups opposed to the fox hunt) in 1824. Darwin first published his theory of natural selection in 1838, and there were a number of anti-vivisection movements common across the Victorian era until the turn of the twentieth century. These movements shared the relatively conservative views of the contemporary animal welfare movement, which favours the regulation of animal use, not its abolition. They did not share the radical nature of the contemporary animal rights movement, which frequently calls for the end of all animal use per se. As a discourse, animal rights began to emerge with Richard Ryder's first use of the term speciesism in 1970 to describe irrational prejudices towards animals. Precise definitions of speciesism have varied since. Shortly thereafter, Peter Singer's *Animal Liberation* was published in 1975, in which Singer first formalizes the argument that human beings should weigh animal interests equally with their own. The Animal Liberation Front (a group dedicated to freeing animals) was founded in the UK in 1976. People for the Ethical Treatment of Animals was founded in 1980, and is now a multi-million dollar NGO. Carol Adams' *Sexual Politics of Meat*, a book that connects the use of women with the use of animals was published in 1981. Tom Regan's *The Case for Animal Rights*, one of the first books to argue that animals have rights because they are the subjects of their own lives, was published in 1983. The Animal Rights Militia, a group that proposes violence towards human beings in defence of animals emerged in 1985, as does the Justice Department, a similar group, in 1993. Gary Franciones' *Animals, Property and the Law* was published 1995, which represents the first political economy of animal rights, in which Francione argued that the root of the problematic nature of our relationship with non-human animals is their property status, and his *Introduction to Animal Rights: Your Child or Your Dog* published in 2001 remains one of the definitive works of the animal rights movement. A number of books have been published since, generally taking up Singer's, Regan's[7] or Francione's views, that, respectively, cover how animals are treated when they are used,[8] or that animals are used at all, reflects the pressing moral dilemma. More closely associated with Singer, the animal welfare view represents a mostly conservative echo to the Bands of Mercy and the anti-vivisection societies of the pre-WWII period. In essence, the animal welfare movement would return the relationship between human beings and animals to one of careful husbandry. More closely associated with Francione and other thinkers, the animal rights movement, however, takes a more radical approach.

Although cultural critics have paid much attention to the debate about human duties to animals in continental philosophy,[9] North American animal rights theory and the animal rights movement in general have received remarkably scant attention by comparison. Without dismissing the importance of the broader animal advocacy movement, what is of particular interest within a Darwinian context is the view that biology provides a basis for rights-holding. Where eugenicists believed that biology provided an exclusionary basis for rights, many animal rights advocates such as Gary Francione, Roger Yates, Bob Torres, Steven Best and others argue that biology (to be specific, sentience) provides a basis for including non-human animals within a broader moral framework of rights. That is, "like [human beings], sentient animals have an interest in not experiencing pain and suffering" (Francione, 2000, *Intro*, xxiii), and as a consequence, deserve moral consideration. In particular, the abolitionist perspective argues that "all sentient beings, humans or nonhuman, have one right: the basic right not to be treated as the property of others", and that, as a consequence of recognizing this right, "we must abolish, and not merely regulate, institutionalized animal exploitation—because it assumes that animals are the property of humans" (Francione, 2007, *Abolitionist*, para 2).

As a discourse, the prospect that all animals, human or otherwise, share a basic right to be ends in themselves because they share common genetics functions in a number of ways. First, it forces us to reconsider the ethical implications of emerging genetic technologies. Gene splicing, cloning and other technologies become more ethically problematic when we consider that the beings these techniques have been and continue to be developed on have a right not to be used in this way. Second, it reconstitutes the figure of the animal as something that is distinguishable from "nature". They are beings to whom we have direct duties and not just an indirect duty of stewardship. Third, it also draws us to a reconsideration of human subjectivity as something that cannot be understood as rigorously different from animal being, undermining a number of historically contiguous roles that the Animal has played in literature. That is, it undermines a traditional binary opposition between human and non-human animal being, in much the same way that other discourses of subjectivity have undermined similar binary oppositions between white/black, straight/gay, man/woman, and so on. Fourth, as a consequence, it provides us with a new opportunity to theorize how animals function in popular culture, complicating previous theoretical work with respect to how animals function in figural terms, in much the same way that feminist, post-colonial and race theory complicated our understanding of cultural work that addresses themes of gender, race and empire.

The prospect that animals are or should be rights-holders poses clear questions about how we imagine animals to be, and how that props up our own sense of subjectivity. It also poses questions in regard to the processes that regulate a system of domination in which to be Human and to be Animal function in ways that are similar to the authorized subjectivities of other regimes of power.

Finally, having rendered these processes into language and representation, the broader discourse of species has rendered anthropocentrism fragile, and animal rights as a reverse discourse draws our duties to other beings, human or otherwise, and to the ecology on which we all rely. That is, animal rights discourse draws out the discursively constructed nature of species difference and in so doing, threatens reifications of human subjectivity, in an effort to foster a shift from a subjective, arrational and passive attitude with respect to other animals, human or non-human, toward an objective, rational and active one; animal rights discourse moves us from a cathartic and anthropocentric pity toward a more egalitarian and dialogic relationship. All of which, in part, is driven by Darwin's ushering into popular discourse the prospect that humans and animals share a common genetic history.

In conclusion, *On the Origin of Species* initiates a discourse of species that establishes biological links between human and non-human animals. Appropriating Darwin, eugenics, as a discourse, puts the (alleged) utility of the group ahead of the rights of the individual. It calls on us to understand biology as a destiny and ethics as a rational calculation. The improvement of the species, at the expense of the individual, was the driving force of both eugenics and its enormities as well as a great deal of more harmless seeming twentieth century discourses around human and non-human animals. In reply, animal rights discourses stake a claim to equal treatment for equal interests based on the shared genetics and biology of non-human animals, their sentience, and therefore, their right not to be used. It is this reverse discourse that Darwin set in motion when he not only argued that humans descended from other animals, but questioned the morality of human domination more than 150 years ago.

Notes

1. As Tim Radford eloquently describes it, "much of the hostility and alarm came not overtly from religion, but from within science. The book was hailed, applauded, challenged, questioned, condemned, cruelly dismissed and, rather astonishingly, ignored: the president of the Geological Society of London in 1859 managed to give Darwin a medal of honour for his geological observations in the Andes and his stunning four-volume study on barnacles, without mentioning his seminal paper with Alfred Russel Wallace, or the forthcoming book" (2008, para 3).
2. In his *Treatise on Human Nature*, Hume argues that "[n]ext to the ridicule of denying an evident truth, is that of taking much pains to defend it; and no truth appears to me more evident, than that beasts are endow'd with thought and reason as well as men" (1739, para 1).
3. "By this method also we put an end to the time-honoured disputes concerning the participation of animals in natural law: for it is clear that, being destitute of intelligence and liberty, they

cannot recognize that law; as they partake, however, in some measure of our nature, in consequence of the sensibility with which they are endowed, they ought to partake of natural right; so that mankind is subjected to a kind of obligation even toward the brutes" (Rosseau, 1755).

4. "The French have already discovered that the blackness of skin is no reason why a human being should be abandoned without redress to the caprice of a tormentor. It may come one day to be recognized, that the number of legs, the villosity of the skin, or the termination of the os sacrum, are reasons equally insufficient for abandoning a sensitive being to the same fate. What else is it that should trace the insuperable line? Is it the faculty of reason, or perhaps, the faculty for discourse?. . .the question is not, Can they reason? nor, Can they talk? but, Can they suffer? Why should the law refuse its protection to any sensitive being?" (Bentham, 1789/2005)
5. Patterson provides an excellent assessment of the influence of Chicago's stockyard system on Ford's system of mass production (p. 72), and the influence of Ford's mass production on the Nazis' system of mass destruction (p. 73). Both a financial supplier to, as well as a source of inspiration for, the Nazis, Ford's life-sized poster was hung by Hitler in his office (p. 75). Ford's reissue of the *Protocols of the Elders of Zion*, perhaps the world's most infamous anti-Semitic text, was so well-edited and updated at Ford's instruction that it became the international standard (p. 75). For all his efforts, "Hitler spoke of Ford in glowing terms" (p. 75) and in reward, he was also one of only four foreigners to be awarded the Grand Cross of the Supreme Order of the German Eagle (p. 79).
6. Unfortunately, I cannot treat all of these works at length. However, the work of ecofeminists (e.g. Josephine Donovan, Carol Adams, Patrick Murphy and others), some post-colonial thinkers (e.g. Helen Tiffin's *Five Emus to Siam: Environment and Empire* among others), and even some class critics (e.g. Charles Patterson's *Eternal Treblinka,* David Nibert's *Animal Rights, Human Rights*, Gary Francione's work, Giorgio Agamben's *The Open: Man and Animal*, Bob Torres's *Making a Killing: A Political Economy of Animal Rights* and others) represent a small but growing body of literature. This budding interest has been mirrored in other disciplines, particularly in the natural sciences (with the works of zoosemioticians and cognitive ethologists) and in philosophy (in Anglo-American analytics with Peter Singer, Tom Regan and others and on the continent with the work of Derrida, Heidegger and Levinas).
7. In *The Case for Animal Rights*, Regan (2004) argues that "the basic moral right to respectful treatment places strict limits on how subjects-of-a-life may be treated. Individuals who possess this right are never to be treated as if they exist as resources for others" (xvii). In this respect, his views are generally shared by other rights advocates. However, drawing on Kantian understanding of rights, Regan (2004) argues that only "some nonhuman animals resemble normal humans in morally relevant ways. In particular, they bring the mystery of a unified psychological presence to the world" (xvi). He does not call for rights for all animals, but rather those that cognitively resemble human beings, "mentally normal mammals of a year or more" (xvi). This differs somewhat from the views of other rights advocates who base rights on sentience alone.
8. For Singer, *animal rights* is actually a "convenient political shorthand" (1975, 8) that addresses several more complicated issues like interests, equality and sentience. According to Singer, "if a being suffers there can be no moral justification for refusing to take that suffering into consideration" and "no matter what the nature of the being, the principle of equality requires that its suffering be counted equally with" the suffering of any other sentient being (*Liberation* 8). The "real weight of the moral argument" for the equality of animals "does not rest on the assertion of the existence of the right" but rather on the "the possibilities for suffering and happiness" (8). In short, Singer argues that animals can suffer (or be happy), that they have an interest in not suffering, and that these interests should be taken into account in any ethical consideration. He does not agree, however, that animals have a basic right to life or not to be used as property—or any basic rights in a deontological sense of the term.
9. In continental philosophy, Levinas, Heidegger and Derrida all pose a reconsideration of "humanism", in the sense that humanism stands for a universal understanding of human subjectivity, what that subjectivity entails and what ethical duties and rights it may accord. For Levinas, the "absolute Other is the human Other" (1961, 12–13), and the face is metaphysical, not culturally mediated, a view that erases animals from most ethical consideration (1961,

p. 9). His argument is a response to Heidegger's that animals, unlike human *Dasein*, have a Being that is poor-in-the-world, "capable of only a limited range and depth of experiential relationships", whose treatment reflects ethical choices of the Subject, but not ethical obligations to an Other (Turner, 2003, para 31–8). In response to Levinas' anthropologocentric position, Derrida defends Heidegger's notion of the ontological differences between human and non-human beings, but criticizes Western anthropocentrism (Turner, 2003, para 45–51).

References

Bentham, J. (1789/2005). An introduction to the principles of morals and legislation [Electronic version]. Adamant Media Corporation.

Bradshaw, G. A & Sapolsky, R. M. (2006, December). Mirror, mirror. American scientist online. Retrieved on May 22, 2008 from http://www.americanscientist.org/template/AssetDetail/assetid/54077

Darwin, C. (1837–8). Notebook B [Electronic version]. *Charles Darwin's notebooks, 1836–1844: geology, transmutation of species, metaphysical enquiries*. British Museum, Cambridge: Cambridge University Press.

Darwin, C. (1859/2003). *On the origin of species*. New York: Penguin Books.

Darwin, C. (1872/2007). *The expression of emotions among men and animals*. New York: Filiquarian Publishing.

Deleuze, G. & Guattari, F. (1987). *A thousand plateaus: Capitalism and schizophrenia*. Minneapolis: University of Minneapolis Press.

Foucault, M. (1977). *Discipline and punish*. New York: Vintage Books.

Foucault, M. (1978). *The history of sexuality: An introduction*. New York: Vintage Books.

Foucault, M. (1997). *Society must be defended*. New York: Picador.

Francione, G. L. (2000). *Introduction to animal rights: Your child or your dog*. Philadelphia, PA: Temple University Press.

Francione, G. L. (2007). The six principles of the animal rights position. *The abolitionist approach*. Retrieved May 26, 2008 from http://www.abolitionistapproach.com/?page_id=86.

Francis, K. (2006). *Charles Darwin and the origin of species*. Westport: Greenwood Press.

Galton, J. (1904) Eugenics: its definition, scope, and aims [Electronic version]. *The American Journal of Sociology*. *X*(1), 1–25.

Harrison, P. (1992). Descartes on animals. *The Philosophical Quarterly*. 42(167), 219–227.

Heidegger, M. (1954/1993). *The question concerning technology and other essays*. New York: Harper Collins.

Levinas, E. (1961). *Totality and infinity*. Pittsburg: Duquesne University Press.

Lorenz, K. (1952). *King Solomon's ring*. London and New York: Routledge.

Morgan, C. L. (1894). *An introduction to comparative psychology* [Electronic version]. London: W. Scott.

Patterson, C. (2002). *Eternal Treblinka: Our treatment of animals and the Holocaust*. New York: Lantern Books.

Radford, T. (2008, February 9). *On the origin of species: The book that changed the world*. The Guardian. Retrieved on May 2, 2008 from http://www.guardian.co.uk/science/2008/feb/09/darwin.bestseller

Regan, T. (2004). Preface to the 2004 edition. *The case for animal rights*. Berkeley: University of California Press.

Rousseau, J. (1755). *A discourse upon the origin and foundation of the inequality among mankind* [Electronic version]. London and Toronto: J.M. Dent and Sons, 1913.

Ryder, R. (1970). *Speciesism* [Electronic version]. Retrieved on May 19, 2008 from http://www.richardryder.co.uk/speciesism.html

Singer, P. (1975). *Animal liberation*. New York: Harper Collins.

Skinner, B. F. (1953). *Science and human behavior* [Electronic version]. Toronto: Pearson Education.

Torres, R. (2007). *Making a killing: The political economy of animal rights*. Oakland: AK Press.
Turner, D. L. (2003). The animal other: Civility and animality in and beyond Heidegger, Levinas, and Derrida [Electronic version]. *disClosure* 12.
Wallerstein, I. (1974). *The modern world-system, vol. I: Capitalist agriculture and the origins of the European world-economy in the sixteenth century*. New York/London: Academic Press.
Watson, J. B. (1913). Psychology as the behaviorist views it. [Electronic version]. *Psychological Review* 20, 158–177.

Intimate Strife: The Unbearable Intimacy of Human–Animal Relations

Beth Carruthers

Abstract As a direct result of our cultural practices, each day species go extinct. Each day, significant habitat is lost or poisoned. Countless non-humans suffer and die in industrialised farming and laboratories. Life forms also are used by artists as media. The freedom to utilise whatever we like in our own interests is largely unquestioned. But one must ask: are all, or any, of these practices desirable or even acceptable? And if they are acceptable, why do we find them so? Also largely unquestioned is the status of human as subject, and of all else as object—as a "what" instead of a "whom", the value of which is determined by its utility to the subject human.

Discussion on how to better human relations with non-human animals is not new. The extension of human-centred ethics or moral standing to non-human animals has resulted in some significant wins in this struggle. However, my overall goal in this chapter is to look elsewhere for revision and change in human–non-human relations (and in the non-human I include not only what we consider the animate world, but all of nature as we think it). To that end, I argue that an alternate ontology is what is required for radical and lasting change. I also briefly consider some ideas that I think might aid us in locating such an ontology.

Keywords Ontology · Human–non-human relations · Art practices · Ethics

Introduction

> . . .the development of a radically different ontology. . .would mean a profound transformation in our conception of alterity and, in keeping with that, a fundamental alteration in our relations with the non-human other. (Langer, 1990)
>
> They are called Beasts because of the violence with which they rage, and are known as "wild" (ferus) because they are accustomed to freedom by nature and are governed (ferantur) by their own wishes. They wander hither and thither, fancy free, and they go wherever they want to go.
>
> *Medieval Bestiary* (as cited in Salisbury, 1994)

B. Carruthers (✉)
Independent Scholar, Galiano Island, BC, Canada
e-mail: beth@bethcarruthers.com

C. Gigliotti (ed.), *Leonardo's Choice*, DOI 10.1007/978-90-481-2479-4_3,

Carol Gigliotti (Baker and Gigliotti, 2006), in conversation with Steve Baker about animals, humans and technology, commented that "The idea of unfettered creativity is the holy grail, not only in the arts, but in the sciences and society at large" (p. 37). Certainly in Western culture this freedom to go wherever we want to go in pursuit of our own interests is seductive, considered almost a right. Rights and freedoms are generally thought of as applying to persons, as are morals and morality in action, or ethics.

That one cannot exercise even simple rights and freedoms without interacting with other beings and with the world seems clear. Yet whether ethical considerations apply to this exercising of freedoms varies depending upon the others affected. In the case of non-human animals, ethical considerations tend to become confused, an on-again, off-again proposition, or they are simply absent. There is no doubt that we require ethical, or moral systems—but what should these systems be and on what should they rest?

In the worldview of Western culture, ethical behaviour is considered specifically in light of the human, on human terms. In this configuration that which falls outside the human becomes fair game in almost every sense of the word.

I believe that the human-centred system of ethics of the Westernised world can only be inherently confused and flawed in respect to the other-than-human—that in fact, it is essentially blind in this regard. Applications or extensions of moral standing or significance to the non-human, while effective to some extent, must be seen, when they are effective, as a form of triage, providing interim assistance. So long as our ethics rise from and within an ontology that separates humans from world and from the family of animals, they cannot be inclusive or comprehensive in regard to the non-human world. My interest here then is in tracing and exploiting some flaws in this foundational ontology in respect to human–non-human relations. I then spend some time considering what an alternate ontology—a shared ontology, for want of a better term—might include.

In doing so my strategy is to destabilise a ubiquitous human centrism by exploring the construction of what is human and what is other; who has the status of moral significance, and who, or what, does not, with an intention of revealing a quagmire of cultural assumptions underlying our practices. I have argued elsewhere (Carruthers, 2006) that although the constructions and assumptions of what is a human or what is an animal are taken as fact, they are rather cultural myths, stories requiring regular revision and maintenance. Even in their status as "givens" dominant ideologies are far from secure. Cultural "truth" is subject to change.

Were boundaries of humanness and otherness to shift, distinctions to change, the bounds of relationship and ethics would have to take on new configurations. Freedoms, creative and otherwise, and the behaviours associated with them, would likewise require reconsideration.

Grail

Was I too dark a prophet when I said
To those who went upon the Holy Quest,

> That most of them would follow wandering fires,
> Lost in the quagmire?
> *The Holy Grail* (Tennyson, 1869)

Although intended by Gigliotti in a pejorative sense, her framing of the Holy Grail as a metaphor for creative freedoms serves as an intriguing point of entry into this discussion, which deals with questions of morality, virtue—and above all, the quest for an alternate ontology.

Tennyson, in his poem *The Holy Grail* (1869), provides us with a view into both the single-minded passion and devotion inspired by the Grail, and the virtue required if one is to be worthy of both the Grail quest and its successful completion.

Central to the character of the Grail of popular legend is its mystical power of transformation. It is both the original chalice from which Christ drank at the Last Supper and later the vessel of the ritual-miracle of transubstantiation, where wine (spirit) becomes blood (body). The Grail moves between, and has powers to bring its seeker across worlds—the refined world of spirit and the gross material (animal, natural) world. It is tenaciously sought, and while it may be found, is never obtained, in the sense of ownership or control. In this sense the Grail itself remains eternally free, an incomprehensible mystery. Its seekers are configured as heroic while they encounter and overcome great trials and obstacles in the Grail quest, often failing these many times before they move forward.

Bound by vows and moral conditions, the seeker who transgresses these is a seeker to whom the Grail is lost, for the Grail will be revealed only to the truly virtuous, whose hearts are pure. The quest for the Grail must then be, at its core, an ethical quest, and as such, it must be bound by moral obligation. In addition, the Grail itself embodies sacrifice—being the vessel which contains the sacrificial blood of the Lamb of God—and sacrifice is required in its seeking.[1]

As representative of an unfettered creativity, the Grail makes an interesting metaphor. It implies a purity of motive, of ends and process. It offers the possibility of transformation. And while in our contemporary creative quests we are assured that personal sacrifice—at least of the bodily kind—will not be required, this is not to say that sacrifice is *not* involved.

Freedom and Sacrifice

> Our large, industrial society seems to work toward the goal of obscuring the sacrifice that makes our existence possible. (Steeves, 1999, p. 166)

The sacrifice that I want to consider first is the sacrifice of other beings in the pursuit of various human quests—quests for cures for human diseases, for food security, for clothing, for vanity and even simple curiosity. I presume such sacrifices to be unwilling, as I don't imagine, for example, that any being would truly—to paraphrase a mid-twentieth century advertising jingle—"want to be an Oscar Meyer wiener", or for that matter, a designer handbag, nor thrive on being imprisoned in order to discover the potential toxicity of substances.

The question that is commonly raised, of whether other beings have wants, relationships, or are sufficiently intelligent, sentient—or whatever descriptor you like—to know what is at stake for them, seems to remain open. This being the case, the question then following on the heels of this first is whether we should take the interests of non-human animals into consideration at all. What these questions really ask is, how much like a human (as we define the human at this point in time) is this other? At the same time, were non-human animals not sufficiently like humans, the results of tests undertaken on them in order to gain knowledge about humans would be completely unhelpful.

We like to view ourselves as moral beings; we want to do the right thing, whatever that may be, and in the case of human–non-human relations, when we go in search of it, somehow that right thing seems always to reflect back as an image of self-interest (solely, or primarily human) as paramount and separate from the interests of other beings.

Contemporary Western society exemplifies the kind of lifestyle to which much of the human world seems to aspire. In this society it is not only creativity that we want unfettered, it is also our quest for the fulfilment of every desire—life, as we would like to know it. In this pursuit, it is almost always non-human others that pay the real price—from battery hens to the rapid extinction of species due to habitat loss.

Although commonly descried in contemporary society, and defying both common sense and lived experience, the legacy of the Descartes' era of a mechanistic world of animal automatons available for our use and amusement survives in daily practice. Non-human animals, and the other-than-human world in its entirety, continue to be utilised at no moral cost. It is no surprise then, that non-human beings should be readily perceived as material, or media, in creative practices.

Examples of recent works utilising living beings include transgenic artworks, artists working in collaboration with scientists where species specimens are gathered and exhibited as a part of the artwork, and works incorporating live non-human animals. An example of this last is artist Huang Yong Ping's *Theatre of the World.*[2] This work, when exhibited at the Vancouver Art Gallery in Canada, was closed down by the artist and the gallery in protest after the local Society for Prevention of Cruelty to Animals insisted that modifications be made to the work in order to accommodate the non-human animals that formed part of the exhibit.

Explanations given by artists in support of these works and/or processes vary from the need to encourage through these practices debate about the ethical use of life forms and transgenic technologies, to the need to better understand effects on life forms of ecosystem contamination.[3] Ping's defence of his work was based on cultural tradition, and on the customary use of non-human animals as metaphor for the human. He refused to agree to the request to make changes to the work on ethical grounds, claiming that the request was tantamount to censorship and as such must be resisted (Ping, 2007).

It is not my intention here to undertake a lengthy discussion of these and other works, but to stress that in each case these works are defended within an unquestioned acceptance of the supremacy of human interests and cultural priorities, whatever these might be at the time. The attitude of exaggerated parental stewardship Eduardo Kac appears to bring to his relationship with his artwork, Alba the GFP bunny,[4] stresses his human mastery over both Alba and the creative processes of life itself. The human centric framework also allows concerns over censorship of an artist's work to take precedence over concerns about living beings—as in the case of Ping's installation at the Vancouver Art Gallery. It is interesting to note however, that the spectacle of Ping's *Theatre of the World* at the Vancouver Art Gallery had the unintended result of revealing the existing ambiguity in social relations among human and non-human animals.

A more dramatic revelatory example of this type might be the 1990 protest by 300 people—also in Vancouver—against a public "performance work" by artist Rick Gibson, who announced his intention to squash a brown and white pet store rat called "Sniffy" between canvas sheets and concrete blocks. Gibson was mobbed and chased by angry protestors and eventually police had to rescue him from the crowd. While it remains unclear as to whether Gibson would actually have snuffed Sniffy, it is the response of the crowd that I wish to note. People were yelling threats that the artist himself should be squashed. Gibson, in the midst, was explaining that Sniffy would have a far better end being squashed than he would have were he to be sold for live pet (reptile) food—his otherwise likely fate.[5] These instances raise multiple questions that have nothing to do with censorship of artworks, but instead foreground a problematic and confused human–non-human relationship.

When did Sniffy attain moral significance in human eyes? It was unlikely to be at the pet shop prior to his celebrity, where he was viewed as animated pet food. Did he attain this status through public attention being drawn to his individual being? Would a similar human response be repeated in an abattoir, among the pigs, cattle and horses? Perhaps this is the fear of the meat industry, since tours are discouraged or not allowed.

In our contemporary world the bothersome question of the sacrifice of the other is shifted to a shadow-world of abattoirs, factory farms and laboratories, while society at large focuses on the pleasurable fruit of such enterprises. The spectre of suffering and death would be an unwelcome taint on a lifestyle that we have configured as safer, cleaner and more ethically sound than any other. That the denial of these practices should collectively trouble us to the extent that they must be hidden reveals a need to engage and come to terms with a disturbing reality.

The landscape of our ethical and cognitive dissonance in this regard is easily revealed as an unstable moral quagmire full of large indeterminate grey zones. People who think nothing of the breeding of designer dogs or race horses for human pleasure will extend (albeit selectively) the embrace of moral significance to very particular non-human others. Those who purchase factory-farmed shrink-wrapped ribs at the supermarket may also contribute to their local SPCA. People eating horse may be horrified at the thought of consuming dog.

We have become collectively adept at splitting or fragmenting our experience of reality, symptomatic of the splitting of the human from an environing world and from the family of animals.

Fig. 1 Baboon readied for xenotransplant (credit: Courtesy of ITV/Carlton)

The Moral Quagmire

> My goal as a scientist, then, is to reveal our moral thinking for what it is: a complex hodgepodge of emotional responses and rational (re)constructions, shaped by biological and cultural forces, that do some things well and other things extremely poorly. (Greene, para. 3)

The most common response to human discomfort with the treatment and utilisation of non-human others is to extend human-centric rights, freedoms and moral standing to other beings. Val Plumwood (2002) referred to this system as "the moral extensionism of contemporary philosophy of human and lesser, or human and similar" (p. 170) whereby an "other" attains special status through some argued or proven similarity to the human, or, conversely, has no status by virtue of being shown as unlike the human within currently accepted terms.

It is not my intention here to undertake a lengthy discussion about the efficacy of the extension of rights or moral standing, but rather to point out that within a Cartesian dualist ontology the non-human must always remain inferior, and that this precept is embedded within the application of our ethics to the non-human other. It might be helpful to think of moral extensionism as similar to "reform ecology", which is exemplified by confining "all intrinsic value to humans, typically regarding them as the apex of evolution" (Langer, 1990, p. 116). While there have been good arguments made for the intrinsic value of other beings and of the non-human world, it is difficult to see how this might be embodied in practice, in the everyday life of our culture, as a "given", without a radically different ontology. In short, I do not believe that extending moral significance is radical or lasting enough to preserve the other-than-human world from the ravages of Western industrial culture.

In order to qualify as morally significant, beings must run an obstacle course of ever-changing and ever-debated criteria. The first set of criteria, and the most easily recognised, is what may be called the terms of moral significance. The second set of criteria, also constructed and evolving over time, is the definition, or terms, of being human, rather than animal, or nature. This latter set of terms rises from an ontological foundation of dualism that splits and insulates the human from the wider world of being among beings.

The terms of moral significance we have created are always open to revision, and this means that while they can be revised in such a way as to make life better for non-human others, there is no guarantee that the well-being and interests of non-human others deemed morally significant now will trump corporate industrial interests when push comes to shove. Non-humans may be only temporarily or conveniently morally significant, or partially so, depending upon the interests of the dominant culture. So long as they rise from the ontology of human as separate from world, of human as apex, conditions of moral significance must remain unstable for non-humans in important ways. The conditions can be configured or reconfigured so as to justify exceptions or revisions to accommodate current cultural desires, which will always have primacy.

But this human that we have configured as so significant is in fact a construct, the form and conception of which is flexible and which must be culturally managed in order to maintain its own otherness from the animal world. If the status of the human–other relationship remains ambiguous, this may be due in part to the ambiguity of the human itself, a constantly reconfigured imaginal form, based on little other than a series of exclusions and tenuous definitions of difference.

Being (In)Difference

> ... in our culture man has always been the result of a simultaneous division and articulation of the animal and the human, in which one of the two terms of the operation was also what was at stake in it. (Agamben, 2004, p. 92)

Linnaeus, "the founder of modern taxonomy" (Agamben, p. 23), said that he found little to differentiate hominids from other animals, and concluded that a human is an animal that recognises itself as human. Obviously, we could not look to biology to make us human. Man is the animal that knows itself to be man, it is self-aware, if you will. The human then, is configured as an animal with a special something else. For Descartes, it was a rational soul, or mind, that differentiated human from beast. From time to time it has also been tool use, brain size and shape, language, and even morality that formed a boundary between animal and human. And while we may (since Darwin) readily admit to being animals, this does not seem to mean being *with* other animals, in any sort of reciprocal relationship with shared interests.

Would an admission, or even a welcoming of such an animal being within humans suggest that we should "regress", return to the cave, as some critics of environmentalism and deep ecology insist would be the case—as if there are only two alternatives; the kind of world we have and a world of darkness? I think not, and suspect that such questions stem from those who fear the animal, their own beast within, a capitulation to which in any form is anticipated as a loss of self-aware being, and a life stripped bare of all culture, pleasure and intellectual pursuit; a life, to those who fear the animal, that could only be, in the famous words of Thomas Hobbes, "nasty, brutish and short".[6] All this as if indeed Descartes were always completely correct, in that only humans have mind, all that we truly are is mind, and all else, including our bodied being, is Other, to be transcended and at all costs not to be mistaken for the self.

One must admire the longevity of a concept that has proved the most successful exclusionary prescription in Western history. Prior to this, distinctions between animal and human moved around a great deal. Joyce Salisbury (1994), in her book *The Beast Within*, shows how these boundaries shifted significantly even from the early to the late Middle Ages, so that by the late Middle Ages there was even less distinction between the two. Non-human animals became exemplars in moral tales and were even held to account for their behaviour in ways we now think of as applying only to the human, with the same punishments meted out. These included being jailed for viciousness, or hung for murder. Interspecies values are also reflected in a chart of Animal Values from British Law of 991-1016, which shows the fines for the killing of a horse to be 30 shillings, and 1 pound for killing a man (p. 181). Tim Ingold (2000) in *The Perception of the Environment*, cites other early examples of the lack of distinction between species:

> The Romans, for example, classified slaves and cattle, respectively, as *instrumentum genus vocale* and *instrumentum genus semi-vocale*, while Vedic texts, according to Benveniste (1969: 48), have a term *pasu* for animate possessions that admits two varieties, quadrupedal (referring to domestic animals) and bipedal (referring to human slaves). (p. 73)

The blurring of human–animal boundaries in the late Middle Ages had the effect of shifting the definition of humanity away from what one physically is, and onto how one acts. In large part this had to do with countering our bestial behaviour through showing compassion toward other beings, including non-humans. The beast is always within us and we can easily recognise familiar human behaviours in the attributes assigned to define the beast in the *Medieval Bestiary* (Salsibury, 1994). This wild fierceness was symptomatic of the loss of Eden, and compassion meant (re)finding that harmony among beings, where the lion lays down with the lamb.

We work hard to police the boundary we have drawn between human and beast, self and world—and we have to stay on our toes to do so. The boundary, after all, is an idea, or a system of belief. Were this boundary to shift, or even vanish altogether, and along with this shift should we still desire to be moral beings, then accommodations would have to be made, assumptions altered, behaviours reassessed. We would need, in effect, a different ontology. The one we have, which insists on hard boundaries, discrete entities and the primacy of an internal self, or I, may be remarkable for its longevity, but it also remarkable for the carnage and violence it has encouraged. The entire other-than-human world, along with those humans who still dwell within different lifeworlds, suffer from Western industrial culture's longstanding indifference to their lives and flourishing. Morality still has to do with compassion, and the Cartesian worldview in essence disallows any such engagement with anything outside the disembodied rational mind of the conceptual being we call human.

Is this then, all that we are, or are content to be—in the words of Canadian Northrop Frye—"Cartesian ghosts caught in the machine that we have assumed nature to be" (Frye, 2003, p. 478)? Surely, if we are imagining ourselves, we can come up with something better for both ourselves and for the ecological community of which we are, inescapably, a part.

Humility and Sacrifice

0 son, thou hast not true humility,
The highest virtue, mother of them all. . .
The Holy Grail (Tennyson, 1869)

What then might we be and how should we live? How might reopening ourselves to being a part of the family of animals change our ways of being in the world? In seeking answers to these admittedly daunting questions, I briefly consider some of the ideas of Val Plumwood and anthropologist Tim Ingold, also along the way drawing in some of Peter Steeve's thoughts. This leads me to Maurice Merleau-Ponty's notions of Flesh and intertwining and, eventually, Giorgio Agamben's idea of a post-human world.

Much of this part of the discussion reflects on the human—non-human relationship via the question of food, of eating and being eaten. This is the most obvious use made of non-human animals by humans, and the contemporary industrial meat machine views non-human beings as units, moving meat, who are denied any self or agency whatsoever, manipulated and killed in ways and in numbers that most consumers would find incomprehensible. Meanwhile, in a twenty-first century parody of Wells' Dr. Moreau, scientists create legless pigs, featherless chickens and other monsters of their fancy—the better to eat you, my dear.

But the real reason that being (as) food seems to lie at the centre of this debate is fear—the dual fears of death and of being eaten. These fears seem particularly strong in Western culture, or perhaps this is because in this culture we have done so much to deny, disarm, cheat, or otherwise manipulate natural systems with a goal of eventually never experiencing death, and certainly of never being eaten. From Plato's (and of course later, Descartes') denial of being the body, through the Christian idea of the Resurrection of the body, to cryogenics and the death-avoidance industries of the contemporary world, we must both defeat death and save ourselves from being food, whatever the cost.

Plumwood (2003), in her paper *Animals and Ecology: Towards a Better Integration*, argues against a problematic moral extensionism and for a position that she calls Ecological Animalism. In taking up the problem of how to better integrate human and non-animals, she does not, in her own words, seek a "unified theory", but "a more contextual and less conflicted" one that takes into consideration the lifeworlds of non-Western human cultures (p. 3). She begins by stressing points of overlap and continuity among human and non-human animals—in particular, the reality of being food. Life feeds on life, and we humans are in no way exempt from this system, although we have configured our contemporary lives in such a way as to conceive of ourselves as consumers who are never consumed. We cannot live without consuming, or making use of, others and within this world humans too are food for bacteria and hungry cougar in the same way a potato is food for the human. Being eaten remains a central terror within Western culture, but, as Plumwood reminds us, we are all food, part of "a chain of reciprocity" (p. 4). She also, importantly, reminds us that in this sense of participating in an ecological economy of mutual use, being perceived as food is neither pejorative, nor reductionist. Part of dwelling within such a system of reciprocity is that all life, all living beings, are also always more than food, more than their potential use to another being. The idea is that in acknowledging this, as well as its corollary that we are all part of an ecological community, dualisms are overcome (at least in part), with all life becoming suitable for respect and care.

Although I quite like this idea, as H. Peter Steeves (1999) points out in his essay *They Say Animals Can Smell Fear*, it is human beings who are doing most of the consuming and non-humans most of the sacrificing; and it is difficult to see how this might change any time soon, with eight billion humans and counting in the world.

Steeves himself approaches the question of animal–human engagement through the embodied relational, exploring that to *know* is more than information about, it is encounter, an intimate reciprocity of being-with, an embodied and acute recognition of the other who may also at times be a dangerous other, an other that could kill and eat you as easily as you could kill and eat him or her. A non-human other can also be a friend, as easily as can a human. *Knowing* what it is to be-with is also expressed in the simple experience of reaching out to stroke, and the cat arching her back to meet and reciprocate—the reciprocity of bodily being-with.

It is to this intimacy, this *knowing*, which he turns in search of an alternative to Cartesian dualism and its abuses. But if what is required is experience—real, embodied experience—in order to *know*, then the likelihood of a wide embrace of such relationality seems, well, unlikely. In the city world of today, where most of us spend our lives, we have so little to do with living non-human others, except for "pets", that meaningful encounters of the sort required to come to *know* others are at a premium, to say the least. And while our world may be shared with other beings, I am struck by the cultural lack of awareness of these others. A colleague recently told me that on asking the question of his college class "is a bird an object?" the students without exception responded that yes, indeed, a bird is an object.

One could say that no relationship could be more intimate than that of eating another, but the backgrounding in any meaningful, lived, way of the origin of what, or whom, one is consuming means that while one may have the *information* that what one is consuming is another being, one cannot really *know* this to be so. Shrink-wrapped bits of muscle in no way resemble, for example, my equine friend, Desi—although a meat-eating human could, conceivably, eat a friend (of any mammalian species) without knowing it. As Steeves put it, "a muscle is a muscle, is a muscle"

Fig. 2 Dogs readied to be slaughtered for food (credit: Courtesy of http://www.all-creatures.org/)

(p. 165). Would a *known* being become morally significant? Would it then follow, if this were the case, that one could not, would not, or should not, eat this other being?

Meat-eating humans frequently point out to those of us who eschew animal flesh that if we consume carrots and potatoes, we are taking the life of several plants. Although much of the perceptual world of plants can only be beyond our ken, we recognise increasing evidence that plants communicate with one another and have more complex lives than we previously thought.[7] While not undertaking a thorough discussion of non-animal life here, I will remark that this readily leads one to question whether the broad acceptance of the complete ownership and manipulation of plant-based life forms for human consumption is so very different from, or more ethically sound than, the manipulation of any other life forms. I am not asserting that it is, but I am interested in the question, since it is the entire living world that we objectify, not just non-human animals.

Only if all beings are somehow morally significant, if life itself is morally significant, and then only if it is given that life must feed on life, and *this* is also morally significant, can the unbearable intimacy of *knowing* we are feeding on life and particular others become an acceptable part of our daily awareness. And along with this must come awareness of the cost—of the price paid by beings who become food for other beings, of the weight of humans upon the earth, and of ourselves as consumed as well as consuming.

Compassion and humility can arise from this shared being in the world, the knowing that like everything living, I too am finite, I too am food, to be used in turn by others. Inequalities, if we want to perceive them as such, certainly exist. Differences exist. A cougar is much more equal to the task of surviving in the forest on her own than am I. I am more likely to be eaten in this scenario than is the cougar, since I am poorly equipped for the task of survival by virtue of being differently bodied and very differently cultured. I am not the cougar's equal on these terms, nor is she mine, even as we share the same world and the equality of being mammals similarly bodied, with similar needs.

As an example of the kind of awareness that I am aiming at expressing here, anthropologist Tim Ingold (2000), in *The Perception of the Environment*, speaks of the relations between hunter and hunted in the alternate lifeworlds of hunter-gatherer cultures, such as the Kyukon, the Cree and the Ojibwa. While it is very difficult to see outside our own lifeworld when it comes to such different human ontologies, Ingold explains it as an entire "sentient ecology" (p. 25), maintaining that the hunter–prey relationship within this lifeworld is "essentially non-violent", a "relationship of trust", as opposed to domination. While this assertion may raise the eyebrows of some, it is important to remember, as Ingold points out, that to judge these cultures from our position well outside such a lifeworld is "profoundly arrogant. It is to accord priority to the Western metaphysics of the alienation of humanity from nature" (p. 76). To those who might assume that non-Western cultures are being viewed romantically, or idealised here, I point to Ingold's comment above, and add that such accusations can be understood as offering an easy way out of the discomfort of challenging one's own worldview. While we experience a world according to an ontology that alienates us from that world, and frame our ethics accordingly, we cannot defend, with any certainty, superiority for either in this case.

Neither I, nor Ingold, are proposing that we should, or could in any sustainable way, take up hunting and gathering. Rather, the proposition is that if we acknowledge the value of learning, and want to learn to *know* and include the other-than-human in ways we have lost within our own culture, we would be best served to seek out those who dwell within complex reciprocal relations in a more respectful and sensitive manner; in short, those dwelling within an alternate lifeworld to ours. We are fortunate that opportunities to do so still exist, since we need help in escaping our perceptual framing so as to see anew, to allow other possibilities for being in the world to present themselves to us. In Ingold's words,

> Were we to rewrite the history of human-animal relations, taking this condition of active engagement, of being in the world, as our starting point, we might speak of it as a history of human *concern* with animals, insofar as this notion conveys a caring, attentive regard, a being-with. And I am suggesting that those who are "with" animals in their day-to-day lives, most notably hunters and herdsmen, can offer some of the best possible indications of how we might proceed. (p. 76)

The world is still "inhabited by beings of manifold kinds", (p. 5) only one kind of which is human, and I contend that while our relations with other beings within Western culture may exist for the most part as background, they nonetheless inform who we are. An ecology is, by definition, a web of intertwined, interdependent, relationships, no matter how we might otherwise think of it. This intertwining is not something we foreground, but is, rather like breathing, not an optional aspect of being alive.

When it comes to the human–world relationship, other, perhaps differently challenging, thoughts are offered by philosopher Maurice Merleau-Ponty. To fully explore the possibilities of these would tax the limits of this essay, but I offer the following to this discussion.

According to Monika Langer's (1990) interpretation of Merleau-Ponty's ideas, "value and signification must be considered intrinsic determinations of the organism and these are accessible only to a new mode of comprehension" (p. 122). This "new mode of comprehension" involves awareness that the perceiver, rather than being a discrete and disembodied consciousness, constitutes a world together *with* the perceived. One can never be a detached observer, outside the world. We are, instead, "inextricably intertwined within a world we inhabit" (p. 124).

In his later work, Merleau-Ponty introduced the idea of "flesh", which offers the possibility for a radical change in human–world relations. Never thoroughly developed by him due to his early death, and somewhat difficult to grasp within the dualist framing of our thinking, this idea of flesh is described by Isis Brook (2005):

> This idea, flesh, arises from an attempt to picture reality in a way that reflects how it is; not just for us, as experiencing subjects seen as somehow separate from an objective world, but also for the world... [Flesh is] a way of reconfiguring the joining of subject and object such that they are not joined because they are the held and the holding in our consciousness, but because both are grounded, prior to any conceptual division, in the same stuff. (p. 353, 356)

Flesh is not the stuff of bodies, at least not in particular; nor is it the stuff of the earth, except in part, neither is it the stuff of mind, as we know it. Rather, it is

> ...the total embrace in which what we think of as us and what we think of as the world are held... .a sharing that breaks down a solitary self-enclosedness, both between me and other humans and between me and non-humans, and even between me and the inanimate. (p. 357, 361)

In this we are not one with things and beings in a mushy, indeterminate kind of way, nor in a chunky I-am-the-mountain-and-the-mountain-is-me kind of way. Rather, we *are* deeply intertwined relationship, within a sharing so intimate that we, or everything, must always be affected by everything else. Within this idea of flesh, then, lies a compelling argument for responsible relationship, as evidenced in caring action. "The world stops being a mass of material and opens up" (Langer, p. 123) revealing "a kinship and participation such that it is impossible to say that nature ends here and that man or expression starts here" (p. 126).

So far, much of what has been presented here suggests that radical change of the sort we need means entering into an immediate engagement with the world, face and feet first, owning up to who we are and what we do in ways that I think must transcend any concept of ethics as we presently understand them. Indeed, what we may need to consider could be thought of as a kind of post-human and post-ethics being-in-the-world, since our current ethical system is rooted in a Cartesian dualism, which one hopes we would be abandoning. Were we not to relinquish it, we would likely never be in a position to offer either care or trust to, or embody the responsibility of, a reciprocal relationship with other beings in the world.

Perhaps we would have a new ethics that would not be ethics as we know it, or a world somehow beyond ethics as we are presently able to think it. Perhaps we ourselves would become something outside what we can now think from where we are and who we allow ourselves to be.

Outside the Machine

> The disappearance of Man at the end of History is not a cosmic catastrophe: the natural World remains what it has been from all eternity. And it is not a biological catastrophe either: Man remains alive as animal in harmony with Nature or given Being. What disappears is Man properly so-called—that is, Action negating the given, and Error, or, in general, the Subject opposed to the Object. (Kojève as cited in Agamben, 2004, p. 6)

Giorgio Agamben begins *The Open* with an image of the animal-headed righteous on the last day of history taken from the vision of Ezekial in a thirteenth century Hebrew bible. He proposes that this suggests that on the last day of humanity, "the relations between men and animals will take on a new form, and that man himself will be reconciled with his animal nature" (p. 3).

Agamben stops short of discussing what this reconciliation might propose in terms of changes to life as it is or might be lived, systems of morality and technology. He brings forward Benjamin's "between" as a space where man neither masters nature, nor nature man. It is rather the relationship itself that is mastered through a kind of balance, or stasis, being achieved by the stopping of the anthropological machine. He explains it thus:

> The anthropological machine no longer articulates nature and man in order to produce the human through the suspension and capture of the inhuman. The machine is, so to speak, stopped. . .and, in the reciprocal suspension of the two terms, something for which we perhaps have no name and which is neither animal nor man settles in between nature and humanity and holds itself in the mastered relation. . . (p. 83)

He tells us that attempting to "trace the no longer human or animal contours of a new creation. . .would run the risk of being equally as mythological as the other" (p. 92). Although he invokes the image of the animal-headed righteous as a clue, this being is mystery. It *is,* without definition, and whether embracing morality or ethics remains unknown, since these are part and parcel of the human being as we now know it, as contained within this being as articulated by the anthropological machine, which we cannot see beyond without in fact being beyond. What Agamben does evoke is a state of harmony, of blessedness. The animal-headed are the righteous; they are the blessed.

Virtue

> With the exception of a few saints, we are not pure. . . But I still struggle. I still yearn. And I am comforted by the Taoist aphorism, "water that is too pure contains no fish."
> (Patsy Hallen, 2003, p. 60)

We see that it is our dualist ontology that supports a purely human-centric restructuring of life, world, other and any moral terms under which we engage these. This structure tells us that it is right and good to undertake the betterment of man, the human, while non-humans are relegated to objects at the disposal of the human in this quest. Our ontology informs both the terms of enquiry and our outcomes. Making things better for humans, in this ontology, assumes that humans are always pitted against an uncaring and brutal nature, of which they are not a part, and over which they must extend and exert control. Everything that is not included in the human is a part of this nature. In the pursuit of making things better for humans, everything that is not human can and should be utilised to the extent of human imagination and creativity. And because this nature and these beings are not human, one need exercise no restraint, none of the care or compassion one calls upon in dealing with humans; after all, why have this kind of compassion for an object whose ultimate raison d'etre is to be of use to the human? This quest to make things better for humans is framed as heroic, virtuous and highly moral as we gird our loins and go forth against the forces of nature.

In practice, undertaking the quest to make life better for humans within the confines of our current ontology gives us the world we increasingly have—a world of clearcut forests, poisoned waters and air, disappearing species, devastating wars over "resources", factory farms, mechanised killing, GM crops, legless pigs, and laboratories filled with caged beings forever at our disposal in any kind of research that takes our fancy. This, then, is what the war against nature, that we believe is also the war against death for the human, looks like, and it will never stop, because there will never be victory.

Artists who participate in this war by adopting the methodologies of control, embodying in their practices typical cultural dogma so far as the objectification and use of non-humans, are not critiquing the system, but supporting it. There is nothing truly radical, moral, or vital about such works. They are tucked into a well-worn ideological niche, supporting the status quo and primarily signifying human self-indulgence. Artists may present themselves as foregrounding issues and problems, but must reflect on whether in doing so they themselves risk becoming a part of the problem.

It may be objected that one should not have expectations of art and artists *doing* something, potentially raising the spectre of an art that may be *for* something other than its own sake. I am not advocating a position that all art must be enrolled in the service of a quest to make change, but art does do something in the world, it has effect, and this effect can, I maintain, be profound. We are engaged, opened, by art, so that an entire experience is offered, in a way that can by-pass the cognitive and the familiar framing of belief and reason we otherwise apply. Artist David Haley (2003) reminds us of Einstein's Theory of Relativity, which "made it apparent that observers inextricably participate in the scenes they are viewing" (p. 146). In the words of Merleau-Ponty (1964), "...it becomes impossible to distinguish between what sees and what is seen" (p. 167) and of Langer (1990), "the seer is inherently part of the seen and coils back over it in an embrace that lets a new meaning emerge" (p. 126).

Maybe in this way, transgenic works could be a potentially transforming blow to the gut, a wake-up call. But for the most part, this would be a call heard only by a few. The cost of making such works is far too high and I very much doubt their ability to draw us into any truly meaningful experiences and conversations toward bringing about change. Real creative challenge lies in finding new ways to do and be. It is far more radical to embody in one's life, and to have one's works embody and therefore present the shapes of a possible, an alternate, lifeworld for the see-er (or viewer) to enter.

David Haley (2003) presents his definition for ecologically sensitive art as "worthy of the name implied by the root of the word art. *Rta*, an ancient term from the *Rig Vedas*, [which] refers to the virtuous, continuing creation of the cosmos" (p. 143).

In virtue ethics, stepping up, or speaking up, to help another is a moral imperative, based on extending care and compassion. Rather than a system of ethics based on a detached reason and duty, I think of this as relational, as more a matter of heart, empathy and engagement. Compassion, as I presented earlier in the discussion, is what we were, in the late Middle Ages, called upon to have in regard to other living beings in the world. We are here together, intimately intertwined as bodied beings together in an embrace with other bodied beings, eating and being eaten, part of a continuous ecological community. Compassion, virtue, care, empathy, love, respect, all make bearable this intimacy we in particular find so terrible and unbearable as to live in a perpetual internal world of separation and denial.

It is through our senses that we can open ourselves to empathy and compassion. It is through the engaged senses that we are opened by art. Ingold tells us that knowledge, as opposed to information, is arrived at through being *shown* [my emphasis] something—literally shown—so that we engage with it (Ingold, 2000). We come to *know*. And Merleau-Ponty (1964), in his essay *Eye and Mind*, tells us, "Immersed in the visible by his body, itself visible, the see-er does not appropriate what he sees; he merely approaches it by looking, he opens himself to the world" (p. 162).

We are not, as Patsy Hallen points out, pure, and purity has become a problematic term. But we may partake of virtue, as an intertwined part of the virtuous, continuous, creation of the cosmos, in part through the eyes of the heart.

> CaNte Ista. Those are the words used to describe a way of seeing that is good and true. . .the true place of the heart is—in that circle where all things are connected. . . CaNte Ista, through the eye of the heart.
> Joseph Bruchac (2001, p. 85)

Notes

1. Various sources, including: popular culture, Alfred Lord Tennyson (1869) *The Holy Grail* and Bryant, Nigel (2006) *The Legend of the Grail.*
2. Images of this work may be found online here: http://visualarts.walkerart.org/oracles/details.wac?id=2227&title=Works&style=images
3. For example, see the works of artist Brandon Belangee on malformed amphibians here: http://greenmuseum.org/content/work_index/img_id-371__prev_size-0__artist_id-19__work_id-86.html
4. For Kac's comments and more on this work, please see his website: http://www.ekac.org and the page on the GFP bunny: http://www.ekac.org/ gfpbunny.html
5. See CBC news archive online: http://archives.cbc.ca/arts_entertainment/visual_ arts/clips/300-1604/
6. Hobbes famously declared in *Leviathan*, that life in "a state of nature" (prior to the formation of a civil society and overseeing state) was "solitary, poor, nasty, brutish and short".
7. There is some recent research enquiring into plant neurophysiology, communications, emotions and interactions with non-plant beings as well as other plants. See for example Brian J. Ford

(1999) *Sensitive Souls: Senses and communications in plants, animals and microbes*. London, UK: Little, Brown and Co., (2006) Baluska, Frantisek; Mancuso, Stefano; Volkmann, Dieter (Eds.) *Communication in Plants: neuronal aspects of plant life*. Springer.

References

Agamben, G. (2004). *The open—man and animal*. Stanford, California: Stanford University Press.

Baker, S. & Gigliotti, C. (2006). We have always been transgenic. *AI & Society 20* (1): 35–48.

Brook, I. (2005). Can Merleau-Ponty's notion of 'flesh' inform or even transform environmental thinking? *Environmental Values 14* (3): 353–62.

Bruchac, J. (2001). The place of the heart. *Parabola: the Journal of the Society for the Study of Myth and Tradition 26* (4): 85–89.

Bryant, N. (2006). *The legend of the grail*. Woodbridge, UK: Boydell & Brewer

Carruthers, B. (2006). *Dreaming human being animal: A shared ontology*. Unpublished paper, Lancaster University, Lancashire, UK.

Ford, B. J. (1999). *Sensitive souls: Senses and communication in plants, animals and microbes*. London, UK: Little, Brown, and Co.

Frye, N. (2003). Haunted by lack of ghosts. In O'Grady & Staines (Eds.), *Collected Works of Northrop Frye* (Vol.12) *Northrop Frye on Canada*. Toronto,Canada: University of Toronto Press.

Greene, J. Personal website: *The Moral Significance of Moral Psychology*. Retrieved from http://www.wjh.harvard.edu/~jgreene

Haley, D. (2003). Species nova [to see anew]. *Ethics and the Environment 8* (1): 143–150.

Hallen, P. (2003). The art of impurity. *Ethics and the Environment. 8* (1): 57–60.

Hobbes, T. (1660). *The Leviathan, Chapter VIII: Of The Natural Condition Of Mankind As Concerning Their Felicity And Misery* [Electronic Version] Retrieved May 2008 URL: http://oregonstate.edu/instruct/phl302/texts/hobbes/leviathan-c.html

Ingold, T. (2000) *The perception of the environment: Essays in livelihood, dwelling and skill*. London, UK: Routledge.

Kac, E. Artist's website: http://www.ekac.org

Langer, M. (1990) Merleau-Ponty and deep ecology. In G. A. Johnson & M. B. Smith (Ed), *Ontology and alterity in Merleau-Ponty*. Evanston, Illinois: Northwestern University Press.

Merleau-Ponty, M. (1964). Eye and mind. In *The primacy of perception*. Evanston, Illinois: Northwestern University Press.

Ping, H. Y. (2007). Artist's statement, Vancouver Art Gallery press release. April 17 2007 [Electronic version] Retrieved May 2008. http://vanartgallery.bc.ca/press_releases/pdf/Theatreoftheworld_PR.pdf

Plumwood, V. (2002). *Environmental culture: The ecological crisis of reason*. London, UK: Routledge.

Plumwood, V. (2003). *Animals and ecology: Towards a better integration*. [Electronic version] Retrieved February 2008 from Department of Social and Political Theory, Research School of Social Sciences, Australian National University http://hdl.handle.net/1885/41767

Salisbury, J. (1994). *The beast within: Animals in the Middle Ages*. London, UK: Routledge.

Steeves, P. (1999). They say animals can smell fear. In H. P. Steeves (Ed.), *Animal Others: On ethics, ontology and animal life* (pp. 173–178) Albany, New York USA: SUNY Press.

Tennyson, A. (1869). *The holy grail*. [Electronic version] Retrieved May 2008 from http://www.lib.rochester.edu/CAMELOT/idyl-grl.htm

Part II

Leonardo's Choice: The Ethics of Artists Working with Genetic Technologies

Carol Gigliotti

Abstract Working with current methodologies of art, biology, and genetic technologies, the stated aims of artists working in this area include attempts both to critique the implications and outcomes of genetic technologies and to forge a new art practice involved in creating living beings using those technologies. It is this last ambition, the development of a new art practice involved in creating living beings, this essay will particularly take to task by questioning the ethics of that goal and the uses of biotechnology in reaching it.

Keywords Animals · Biotechnology · Ethics · Animal rights · Bioart

Although its source may well be apocryphal, the following quote has been attributed to Leonardo da Vinci, "I have from an early age abjured the use of meat, and the time will come when men such as I will look upon the murder of animals as they now look upon the murder of men" (Preece 2002, 93).[1] While da Vinci may not have said exactly that, his compassion for animals is well documented in his notebooks and several sources cite his vegetarianism (Clark 1977, 45). The notebooks contain numerous references to his shock and disdain for man's deliberate choice in abusing the other animals, many of whom provide him with food and labor:

> Of candles made of beeswax
> [The bees] give light to divine service—and for this they are destroyed.
> Of asses
> Here the hardest labor is repaid by hunger and thirst, pain and blows, goads and curses, and loud abuse.
> Of a fish served with its roe
> Endless generations of fish will be lost because of the death of this pregnant one.
> Of slaughtered oxen
> Behold—the lords of great estates have killed their own laborers
> (da Vinci, as quoted in Kellen 1971, 78–79).

C. Gigliotti (✉)
Dynamic Media and Critical and Cultural Studies, Emily Carr University, Vancouver, BC, Canada
e-mail: gigliotte@ecuad.ca

C. Gigliotti (ed.), *Leonardo's Choice*, DOI 10.1007/978-90-481-2479-4_4,

Vasari in describing Leonardo's exemplary character tells us how Leonardo's compassion for animals was such that he bought caged birds merely to set them free (Turner, 1993, 62). Vasari, however, has a very young Leonardo composing a painting of a monster (possibly a Medusa) modeled on dead "lizards, grasshoppers, serpents, butterflies, locusts, bats, and other strange animals of the kind" he had brought to his room. An additional story by Vasari of a much older Leonardo in Rome, describes how the artist spent his time, much to the chagrin of Pope Leo X:

> To the back of a very odd-looking lizard that was found by the gardener of the Belvedere he attached with a mixture of quicksilver some wings, made from the scales stripped from other lizards, which quivered as it walked along. Then, after he had given it eyes, horns and a beard he tamed the creature, and keeping it in a box, he used to show it to friends and frighten the life out of them. (as cited in Turner, 62)

Much of Vasari's information about Leonardo is second hand and some of it is more than likely to have been invented (Turner, 55–68), but, together with Leonardo's notebooks, these tales of Leonardo give us some appreciation of the conflicting priorities that may have existed in Leonardo's attitudes toward animals. He was compassionate toward the plight of animals used solely for human purposes, while at times using animals himself for his own purposes. A painter, a scientist, a naturalist, a technologist, a prophet, Leonardo was both an exemplar of his time and ahead of it.

These preoccupations of Leonardo are reflective of ethical issues brought up by artists working with genetic technologies involving bacteria, plants, and animals. Many of these artists are seen or see themselves as descendants of Leonardo and his abilities to cross the disciplines of art and science. It is no accident that the leading publication of such crossover activity for the last 36 years has been the influential *Leonardo: Journal of the International Society of the Arts, Science and Technology*. Held (2002) Curator of Gene(sis): Contemporary Art Explores Human Genomics, an exhibit traveling from 2002 to 2004, introduces the exhibit in this way:

> As artists take up the tools and materials of genetic and genomic research, their experimental reflections are changing our notions of artistic practice. Many artists function as researchers, engaged in nonhypothesis-driven, open-ended investigations. Their studios are laboratories for this experience-based inquiry. In addition, artists such as Eduardo Kac, Critical Art Ensemble, Paul Vanouse, Joe Davis, Tissue Art and Culture, Jill Reynolds, Iñigo Manglano-Ovalle, and Justine Cooper, to name just a handful, regularly use biological materials in their work. In some cases, artists are creating new life forms and releasing them into the environment. (p. 4)

Working with current methodologies of art, biology, and genetic technologies, the stated aims of artists working in this area include attempts both to critique the implications and outcomes of genetic technologies and to forge a new art practice involved in creating living beings using those technologies. It is this last ambition, the development of a new art practice involved in creating living beings, that this essay will particularly take to task by questioning the ethics of that goal and the uses of biotechnology in reaching it. Doing so has proven to be a contentious activity, involving as it does discourse about both artistic and scientific practice, each bringing along its own linguistic and conceptual assumptions, metaphorical, and otherwise. Participants in this debate have come from what some may consider to be both "inside" and "outside" the art world: the artists themselves, curators, critics, art theorists, philosophers, cultural and political critics, theologians and scientists, and the general public. Many of these participants, wherever their disciplinary reference-point might be, see art as one of the few environments left where uncensored thought is not only condoned, but also encouraged. As Efimova (2003), Associate Curator of the Berkeley Art Museum presentation of Gene(sis), says in her introductory essay, "…experimental art remains one of the few enclaves where imaginative, impractical, non-mundane thinking is still tolerated" (p. 1).

Is Thinking in Art Always Radical?

The idea that art is a last bastion for radical thinking has frequently been used as a rationale for this new art practice. In an essay included in *The Eighth Day: The Transgenic Work of Eduardo Kac*, Machado (2003) argues this point. He contends that critiques of biotechnologies tend to take a "conservative bias" or "even dogmatic interdictions of religious order." He sees "The more experimental and much less conformist sphere of art—with its emphasis on creation, by means of genetic engineering, of works which are simply beautiful, not utilitarian or potentially profit making…" (p. 94) as conducive to more sophisticated discussion about genetics as well as science and technology in general. He adds, however, one of the benefits of Kac's work, *The Eighth Day*, is to develop science and technology "away from the unproductive dichotomy of good and bad, right or wrong, and toward a confrontation of the whole of its complexity" (p. 95).

Two assumptions are at work in these statements by Machado and in much of the writing by both artists and critics about artists working with genetic technologies. The first assumption is that thinking in art is consistently experimental and non-conformist. While one may assume that to be true, based on some historical precedents, the assumption does not insure that all thinking emerging from art is necessarily radical. The second assumption concerns the idea that a confrontation with the complexity of a topic or issue precludes the necessity of confronting ethical choices embedded in that complexity. On the contrary, one of the main reasons for understanding complexity is the insight it may offer to ethical choice. I highlight the limits of these two assumptions because their uncritical acceptance muddies the discussion of two aspects pivotal in discourse surrounding the ethics and aesthetics of a new art practice involving living beings.

These two aesthetic aspects are most clearly delineated by art critic Bureaud (2002) in an issue of *artpress* that included seven essays and a "dossier" on "art bio(techno)logique." She summarizes the discourse around "biological art" as ranging from technical practices involving biological and biotechnological methods to contexts dealing with related human-centered social, political, environmental, and ethical issues to human-based perspectives on immortality. Bureaud makes the point, however, that while these perspectives are essential to understanding and evaluating these works, "…analysis often fails to get as far as their artistic or aesthetic aspects" (p. 38). While this implies that artistic or aesthetic aspects will be discussed separately from the ethical aspects of this work, Bureaud describes, among the seven aesthetic aspects she has observed in this growing body of work, two characteristics or orientations one can only see as inextricably entwined with ethical perspectives.

The first is an approach she calls the "anti-anthropocentric art of the continuum" including the semi-living (as in Tissue Culture and Art Projects) and transgenic organisms (as in projects by Eduardo Kac). She describes this approach as emphasizing the "permeability of the frontiers between species, the continuity that goes from the non-living to the different degrees of complexity in life forms" (p. 38). Catts and Zurr (2003 & 2004), artists involved in Tissue Culture and Art Projects, utilize the idea of "a continuum of life" as oppositional to an anthropocentric worldview:

> . . . we argue that the underlying problem concerned with the manipulation of life is rooted in the perceptions of humans as a separated and privileged life form, a perception inherited in the West from the Judo-Christian-and Classical worldviews. This anthropocentricism is distorting society's ability to cope with the expanding scientific knowledge of life. Further this cultural barrier in the continuum of life between the human and other living systems prejudices decisions about manipulations of living systems. (p. 2)

This confusing statement can be read as both critique and support of the historic values of anthropocentrism leading to the instrumentation and destruction of living systems. Catts and Zurr, themselves, point out that much of what they are saying and doing is contradictory and this statement and much of the rest of the essay, from which it is excerpted, is evidence of this. Although they claim that their hope is to challenge "long held beliefs" about the perceived barrier between humans and other living beings, they see their involvement in actually manipulating life as "highlighting the inconsistency of the still prevalent view of the dominion of man" (p. 3).

They acknowledge the paradoxical quality of their position:

> . . .on one hand we attempt to break down specism and make humans part of a broader continuum. On the other hand, we artists-humans, are using (abusing?) our more privileged position to technically manipulate an aesthetic experiment. (p. 17)

They insist, however, ". . .only when humans realize that they are a part of the continuum of life will manipulating life not be as alarming as it now seems" (p. 17). The absurdity of this claim is hard to miss. Humans have been manipulating animal life with impunity for thousands of years. Most do not find it alarming, but customary. If, as Zurr and Catts claim, their goal is to encourage people to understand the distortions a human-centered view causes in recognizing the continuum of life, more manipulation of life forms will most certainly not contribute to that project, but only serve to reinforce it.

Kac (2000) references an anti-anthropocentric approach in his essay, "GFP Bunny:"

> Rather than accepting the move from the complexity of life processes to genetics, transgenic art gives emphasis to the social existence of organisms, and thus highlights the evolutionary continuum of physiological and behavioral characteristics between the species. (p. 111)

Fig. 1 Rabbit with skin removed (credit: none)

Inherent contradictions appear when reading these quotes in the context of Kac's transgenic work and his stated goals of "...a new art form based on the use of genetic engineering to transfer natural or synthetic genes to an organism, to create unique living beings" (p. 101). Others have questioned these contradictions in essays on Kac's work. Hayles (2000) asks about Kac's Gene(sis):

> Does Kac's intervention in the genetic sequences of bacteria contest the notion that humans have dominion or reinforce it? The ambiguity inheres in any artistic practice that uses the tool of the master to gain perspective on the master's house. (p. 86)

Hayles sees the usefulness this approach might have for imagining that same drive for domination and control executed upon the future of the human. But what of the animals who currently exist under that drive? Seen through the lens of transgenic art, what future can we imagine for them?

Baker (2003), in one of the most thoughtful essays on Kac's work to date, enlists Derrida's investigations of the human responsibility to the non-human animal to help in understanding Kac's work. Derrida (2002), in "The animal that therefore I am," relates how in the last 200 years

> ...the traditional forms of treatment of the animal have been turned upside down by the joint developments of zoological, ethological, biological, and genetic forms of knowledge and the always inseparable techniques of human intervention with respect to their object, the transformation of the actual object, its milieu, its world, namely the living animal. (p. 394)

Derrida describes this as "violence that some would compare to the worst cases of genocide." While Kac judges the procedures he uses to be safe because they have been regularly employed on mice and rabbits since 1980 and 1985, respectively, Baker (2003) says, "...that is precisely the technology that has led to an increase in the numbers of animals currently subjected to laboratory experiments" (p. 36).

In fact, the best estimate of current use of animals used in research in the US is 20 million, and about 2 million in Canada (Mukerjee, 1997). According to more recent sources, however, worldwide animal use was estimated to be between 60 and 85

Fig. 2 Dead animal refrigerator (credit: ©Brian Gunn/IAAPEA)

million animals in the early 1990s (Rowan, 1995). And though the use of animals in experimentation has decreased slightly over the last 40 years due to the diligence and commitment of a vast network of animal welfare and animals rights organizations, "...the impact of genetic engineering on animal use should be carefully monitored, given its potential to reverse the decreases in animal use seen during the 1980s and 1990s" (Salem and Rowan, 2003).

How, then, do these artists see their goal of a new art practice of creating life forms as part of a world view that is anti-anthropocentric, when in fact it continues along the very traditional, conformist, and conservative paths along which are littered the bodies and lives of millions of animals?

Anthropocentrism, Ecocentrism, and Animal Life

A more thorough examination of the original ideas and goals of an anti-anthropocentric worldview and associated sciences, rather than supporting the new art practice of creating life forms, instead allows the weakness of these artists' arguments and practice to come into view. In addition, in at least these instances, standard ideas about the dominant role played by humans are being reinforced while the truly radical notions of bio-centricism are finding support in other places.

While Kac references H.R. Mantura, and Zurr and Catts mention the work of Lynn Marguilas and James Lovelock, the wealth of material on environmental ethics from which ideas about bio-centric worldviews emerge, both in philosophy and practical ecology, have increased tremendously in the last 30 years. An important component of these truly radical worldviews exists in thinking of the telos of an organism, in the Aristotelian sense of the term, and can be roughly understood as the fulfilled state or end or goal of the organism. A helpful approach is to concentrate on the distinction between instrumental value, which is what we traditionally see the natural world as possessing, and intrinsic value. Taylor (1986) in his book, *Respect for Nature: A Theory of Environmental Ethics* develops a bio-centered or life-centered, as opposed to anthropocentric or human-centered, environmental ethics, as well as arguing that all living organisms possess intrinsic worth. What Taylor and others like him are advocating is a "...world order on our planet in which human civilization is brought into harmony with nature" (308). This goal, however, is built upon cultures in which "...each carries on with its way of life within the constraints of the human ethics of respect for persons" (308).

Unlike some arguments based on a "deep ecology" in which a conception of individual organisms having inherent worth is not included, this understanding of environmental ethics makes integral links to necessary changes in social, political, and economic justice on a planetary scale. These changes would include the elimination of the use of all sentient beings in any form of experimentation. As Taylor admits, these changes would require a profound moral reorientation, and the first step would be an "inner change in our moral beliefs and commitments." Respect for and consideration of the intrinsic worth of individuals of any species, far from being disconnected to these changes in our moral commitments to the human species, is fundamental to altering current compulsions for ways of life inherently devastating to much of the planet and its inhabitants. Shiva (2000) argues,

> The emerging trends in global trade and technology work inherently against justice and ecological sustainability. They threaten to create a new era of bio-imperialism, built on the impoverishment of the Third World and the biosphere. (p. 25)

These trends threaten both ecological and cultural diversity. Shiva cites two root causes for the West's adherence to these obviously negative compulsions:

> The first arises from the "empty-earth" paradigm of colonization, which assumes that ecosystems are empty if not taken over by Western industrial man or his clones. This view threatens other species and other cultures to extinction because it is blind to their existence, their rights and to the impact of the colonizing culture. The second cause is what I have described as monoculture of the mind: the idea that the world is or should be uniform and one-dimensional, that diversity is either disease or deficiency, and monocultures are necessary for the production of more food and economic benefits. (p. 26)

For Shiva, genetic technologies are turning life, or biology, into "capitalism's lastest frontier" (p. 28). She convincingly reasons, however,

> There is . . .one problem with life from the point of view of capital. Life reproduces and multiplies freely. Living organisms self-organize and replicate. Life's renewability is a barrier to commodification. If life has to be commodified, its renewability must be interrupted and arrested. (p. 30)

This is being accomplished by industrial breeding, genetic engineering, and patent and intellectual property rights. Shiva, like Taylor, eloquently calls for a major shift in thought, referring to genetic engineering as based on genetic reductionism:

> A shift in the paradigm of knowledge from a reductionist to a relational approach is necessary for both biological and cultural diversity. A relational view of living systems recognizes the intrinsic worth of all species protects their ecological space and respects their self-organizational, diverse, dynamic, and evolving capacities. (p. 129)

The fundamental goals for manipulating nature, at any level, are always grounded in human interest. Attempts by artists to make the case for biogenetic art involving living matter or beings, by and large, have come from a truly non-radical worldview, one that still posits human beings as the center and rationale of all endeavors. This should not be surprising, since this anthropocentric viewpoint matches that of both past and current genetic research upon which this art is based. Using living nonhumans in experiments or for other human purposes is a generally accepted practice, one not often questioned within the discourses of science or, for that matter, within the discourses of art.[2] The worldview upon which these activities rest sees all of nature as available for human intervention.

Although the stated aims of some artists involved in these discourses are to question the anthropocentric standpoint while at the same time using the tools, methods, and assumed ideologies of biogenetics, the reality of animal use in both biotechnology in general, and in biogenetic art forms specifically, can only highlight in this work a fundamental misunderstanding of what a real commitment to antianthropocentric aims might mean. And it is precisely at this point where artistic practice using living beings falls short of any contribution to those aims.

Theories of Aesthetics and Ethics and the Realities of Animal Life

The second aesthetic aspect Bureaud mentions that is linked to ethical perspectives is "an aesthetics of attention and responsibility." Both Zurr and Catts, and Eduardo Kac, insist that ethical questions are central to their aesthetic concerns. They have discussed at some length how their use of living or semi-living beings is included in that "aesthetics of care."[3] All three, however, also insist that their art practice of creating living beings offers as Kac (2000), claims "important alternatives to the polarizing debate" about genetic engineering, replacing dichotomy with "ambiguity and subtlety" (p. 1). Catts and Zurr (2003, 2004) claim artists involved with this new art form are

> ...manipulating life and "inserting" life into new contexts including the art galleries. By that they are forcing the audience to engage with the living artwork and to share the consequences/responsibilities involved with the manipulation/creation of life for artistic ends. (p. 2)

Some critics, particularly Baker in discussion about Kac's transgenic work, acknowledge the dichotomy between Kac using the techniques of animal experimentation for biotechnological investigations and his emphasis on responsibility that does not "treat an animal as an object, be it an art object or an object of any kind."[4]

Comparing Kac and Derrida, Baker (2003) says,

> Kac, with similarly serious intentions, engages with the animal through techniques that strike many people as meddlesome, invasive and profoundly unethical. (p. 29)

Baker believes that a more detailed reading of the connection between this engagement of Kac's and the concern of Derrida with "how to do philosophy by means of a prolonged and serious meditation on his relationship with the cat who shares his home" (p. 32) is necessary. Baker sees the value of the comparison of the two thinkers in what it may tell us about the "relation of intentions to actions in both ethics and aesthetics" (p. 29). I agree, but for somewhat different reasons. Both Kac and Derrida see the animal, through very thick filters, and both have disconnected their actions from their intentions. The goal for each is not an understanding of the animal for itself, but for a human-centered reason. Derrida sees the animal through the filter of "doing philosophy" and the impossibility of an excessive responsibility to the animal in the present culture, while Kac sees the animal through the filter of an acceptance of the inevitability of a biotechnological future in which his goal of the new art practice of creating living beings makes sense.

Derrida (1991), even as he attempts to outline an "excessive" responsibility that includes the non-human, builds a case for the impossibility of acting on that responsibility:

> [A] pure openness to the [O]ther is impossible—and certainly in this culture. We can no more step out of carnophallogocentrism to some peaceable kingdom than we can step out of metaphysics. Put another way, a violence of a sort, "eating [O]thers," is not an option, but

> a general condition of life, and it would be a dangerous fanaticism (or quietism) to suppose otherwise. The issue is not whether we eat, but how. (p. 115)

Unfortunately, though Derrida's intentions are to deconstruct the general scheme of dominance in Western metaphysics and religion, he disregards something that might be helpful in being able to see animals as other than food, tools, or entertainment. He does not seem to be able to take even the fundamental step of vegetarianism. His reasons for this and arguments against those reasons are the stuff of a much longer essay, but suffice it to say, that a sense of inevitability and an inability to see or imagine outside that scheme of dominance seems to play a role in Derrida's disconnection between his intentions and actions.

While his equivocation toward vegetarianism "seems to rest on the restricted, cautious assessment of its significance; one which would allow vegetarians to buy good conscience on the cheap," he misses the opportunity in a committed ethical vegetarianism, and even better veganism, for what Wood (1999) calls a "motivated possibility of response." As Wood so forcefully puts it,

> Carnophallogocentrism is not a dispensation of being toward which resistance is futile; it is a mutually reinforcing network of powers, schemata of domination, and investments that has to reproduce itself to stay in existence. Vegetarianism is not just about substituting beans for beef; it is—at least potentially—a site of proliferating resistance to that reproduction. (p. 32)

The disconnection between actions and intentions is less overt in the case of Kac. Kac's ambivalence toward the inevitability of a biotechnological future, and in some cases his welcoming of it, clouds our reception of what his inner thoughts might be about his uses of living beings, even as he critiques the Western philosophical canon on which those uses are based. Lestral (2002) touches briefly on why this might be so in general for artists working in this vein:

> We must also take into account the inglorious possibility that these artists are being manipulated—and not necessarily consciously either—by technologists and multinationals; that they are serving to legitimize practices that our cultures otherwise find it hard to accept. (p. 45)

While Lestral backs away from this conclusion, for one that he finds "more reasonable" that sees these artists as doing what artists have always done by exploring practices of their period, he questions if art can play the role of being "critical."

And while Kac's work has been effective in highlighting the complex quality of this society's involvement with biotechnology, his work adds to this complexity in unhelpful ways. His acceptance of a biotechnological future and his use of techniques that objectify the animal frustrate responses seeking to confront ethical choices embedded in that complexity and also frustrate imagining alternatives. Put another way, one cannot say they object to treating an animal as an object of any kind, and then use the very techniques of objectification to create an animal for the purpose of continuing this objectification in the form of a new art practice. What value does "confronting complexity" have if it obscures insight into a "possibility of response?"

Catt and Zurr's work suffers from the same burden, though they appropriate arguments from two of the most important animal-rights philosophers, Regan (1983) and Singer (1975) to support their work with tissue cultures. Taking liberties with both philosophers' positions, they use Singer's utilitarian position that would allow experiments on animals if the experimenters also would be willing to use humans at an equal or lower level of consciousness, rather than Regan's individualist deontological position that makes a case for an end to all experimentation on non-humans. Regan argues that animals are the "subject-of a-life" and thus have inherent value:

> A being that is a subject-of-a-life will: have beliefs and desires; perception, memory, and a sense of the future, including their own future; an emotional life together with feelings of pleasure and pain; preference-interests and welfare-interests; the ability to initiate action in pursuit of their desires and goals; a psychological identity over time; and an individual welfare in the sense that their experiential life fares well or ill for them, logically independently of their utility for others, and logically independently of their being the object of anyone else's interests. Those who satisfy the subject-of-a-life criterion themselves have a distinctive kind of value—inherent value—and are not to be viewed or treated as mere receptacles. (p. 243)

In contrast to Singer's utilitarian interpretation of formal justice, and in this case Catt and Zurr's interpretation as well, Regan argues for an acceptance of the respect principle in relation to those who are subjects-of-a-life:

> We cannot justify harming them merely on the grounds that this will produce an optimal aggregate balance of intrinsic goods over intrinsic evils for all concerned. We owe them respectful treatment ... because justice requires it. (p. 261)

Catts and Zurr (2003, 2004), however, seem to be filtering any understanding of their commitments and "care" for non-human beings through their fascination with the techniques of biotechnology and a confounding of what might constitute intrinsic good or evil for the animals involved in genetic technologies. Their project, "Disembodied Cuisine" investigating "the possibility of eating victimless meat by growing semi-living steaks from a biopsy taken from an animal while keeping the animal alive and healthy" (p. 13) is an example of this.

They believe

> That by the creation of the new class of semi-living /partial life we further shift/blur/ problematise the ethical goalpost in relation to our (human) position in the continuum of life. The discussion that being generated [*sic*] regarding the rights of the semi-living will draw attention to the conceptual frameworks in which we humans understand and relate to the world. (p. 17)

And to this they add: "To manipulate life is to be at home with the other that can be anything within this continuum" (p. 17). Once again, this rhetoric begs the question. One cannot say they do *not* relate to the living or semi-living as an object and then manipulate that living/semi-living being as an object. Why continue to use animals in any form as food if you wish to question the traditional view of the non-human? The continuation of such a program can only make suspect their stated goals of a "humble attempt of ethical consideration which goes beyond the 'I' the 'You' and even the 'Human' (as much as our humanness 'burden' enables us)" (pp. 18–19).

In light of the urgency of the future of the ecosystem's integrity, of which biotechnology is increasingly playing a large role, and the millions of our fellow creatures whose lives we are destroying in that process, it is important to ask, what does art contribute to that future? What responses come from those contributions? The continuation of art practices of creating life-forms through biotechnological means can only serve to implicate these practices, and artists who are involved in them, in contributing to a worldview that still values particular human needs above all else. This worldview, based as it is on the control and manipulation of nature, will continue to blind us to the more radical transformation of acknowledging that we have always been transgenic. The practices of biotechnology are misuses of that knowledge. Far more radical and creative responses to that fact may be based upon a number of increasingly influential ideas from broader areas of thought: the revolutionary idea in cognitive ethology that all living beings are equally gifted with their own worldview or a bio-centered environmental ethics that sees all living organisms possessing intrinsic worth. Appreciating and protecting the biodiversity existing already in the natural and cultural world, or what we have left of it, is part of this learning curve. Additionally, imaginative responses might be based on the enormous amount of thinking now going on in philosophy about the status of the animal. Reminiscent of Regan's *The Case for Animal Rights*, another more recent philosophical argument along these lines is cogently outlined by Italian philosopher, Cavalieri (2003). It is supported by much of the information and research from cognitive ethology and studies of the mind, but also rests on the major point, similar to earlier debates on the status of women and slaves, that

> . . .the shift from the condition of objects to that of subjects of legal rights does not appear as a point of arrival but rather as the initial access to the circle of possible beneficiaries of that "egalitarian plateau" from which contemporary political philosophy starts in order to determine any more specific individual right. (p. 142)

This shift calls for a further reorganization of society, similar to ongoing shifts concerning human rights, requiring the abolition of the status of animals as property or assets and the prohibition of all practices made possible by that status. This would, of course, include the prohibition of animal experimentation of all kinds. Describing her argument as neither contingent nor eccentric but the "necessary dialectical derivation of the most universally accepted among contemporary ethical doctrines—human rights theory," Cavalieri insists the argument demands a commitment to not only avoiding participating in, but also demands a commitment to opposing discrimination. Denying these demands would subvert ". . .not merely what is right, but the very idea of justice" (p. 143). Whether to continue to put energies toward a new art form of creating living beings or to commit to a more radical worldview that responds to the urgent cries of a disappearing natural world is the choice before the contemporary artist.

Notes

1. The quote is actually from a work of fiction written by Merijkowsky in the 1920s Romance of Leonardo da Vinci.
2. There are, of course, exceptions to this statement, including artists Britta Jaschinski, Angela Singer, and Julie Andreyev.

3. This phrase comes from the title "The Aesthetics of Care?" a symposium presented by SymbioticA and The Institute of Advanced Studies, University of Western Australia. August 5 2002 at the Perth Institute for Contemporary Arts. Catts and Zurr and their Tissue Culture and Art Project are hosted by SymbioticA—The Art and Science Collaborative Research Lab.
4. "Interview with Eduardo Kac" Interview conducted online, with questions posted to the Genolog website, July–September 2000 <http://www.ekac.org/ genointer.html>, pp. 3–4.

References

Baker, S. (2003). Philosophy in the wild? In S. Britton & D. Collins (Eds.) *The Eighth Day: The transgenic art of Eduardo Kac.* Tempe, Arizona: Institute for Study in the Arts.

Bureaud, A. (2002). The ethics and aesthetics of biological art. (C. Penwarden, Trans.) artpress 276, 38.

Catts, O. & Zurr, I. (2003 & 2004). The ethical claims of bioart: Killing the other or self-cannibalism. *Australian and New Zealand Journal of Art: Art Ethics* Double Issue 4(2) and 5(1):167–188. Retrieved on January 2, 2009 from http://www.tca.uwa.edu.au/atGlance/pubMainFrames.html

Cavalieri, P. (2003). *The animal question: Why non-human animals deserve human rights.* New York: Oxford University Press

Clark, K. (1977). *Animals and men: Their relationship as reflected in western art from prehistory to the present day.* London: Thames & Hudson.

Derrida, J. (1991). 'Eating well,' or the calculation of the subject. In C. Cadava & J. Nancy (Eds.) *Who comes after the subject?* New York London: Routledge.

Derrida, J. (2002). The animal that therefore I am (more to follow). (David Wills, Trans.) *Critical Inquiry*, 28(2):369–418.

Efimova, A. (2003). Introduction: Gene(sis): Contemporary art explores human genomics. Retrieved January 2, 2009, from University of California Berkeley Art Museum and Pacific Film Archive Web site: http://www.bampfa.berkeley.edu/exhibition/genesis

Hayles, N.K. (2000). Who is in control here? Meditating on Eduardo Kac's transgeneic work. In S. Britton & D. Collins (Eds.) *The Eighth Day: The transgenic art of Eduardo Kac.* Tempe, Arizona: Institute for Study in the Arts.

Held, R. (2002). Gene(sis): A contemporary art exhibition for the genomic age. In *Gene(sis) [electronic resource]: Contemporary art explores human genomics: CD-ROM catalogue.* Seattle: Henry Art Gallery.

Kac, E. (2000). GFP Bunny. In P. Dobrila & A. Kostic (Eds.) *Eduardo Kac: Telepresence, biotelematics, and transgenic art.* Maribor, Slovenia: Kibla.

Kellen, E. (Ed.) (1971). *Fantastic tales, strange animals, riddles, jests and prophesies of Leonardo da Vinci.* New York: Thomas Nelson.

Lestel, D. (2002). The artistic manipulation of the living. (C. Penwarden, Trans.) *artpress* 276, 45.

Machado, A. (2003). Towards a transgenic art. In S. Britton & D. Collins (Eds.) *The Eighth Day: The transgenic art of Eduardo Kac.* Tempe, Arizona: Institute for Study in the Arts.

Mukerjee, M. (1997, February). Trends in animal research. *Scientific American*, 276(2), 86. Retrieved January 2, 2005, from Academic Search Premier database.

Preece, R. (2002). *Awe for the tiger, love for the lamb.* Vancouver, BC: UBC Press.

Regan, T. (1983). *The case for animal rights.* Berkeley: University of California Press.

Rowan, A. (1995). Replacement alternatives and the concept of alternatives. In A.M. Goldberg & L.F.M. van Zutphen (Ed.) *The world congress on alternatives and animal use in the life sciences: education, research, testing.* New York: Mary Ann Liebert.

Salem, D.J. & Rowan, A.N. (Eds.) (2003). *The state of the animals II.* Washington, DC: Humane Society Press.

Shiva, V. (2000). *Tomorrow's biodiversity.* London: Thames and Hudson.

Singer, P. (1975). *Animal liberation*. New York: Avon Books.
Taylor, P.W. (1986). *Respect for nature: A theory of environmental ethics*. Princeton, NJ: Princeton University Press.
Turner, A.R. (1993). *Inventing Leonardo*. New York: Knopf.
Wood, D. (1999). Comment ne pas manger—Deconstruction and Humanism. In H.P. Steves (Ed.) *Animal others: On ethics, ontology and animal life*. Albany: State University of New York Press

We Have Always Been Transgenic: A Dialogue

Steve Baker and Carol Gigliotti

Abstract This dialogue concerns the nature of ethical responsibility in contemporary art practice, and its relation to questions of creativity; the role of writing in shaping the perception of transgenic art and related practices; and the problems that may be associated with trusting artists to act with integrity in the unchartered waters of their enthusiastic engagement with genetic technologies.

Keywords Art practice · Transgenic art · Ethics · Aesthetics · Genetics · Postmodernism

Introductory Remarks

This email conversation, stemming from Steve Baker reading Carol Gigliotti's essay, "*Leonardo's Choice*," and conducted on an occasional basis from November 2004 to March 2005, emerged as a format in which we hoped to be able to explore both our common interest in contemporary artists' engagement with questions of ethics and animal life, and the significant differences in our own approaches to those questions. Both of us were in the middle of writing books considering aspects of these issues in more detail, which may explain why we seemed sometimes to have too much to say, without always managing to say it very clearly. The conversation touches—sometimes only very lightly—on issues that include the following: the nature of ethical responsibility in contemporary art practice, and its relation to questions of creativity; the role of writing in shaping the perception of transgenic art and related practices; and the problems that may be associated with trusting artists to act with integrity in the uncharted waters of their enthusiastic engagement with genetic technologies. Running through much of the conversation is a tension between what is perceived by each party as the adoption of a wide or too narrow focus: is the wider focus to see this area of art practice as something that should be viewed in a similar manner to other forms of contemporary art, where an artist's "proper business"

S. Baker (✉)
Professor Emeritus of Art History, University of Central Lancashire,
Preston, Lancashire, UK
e-mail: sbaker1@uclan.ac.uk

C. Gigliotti (ed.), *Leonardo's Choice*, DOI 10.1007/978-90-481-2479-4_5,

is "curiosity and awareness" (as John Cage once remarked), or is the wider focus to see the ethical implications of this kind of art practice as inextricably linked to the methodologies and goals of the techno-science it purports to critique? We call what follows a conversation because—like any lengthy face-to-face conversation—it seems occasionally to display a frustrating circularity, a failure on both our parts to quite grasp the point the other is making, and a failure to return to certain issues to which we promised to return. Conversations develop at their own pace, and cannot always be hurried, so as the publication deadline loomed there was a sense of much remaining to be said, and of little having been resolved. This reminded both of us of the character in a Martin Amis novel who reflects: "I am taking a good firm knot and reducing it to a mess of loose ends." It is perhaps for readers to decide whether or not any discernible pattern emerges from these loose ends.

Steve Baker and Carol Gigliotti, March 2005

Steve Baker: Let me try this out as an initial question: For me, one of the most striking statements in your "*Leonardo's Choice*" (Gigliotti, 2006) essay is the assertion, towards the end of the essay, that "we have always been transgenic." It struck me as having as much to do with politics as with science, coming across as a kind of rallying-cry, or as the title of an as-yet-to-be-written manifesto. And without really stopping to analyze the reasons, it was a rallying-cry to which I was immediately drawn. Am I misreading what you had in mind?

Carol Gigliotti: I do appreciate your close reading of my essay. It is that phrase, "we have always been transgenic," which made me pause briefly before I decided to, in fact, include it in the essay. My hesitancy about including it had less to do with the meaning behind it, which I shall explain in more detail in a bit, and more to do with what the possibilities are for readers to misunderstand its actual intent. You are correct in intuiting it as a kind of rallying cry, or to be more descriptive of my thoughts when the phrase rushed from my mouth in intense frustration: a rallying scream. This frustration is both personal and representative of a bafflement borne by people from many corners who see the state of the non-human world as so obviously at risk. I worked on this essay for over 6 months and have been deeply involved for over 25 years with many of the very sources that artists I was writing about were utilizing to rationalize their work: cognitive ethology, animal consciousness, animal rights, and ethical vegetarianism and veganism. After, at least I hope, specifically taking to task the misuse and misunderstanding of these commitments in the service of activities so obviously counterproductive to an awareness of either this growing risk or the goals upon which much of these commitments have been undertaken, I must have felt the need for a phrase that might indicate some of this bewilderment at what seemed to me a severe myopia among a number of artists and thinkers. The phrase itself, as you suggest, has to do with both its political and scientific underpinnings, as well as a bit of a nod towards Bruno Latour and Donna Haraway. I wanted to throw the reader, the artist, the writer, the techno-theorist, the student, who appreciated my very specific points in earlier parts of the essay, a metaphoric hook upon which they might begin or continue their own thinking. The fact that there is a vast amount of genetic similarity between organisms, including humans, and we are all related by a shared evolutionary history, is the basis for the idea that we are all transgenic, and the basis, as well, for notions of a bio-centric compassion. What current transgenic technologies are doing, however, is based on a flawed application of this similarity by reducing complex behaviors to single genes completely apart from the context of the formations of those behaviors. The problem with using what might be construed as an ambiguous metaphor is that it, too, might be misread and misapplied. I am happy you read it as a rallying cry and were immediately drawn to it. After all my consternation, I would love to hear why you think you particularly were drawn to it.

SB: Perhaps I will come back to that question a little later. For the moment, I would rather take up your concern with the "problem" of using an ambiguous metaphor. I think ambiguous metaphors can be thoroughly good things. Yes, it is hard to patrol or control the work they may do, but they allow thought in, instead of shutting it out. I have always been struck by a statement made by Sue Coe, who I guess is still the best known artist working in North America to address an explicit animal rights agenda through her work. She said in an interview, with reference both to her own work and to that of other artists she admires, that "the most political art is the art of ambiguity."

Because I agree so thoroughly with that view, I have reservations about using terms like "severe myopia" to characterize the attitudes of some of the artists whose work you describe. Especially when working in relatively new or unfamiliar fields, artists' experimental strategies inevitably run the risk of being open to contradictory readings—and of course to being unsuccessful!

Does that risk (and far more importantly, of course, the risk to animals' well being that their work may also—sometimes inadvertently—entail) mean that such strategies are automatically invalidated? To my surprise, I find it increasingly difficult to separate that question from the question of what art is for, and what kinds of work art can do, and why it is that we believe (as I am guessing we both do) that art matters. This raises all manner of difficulties and contradictions. I suspect, for example, that I am perilously close to arguing that artists should be allowed certain freedoms that scientists should not be allowed! And my interest in pursuing this conversation with you is partly to explore those contradictions in my own thinking. If we believe that art can indeed, in principle, open up productive new perspectives on pressing and controversial contemporary political issues (not least issues to do with animal life), how do we deal with those very many cases of artists who may actively be trying to act responsibly, but whose views on animal life, animal welfare and animal rights may be slightly (or even significantly) different to ours? I am not suggesting, of course, that you and I necessarily hold exactly the same views on these matters, and those differences are also ones that I think we both see as worth teasing out in the course of this exchange.

CG: Ah, Steve. You have hit on the crux of the matter for artists, but not for artists alone, when you allude to how close you might be to "arguing that artists should be allowed certain freedoms that scientists should not be allowed!" The idea of unfettered creativity is the holy grail, not only in the arts, but in the sciences and society at large. In my experience, and increasingly so, as I have not only been researching artists' and scientists' work, but also involved actively in technological development, the push for an unfettered path to the new, the imaginative, the original has gained incredible momentum and force for the last half century in all areas of endeavor. I realize you see the goal behind this freedom as opening up "... productive new perspectives on pressing and controversial contemporary political issues" as a very positive one. Many scientists, however, see this need for freedom of investigation as necessary for creative progress in solutions to problems as well, medical ones, for instance. Granted there are good reasons for progress spurred on by the creative imagination, the freedom to dream, but there is also a darker side to that lust for the new, the original. I will not go into detail here about the repercussions of that drive, but as Ronald Wright (2004, p. 5) points out, "The myth of progress has sometimes served us well—those of us seated at the best tables, anyway—and may continue to do so. But I shall argue in this book that it has also become dangerous. Progress has an internal logic that can lead beyond reason to catastrophe. A seductive trail of successes may end in a trap."

Like you, I have struggled with seeming contradictions between my long-time support of creative freedom (I teach art and design, after all) and my increasing discomfort with the work and attitudes of some of my colleagues in interactive technologies or what is now called New Media. My discomfort and criticism of their work stems not only from their inclusion of the areas concerning animals I mentioned in my last email and my essay, areas I feel are being used for reasons directly opposed to ideas about the intrinsic worth of animals and the natural world. My disagreement with this work also emerges from an ongoing concern over where our increasingly technologized world is taking us, and what we are losing because of the bargains we have made along the way. I do not see these works as "opening up new perspectives," but immersed in existing perspectives, which I, among others, have already found to be not only troublesome, but also disastrous, not just for animals, but for the planet at large. I said very specifically in my essay what I was taking to task was the "new art form of creating life." The drive behind that goal is not altruism.

The book on which I am working (struggling with) is about this very issue. I feel I have gone on too long, but must respond to several other points about metaphor. I think metaphor is a very powerful form of communication. And I think all metaphors can be ambiguous to a greater or lesser extent depending on the context and to what degree people agree on experiences from whence the communicating metaphor has emerged. But while I see metaphor as a form of imaginative communication, I also understand it as neither objective nor subjective, but experiential. As a writer and artist, I believe that when I use a metaphor to communicate, I also have a responsibility for that communication. In the case of my "we have always been transgenic" metaphor, embedding it in the essay allowed me to feel that overall, its meaning would be taken in the context of the essay. My "severe myopia" metaphor is one, that while you may have reservations about, I do not.

Fig. 1 Rabbits in laboratory stocks (Credit: Brian Gunn/IAAPEA)

Severe myopia is an extreme state of nearsightedness, due to the faulty focusing of light rays from objects at a distance. In other words, holding an object close allows the myopic person to see quite well, but what they cannot see or distinguish are distant objects. I think this is a very relevant metaphor for artists who seem unable to look at the larger world in which their art exists. I do not see myself as an art critic, but an artist and writer whose main interest has always been ethics, and in my case, an ethics informed by my direct relationship with the natural world, about which I am determined to speak out. I could go on, of course. I feel I have answered and not answered your questions, but tried to explain how I am answering those same questions for myself. How do we differ in our views on animal life, animal welfare and animal rights, may I ask?

SB: Let me start with that last question. How do we differ? Perhaps (though I may be wrong) in how we choose to approach the relation of our writing on art to our views on animal life. I first got interested in animal rights issues about 20 years ago when I first learned of the work of the work of the British Union for the Abolition of Vivisection (BUAV), and opposition to all forms of animal experimentation has always been the animal rights issue that has engaged me most directly, and about which I have felt most strongly. The first thing I ever published on an animal-rights theme was a very short piece in the BUAV magazine, *Liberator*, in 1986, on the misrepresentation of animal rights activism in RSPCA publicity in the UK.

When my first book, *Picturing the Beast*, came out in 1993, my concern was again with representations of animals and representations of human attitudes to animals—not in contemporary art, but in contemporary Western culture as a whole. I was arguing a point that now seems blindingly obvious: that these representations (no matter how ephemeral) do matter, and that they help to shape our perceptions and our understanding. The final chapter also included discussion of the contested representation of the animal rights activist. The main point of that final chapter, though, was to explore what I provisionally termed "strategic images for animal rights"—images adaptable enough to resist easy incorporation into the culture's dominant meanings, but sufficiently attuned to those meanings to enable them to be recognized within that culture. In trying to identify and to characterize effective strategies for image-making in this field, I had of course to be quite open about my own position. It was a position or undertaking that even then I preferred to describe as political rather than moral; the point was to get beyond what I already, intuitively (which is to say in a thoroughly under-researched manner!), tended to think of as the evident limits of moral philosophy for thinking about animals.

At any rate, it seemed important to declare my hand in terms of where I stood on those issues, simply for that particular chapter to make sense. I am aware that I use writing differently now, and see it doing a different kind of work. I regard writing in much the same way that I regard artworks (even transgenic artworks): as attempts to undertake particular kinds of work. And that is where I think you and I may differ: in the kinds of work we want our writing to do.

CG: First of all it was obvious to me, at least, from reading *The Postmodern Animal* and other writing of yours that your beliefs lay within the sphere of animals rights. I, too, was very involved with the animal rights movement, through PETA (People for the Ethical Treatment of Animals) and Tom Regan's Culture and Animal Foundation. And I am aware of, and have used in the past, writing strategies that do not entail completely showing one's commitments. I sometimes used visual strategies in this manner as an artist since most of my visual work was about our relationship with animals, specifically our use of animals in the laboratory and factory farming. In fact, *Leonardo's Choice* is probably the most overtly committed of my writing up to this point on issues surrounding animals. I have alluded to those commitments in many of my past essays, however.

But, personal experiences coupled with the drift in rhetoric and practices of a number of artists involved with technology over the last five years or so have caused me to rethink the viability of my continuing to use those strategies, especially when faced with the realities of biotechnology. I know Eduardo Kac very well and have been friends and colleagues with him for over ten years, supporting his previous work and curating an earlier work into a Siggraph Art Show. I tell you this because some of my feelings about the importance of not only considering but also clarifying how ethics and aesthetics are so entwined has to do with my intimate involvement with technological art and many of the people who have pioneered artistic technological work.

My dissertation, "Aesthetics of a Virtual World," concerning ethical issues in interactive technological development specifically virtual environments and finished in 1993, was somewhat ahead of its time. After a number of years of attempting to embed these ideas in publications, lectures and projects, I have decided that time is too short, for me, and for the planet, not to speak directly about these issues. I do, however, appreciate what you are doing a great deal! Over time I have learned to appreciate Dorothy Parker's assertion, "If you can't be direct, why be?"

SB: I think it is fascinating that our writing is moving in opposite directions, or that our strategies for writing are taking us in those directions. I wonder why that is happening? I had one idea about this, though I could be quite wrong. In your current work your focus is directly on artists working with genetic technologies, and more specifically with those who presume to explore the creation of new life forms as their art practice. I have no difficulty at all in seeing why you want to take a stand against that work. Why do not I do so too, in an equally direct manner? I think because I continue to want to address the whole spectrum of contemporary art that engages with ideas of the animal—everything, that is, from the animal advocacy that is at the heart of the work of artists like Angela Singer and Sue Coe, to the transgenic explorations of Kac, Adam Zaretsky, and so on. And all points in between, including non-transgenic work with living animals, such as that of Catherine Chalmers or Olly and Suzi, often undertaken for incredibly diverse reasons. And I am very reluctant indeed to identify (or even to try to identify) a point on that spectrum beyond which my disapproval kicks in. (It does kick in, of course, but to say so would I think get in the way of how I am trying to use that writing.) Why? Because my conviction that recent and contemporary art can offer fresh perspectives on ethical questions—and we will have to come back shortly to the reservations you have already expressed about that idea—has led me to the really quite uncomfortable view that it is not only "ethically sound" art, for want of a better term, that can do this. I seem to have ended up taking the view—or more accurately, I am still in the early stages of trying to articulate the view—that postmodern animal art constitutes a kind of fluid sub-ethical practice, and that the integrity of that work is not fashioned out of, and is not best expressed through, the language of morals and ethics.

CG: I agree that our different strategies, and perhaps, directions, are due to the breadth of contemporary art about which we write. At this point, my focus is on technologically mediated work, alife, genetics, interactive installation, augmented realities, computer game design, immersive environments and how and why the non-human, though often seemingly absent or exploited, is fundamental to understanding the meaning of these technologies for the future of both the human and non-human.

SB: That part of writing that you describe where you know things are connected, connectible, but have not yet figured out how to articulate that conviction is, I always think, the most exciting part or stage of writing (or of making art). It has a lot to do too, I think, with my sense of the integrity of how artists think and work. Perhaps my favorite statement by an artist is from Jim Dine in the 1960s, saying "I trust objects so much. I trust disparate elements going together." And even if the artists we are talking about in this exchange are ones who push that perception close to or beyond tolerable limits, I am reluctant entirely to let go of that notion of trust.

I want to try to bring us back on track, because our recent exchanges around writing have perhaps (though perhaps necessarily) diverted us from the direct question of genetic technologies, and of artists whose practice involves "creating living beings" (Gigliotti 2006, p. 2) or "new life forms" (Robin Held, quoted in Gigliotti, p. 5).

To do this perhaps we could look at one particular passage in "*Leonardo's Choice*" (pp. 21–22). There, you effectively distinguish an anthropocentric worldview in which those practices are implicated—and I want to return to that word—from a more radical worldview that would allow that "we have always been transgenic." What I would like to try to think about is the difficulty of disentangling those worldviews. When you first sent me your essay, thinking back after reading it I made a brief note to myself, and misremembering your crucial phrase as "we are already transgenic," I scribbled down an echo of that phrase as "we are already (necessarily) compromised... gloriously." This little Haraway-style celebratory observation, which I came across again the other day, seemed to resonate with your word "implicate" (p. 21),with its intriguingly varied uses (to show to be involved in a crime; to imply; and to entangle).

What made me think about that was a statement by Yve Lomax that I read recently in a student assignment: "The implication of metaphor brings a much richer knowledge of the World. To speak by way of implication is not about standing back and attaining the right distance in order to see things clearly; implication is about being plunged into things, finding oneself in the middle of things" (quoted in Wheale (ed), *The Postmodern Arts*, Routledge 1995, p. 157).Without going back just for now to our earlier brief comments on metaphor, it was that sense of "finding oneself in the middle of things" that I wanted to hold on to. Perhaps this is one of the points at (or from) which you and I articulate the fairly small differences in our outlook quite markedly. You justifiably challenge "the idea that a confrontation with the complexity of a topic or issue precludes the necessity of confronting ethical choices embedded in that complexity" (Gigliotti, p. 3), and conclude that opting for one or the other of the incompatible worldviews you have described is "the choice before the contemporary artist" (p. 12). I think I would simply choose to articulate the negotiation of those choices differently.

This, and my admittedly rather wayward refusal (or disinclination, at least) to engage with the language of ethical choice will take some explaining, so let me at least try to say something plainly: Do I think that artists messing around with animals in laboratories is wrong? Yes. But my disapproval alone is not going to stop them doing it. Guattari (1995, p. 131) (of whom, more later) wrote: "The work of art, for those who use it, is an activity of unframing, of rupturing sense..." And that is what writing is, too. That is why the question we have already touched on, of how we were using our writing, seemed so important. When Eduardo Kac invited me to write about his work, I had some reservations, but when I realized just how seriously his transgenic work was being taken by writers with little apparent interest in its ethical implications, it seemed important to accept the invitation. And of course it was a compromise, with uncertain outcomes on both sides, I think. He can legitimately take the view that I find his work interesting (which I do, in all sorts of ways) because I agreed to write about it in a book devoted to that work, and I managed at least implicitly to put in place some serious criticisms of the work by juxtaposing his words with those of Derrida's forthright remarks about the "hell" of "the imposition of genetic experimentation." It was a fairly small-scale attempt at unframing, at rupturing the sense of what Eduardo was doing. I realize that by writing about it at all, some people may think I condone the work, but the idea—whether or not it works out that way—was to open a dialogue.

CG: I am finding, have found, ideas around creativity as central to finding a direction through these issues, just as you have. I see our differences as one of context, since you seem to be making a case for artists while I am looking at artists and scientists as one increasingly blended field. For me that is a large and very important difference. The realities of genetic technologies and artificial life research and practice are the context I am trying to keep in the forefront, and I think the kind of creativity this blending of art and science has encouraged is key to both a positive and negative future for the human and non-human world.

SB: I want to move on to something slightly different now, though still around those ideas of implication, and of the difficulty of sharply differentiating between particular worldviews. I want to suggest (though you know this perfectly well, of course!) that there is a rather complex contemporary rhetoric around body image, and body images, and to juxtapose a few statements that employ it in various ways in order to see what kinds of work that rhetoric is being used to do. The first is from David Abram (1997, p. 46), from a broadly ecological/anthropological standpoint: "the boundaries of a living body are open and indeterminate; more like membranes than barriers, they define a surface of metamorphosis and exchange."

Next, the British sculptor Anthony Gormley whose work focuses directly on the form of his own body: "Our bodies are on temporary loan from the circulation of the elements in the atmosphere."

Then Donna Haraway (1991, p. 149), whose particular elaboration of the idea of the cyborg, which she terms "my blasphemy" is one that "insists on the inextricable weave of the organic, technical, textual, mythic, economic, and political threads that make up the flesh of the world" (Haraway 1995, p. xii).

Taking it into more fanciful, virtual and dodgily Baudrillard-style imagery is Arthur Kroker (2004) relishing the idea that "skin smoothes out into Flash flesh cut at the speed of syncopation, and the face itself floats away, not into an aesthetics of facialization, but into something more indeterminate, more tentative, more slippery in the codes." And from there it seems but a small step to Catts and Zurr's (quoted in Gigliotti, p. 8) “only when humans realize that they are part of the continuum of life will manipulating life not be as alarming as it now seems.”

Somehow, somewhere, across those five small statements (a random selection) that all seem to share a desire to celebrate rather than to fear the openness of bodies, a slippage—a rather terrible slippage—happens. (The one thing Kroker gets right is the slipperiness.) But—and this is a hard thing to express clearly, so forgive me if this first attempt does not quite get it right—poststructuralism’s decentering of the subject, the integrity with which it refuses the fixity or the propriety of all selves, all bodies, all identities (I am thinking here of Deleuze & Guattari, and of Cixous, and Haraway, at their radical, uncompromising best) is perhaps necessarily difficult—impossible?— neatly to separate from what I think of (excuse the judgemental tone) as a less accomplished playing with or employment of that rhetoric. If I seem to be reducing this body-immediate stuff to aesthetics, I guess it is because, in part, I just cannot see ethics as a higher ground to which to retreat here.

So, to my last main point here. Can we perhaps, no matter how tentatively, distinguish three claims (in relation to the kinds of work discussed in “LC”)? First (and these are not direct quotations), “I make new life”— presumptuous, immodest, oddly uncreative, and somehow not what we want or expect from artists. Second, “I play with materials (sometimes ‘new’ materials) and look on with interest as things happen”—better, and more in the spirit of Jim Dine on trust, at least if we can bracket out the idea that those materials might include animals (but we cannot bracket that out). Third, “I use this (art) play to address issues” (big, important issues of course)—but that is maybe too conscious, too intentional, too self-consciously “responsible.” I am still, pressingly, in the middle of trying to think this through, but it seems to me the second claim that is the creative one, the one that has a space for and of not-knowing, despite the terrible precariousness of that position for animals who somehow get caught up in it.

Why raise this now? Well, partly because “*Leonardo’s Choice*,” Catts and Zurr aside, tends to quote critics rather more than artists themselves (critics who appear to assert the first claim, on behalf of artists). And what the artists say seems to me to matter a great deal, no matter how contradictory it is. Eduardo’s claim both is and is not the “I make new life” claim. His transgenic art is about the making of new life, but he is equally adamant that Alba is not/was not herself the art, the artwork. GFP Bunny, as an artwork, is a dialogue and (of sorts) an ecology of social exchange, not the one-off creation of an out-of-species being.

CG: I would like to take up first your ideas about implication. I agree wholeheartedly with an understanding of speaking, writing, or making art that is not distanced from that which it is about. “Finding oneself in the middle of things” is, more often than not, when ethical dilemmas become apparent. It is because we are implicated in something, that our judgment becomes so important, and carries so much weight. How we choose to be in the middle of things is the important question. Your assumptions about what kind of ethics I am referring to are, perhaps, hindering our discussion of these issues. I would like to come back to that in just a moment, but first, I would like to make it clear I am not saying that artists should hold themselves back from involvement with ideas about genetic technologies.

Fig. 2 Baboon in laboratory (Credit: None)

These technologies will have a huge impact, are having a huge impact, on contemporary life and artists will, of course, be involved in the making of the meanings of that impact. And it is here where I think we may differ in more than our strategies. The description of art in the quote from Guatari is very much a postmodern one. The postmodern sees art as an activity about "an unframing, a rupturing sense." But what about other functions of art: to make sense, for instance, to creatively look for alternatives, to offer connections where none were seen before. My point is that there are many more creative and, in the long run, more imaginative ways for artists to make sense out of today's technologies, to even have an impact on their development, than to merely repeat current technological methods. This is especially relevant when the effects of the much sought after experience of implication fall not upon the artist herself, but another being.

Now back to ethics: I do not see ethics as a higher ground where one can retreat! In fact, I see ethics in the exact opposite way. For me, ethical understanding and action emerges from a common embodied imagination that celebrates and respects "...rather than fears the openness of bodies." In the midst of all those variations of rhetoric you mention and, perhaps more importantly in the midst of the rhetoric of today's corporate techno-science, my interest is in what, specifically, the speaker means. What do they mean, I want to know, not in any abstract theoretical way, but in ways that are "implicated" in real outcomes for specific beings, both human and non-human. It is our embeddedness in the physical world we share with humans and non-humans that allows us opportunities for an empathic understanding of what those outcomes might mean. And through that empathic experience we search for actions to mirror that imaginative understanding in the real world.

Now, on to your three claims. The artists discussed in *LC* and others involved in "creating life" wish to make or are involved in making all three claims. Certainly Eduardo makes the first and means it, though he problematizes that claim as well. And though he may be interested in the other two claims, his interest in "creating life" is long held on his part, as can be seen in his work with and writings on robotics and telepresence. The fact that we cannot bracket animals out of the second claim of playing with materials means, for me, that "looking on with interest as things happen" is not to be trusted in any conclusive way. Your phrasing of the third claim undermines any attempt at responsibility, something I am not sure you meant to do. Permit me to use myself as an example. I, who know through my research and experience that using animals in experiments almost always puts animals in a precarious position, cannot pretend that I do not know this. The artists involved may not know that at the outset or care, but I do, and I cannot pretend that I do not. In this way I am consciously responsible. How else can someone take on responsibility for another, if not consciously?

Your observation that "*LC*" tended to quote critics rather than artists is a bit unfair. I quoted Eduardo four times in this essay and Catts and Zurr seven times. Since this essay was being written almost three years after the installation of The Eighth Day and almost four years after the announcement of GFP Bunny, I thought that Eduardo already had ample opportunities to be heard. More to the point however, I chose to look closely at what Eduardo insists is the real artwork, the surrounding dialogue. He is correct in asserting this and I chose to take him at his word and join in the dialogue. I am interested in what artists have to say, but I am even more interested in all the voices taking part. But one voice was not heard: Alba's. Eduardo sees Alba as part of an "ecology of social exchange," but I completely disagree since no animal used by humans in this way has a choice or a voice. The "social exchange" in this case is entirely a human one, bounded entirely by the desires and needs of the human. My goal was to give Alba a voice.

SB: Right, let me try to work through the points you raise fairly systematically. First, the question of the form or the nature of one's implication in things, one's being in the middle of things. I am not at all sure that it is possible to get this right. Wishy-washy as it may sound, I think that trying to act with integrity in specific circumstances is the most anyone can do, regardless of one's chosen medium. I recently read an interesting and broadly sympathetic review of Steven Wise's *Drawing the Line* (Wise, 2002), the book in which Wise argues that the "probability of conscious awareness" of individual species can be credibly estimated, and that basic legal rights should be accorded to those species that are ranked at anything over a particular score on the numerical scale that Wise adapts from the one used in Griffin's book *Animal Minds* (1992). Wise has indeed thus drawn a line, a line informed by scientific research, and the reviewer's concern (as with some of the more familiar complaints about the Great Ape Project, for example) is that even when erring on the side of generosity, as Wise insists we must, certain species appear to fall on the wrong side of that line, and do not get the protection of legal rights. It reminded me of the point I made earlier in our exchange that in my proposal to trust artists to act with integrity in relation to questions of animal life, instead of drawing a line beyond which they should not go, certain animals like Alba may indeed end up on the wrong side, the unsafe side, of contemporary art's practices. I think the point I am trying to get at is that with or without a dividing line (which is only one of the things one might try to draw in the middle of things), we (merely intellectually) and animals (tragically physically) are left in the same uncomfortable, contradictory, inconsistent pickle. And whether or not that's the pickle of always having been transgenic, I am really not sure. Wise, it seems to me, was looking for a creative approach to furthering the case for animal rights, and the reviewer's question is whether or not the granting of rights is "the only way, or the best way, to secure a better deal for members of other species." But that's the whole point about inventiveness: you do not know if it is going to work. You try things, with the best of intentions, and if they do not work they may still help to nudge the debate forwards a little.

I agree with your idea about artists helping to make the meanings of the impact of genetic technologies, and I think you were right to say that the artists' good intentions will not necessarily enable them to control what those meanings are eventually judged to be. Being "consciously responsible," and acknowledging that responsibility, does not make the outcome predictable. I can see why you object to the incautious phrasing of my "third claim," and read it as undermining "any attempt at responsibility," which was certainly not my intention. It perhaps resulted from my trying to play up the importance of the second claim, which for me has parallels with Sue Coe's promotion of an art of ambiguity. Art's importance is in making statements, shaping meanings and shaping responsibilities, whose forms are distinct from those of philosophy or science, and I think that shaping an ambiguity can be a difficult and a highly responsible action. In an art of indirection and ambiguity, and of rhetorical inventiveness, an animal such as Alba's voice might just resonate all the more strongly, as might the silent "voices" of all the animals represented in Sue Coe's work.

On the ethics question, I am absolutely not suggesting that you see ethics as a higher ground to which to retreat. Absolutely not. But in arguing for the entanglement of aesthetics and ethics, I wanted to challenge the widespread assumption that the ethical is the more secure and important of the two concepts. As an editorial in *Geoforum* (Davies 2003, p. 411) put it a couple of years ago in relation to the "geography of monsters" that include transgenic creatures such as Alba, "there is nothing trivial about the role of aesthetics in the development of these new forms of hybrid nature."

And this brings me to Guattari, whose view of the work of art as "an activity of unframing, of rupturing sense" (p. 131) you see as very much a postmodern one, and one that may be distinct from other art's (or other artists') concern "to make sense" and "to creatively look for alternatives." I read this differently, and see his rupturing of bad sense as much the same thing as your making of good sense—though that's probably far too abrupt a manner of putting this. The complacent, unthinking sense he is trying to rupture might be seen, for example, as the "sense" that informs what Eduardo calls "corporate genetic engineering" (I think that's the term he uses), the "sense" that the GFP Bunny project was an attempt—a sincere but terribly flawed attempt—to rupture.

If you doubt my reading of Guattari's phrase, bear in mind that art is one of Guattari's examples of "incorporeal species"—things that act in the world, and act on the world—his other examples include "love and compassion for others" (p. 120). Our living interdependently among both corporeal and incorporeal species is something that will only work, he insists, accompanied by the development of a "sense of responsibility," and is directly about how we "change mentalities" (p. 119). I have no objection to this being labeled postmodern, if you like, but to my mind this is postmodernism at its responsible, inventive best!

CG: In relation to what you say about acting "with integrity in specific circumstances" as the most anyone can do, on some levels I agree with you. But there is something to be said for thinking things through on a conscious and broadly inclusive level, and based on a great deal of information and experience and thought, deciding to commit to a particular course of action—in other words, what integrity is based on. Gandhi, Martin Luther King, Nelson Mandela and Vandana Shiva come to mind here as examples of people who have done that to great ethical effect. I do think, however, a great many sources "nudge the debate a little bit forward" which is why I chose in this journal issue to bring in authors from a wide range of perspectives. My concern has been, having been in the new media/ technology field for some time, that artists working in this particular field with its intricate connections to science and technology at large, did not have the realities of animal experimentation at the forefront of their concerns. I think it has been a side issue for most artists and in many cases, the viewers and critics have brought up these ideas. This may not be the case for the Tissue Culture and Art Project, interestingly enough, though their rhetoric seems to me to obscure their real intentions in this area. It is up to all of us who see those concerns not as a side issue, but at the center of a larger debate about the whole non-human world and our relationship with it, to speak up as we see fit. Exactly what you did in agreeing to write your excellent essay for *The Eighth Day*. The understanding of "we have always been transgenic" leads to actions such as this. I, too, am arguing for the understanding of aesthetics and ethics as intricately connected practices. While we often can disentangle them conceptually, in practice, they are entwined. That is the whole point of the "*LC*" essay, and all my prior writing on ethics and technology, and this discussion only encourages me to be even clearer about that. Guattari, Derrida, these people have moved us further in understanding ideas about ourselves and with that knowledge one has to decide, on one's own, how to act. I suppose I am seeking to move past the indecision of much of postmodernism. I think it is time to do that. As I have said before, for me at least, the time seems too short not to be direct and clear.

SB: Well, as you know, my own view is that quieter or more indirect strategies may in fact be more effective for animals in the long run—and there is a long run to be considered here, as well as the more immediate issues that rightly concern you. The strategies I have in mind have nothing to do with "indecision." They have to do with recognizing and articulating the considerable difficulties of being direct and clear, and finding ways of acting in the wake of those difficulties. And perhaps they have to do also with seeing contemporary art's experimental drive and its disregard for hierarchical thought as the things that create its potential as the space of a genuinely transgenic practice—in that wider, more metaphorical sense that we have used it here, as in "we have always been transgenic." If this is difficult to articulate at present, it may be because the force of such work would lie (and occasionally already does lie) in its fusing of ambiguity of expression and clarity of purpose.

References

Abram, D. (1997). *The spell of the sensuous: Perception and language in a more-than-human world*. New York: Vintage.
Davies, G. (2003) Editorial: a geography of monsters? *Geoforum* 34(4), pp. 409–412.
Gigliotti, C. (2006). Leonardo's choice: The ethics of artists working with genetic technologies. AI & Society 20(1), 22–34.
Griffin, D. (1992). *Animal minds*. Chicago: The University of Chicago Press.
Guattari, F. (1995). *Chaosmosis*. Sydney: Power Pubications.
Haraway, D. (1991). Manifesto for cyborgs: Science, technology, and socialist feminism in the late twentieth century. In *Simians, cyborgs, and women: The reinvention of nature*. New York: Routledge.
Haraway, D. (1995). Foreword. In C. Hables-Gray (Ed.), *The cyborg handbook*. New York: Routledge.
Kroker, A. (2004) The Transgenic Art of Telebody. Retrieved from http://www.telebody.ws/KrokerOnTELEBODY2.doc
Wheale, N. (Ed). (1995). *The postmodern arts*. London, New York: Routledge.
Wise, S. (2002). Drawing the Line: Science and the case for animal rights. New York: Basic Books.
Wright, R. (2004). A short history of progress. Toronto: House of Anansi Press.

Negotiating the Hybrid: Art, Theory and Genetic Technologies

Caroline Seck Langill

My mother doesn't like me
And neither do I have a sweetheart
Oh why can't I just die
What should I do?
Johannes Brahms, Die Trauernde

Abstract Over the past decade artists have increasingly turned to science in order to investigate technology's effect. The move from hardware-based technologies to live organisms as media, raises ethical issues that the broader art community is addressing. This chapter tracks the history of instrumental disengagement to determine when and how the gradual codification of life contributed to the eventual use of live organisms in art practice.

Keywords Animals · Ethics · Genetics · Hybrid · Instrumental reason · Posthumanity

The above quotation consists of the first four lines of the Brahms lieder, *Die Trauernde*, "*The Mournful One.*" They are also the first words we hear sung by Jesse Norman, in Wendy Coburn's video work of the same name. The image accompanying this sad song is of a young monkey leaping toward, and then hugging, an effigy of its mother, a cloth-wrapped wire body topped with a stark doll-like head. Prior to the title sequence, before we hear the folk song, we see a straight documentary image of a baby monkey swaddled in muslin, stroked by a man. They seem to be in a laboratory of some sort, and we might place the image, based on the scientist's dress, in the 1950s. It is a bright image, almost cheerful. After the title comes up, the soundtrack begins, a slow, sad song sung in German. The image of the monkey and the surrogate mother lie in the lower left-hand corner of the screen. The space beyond them is dark providing good contrast for the subtitles that occupy the top half of the image. The monkey, in slow motion, moves from one side of the frame

C.S. Langill (✉)
Liberal Arts, Ontario College of Art and Design, Toronto, ONT, Canada
e-mail: clangill@faculty.ocad.ca

C. Gigliotti (ed.), *Leonardo's Choice*, DOI 10.1007/978-90-481-2479-4_6,

leaping over to the mother, hugging it, but with an expression difficult to place, perhaps mournful. The narrative goes on to describe someone, who is invisible, has no one, is in pain, waiting to die.

The juxtaposition of the song and the image of the baby monkey caressing the model creates a cautionary artwork that suggests that our relationships with animals, and the trauma they have suffered in order to further Western science, requires reconsideration. Coburn drew the documentary footage from Harry Harlow's experimentation on macaque monkeys. Harlow (1979) created a breeding colony of research subjects at the University of Wisconsin that remained active from 1955 to 1960, the laboratory providing an apparently safer and cleaner living environment. The scientist's statements regarding the lab as a home sound naive and dated:

> Human laboratory mothers or fathers are brighter and better trained than monkey mothers. Human caretakers are motivated to improve their methods, and unfortunately monkey mothers and fathers cannot read. Trained human beings know more about infant monkey diseases than the most highly educated monkey mothers. (p. 7)

Harlow's research on maternal bonding, extrapolated to human behavior, is considered to be a cornerstone of modern psychology, teaching students about bonding between mother and child, although his methods and the images of suffering that we witness through this artwork make one wonder how such an experiment can even begin to broach theories of love and affection. This research occupies a central position in a history of scientific practice that utilizes animals as analogous subjects for human beings. The trajectory from the ordering of the species to the cloning of sheep is hardly a straight path, but the utilitarian attitude that animals are there to service science in order to provide insight into human health has not really shifted. Donna Haraway (2000) recommends the use of animals for experimentation be carefully limited, for it begs the question,

> How much suffering is who bearing and how do I respond to that? I can't quantify such suffering, and an ethical judgement is not a quantitative calculation at root, but an acknowledgment of the responsibility for a relationship. (p. 147)

The intention of this chapter is not necessarily to contribute to the already vigorous anti-vivisection debate, but to explore how scientific disengagement from animals, and other living organisms, enabled biotechnology companies access to all species for manipulation and profit-making. I track certain branches of scientific progress from the 1800s that influenced the eventual coding of all organic matter to the point where the theories and language surrounding current scientific practices have more currency than the subjects themselves. The rhetoric of genetic technologies, of transpeciation, has seduced not just academia in recent years, and its funders, but also artists that one might presume would provide counternarratives to the resulting disruption of life as we have understood it, up to this point in time.

This may sound nostalgic, but the creation of a phosphorescent rabbit produced through transpeciation with a jellyfish gene by the artist Eduardo Kacs has opened up a debate that brings into question the ethical climate surrounding art practice in the postmodern biotechnology environment. In Canada, there are new funding opportunities jointly provided by the Canada Council for the Arts and the National

Research Council encouraging artists and scientists to collaborate, making one wonder about common ground; about how artists might incorporate the delimitations of the scientific method, or how scientists will digest subjective reasoning. Wendy Coburn's *Die Trauernde* is only one of many art works that contest scientific methodology. Jeanne Randolph (1991) in her curatorial essay for the exhibition Influencing Machines asked the question, "Can artworks ever offer perceptions of technology that influence technology's effect" (p. 37)?

I am reviving Randolph's provocative question, for its query frames this discussion of work by artists who challenge the precepts of science, psychology, the origin of the species and genetic engineering. They question our acceptance of science as the ultimate purveyor of truth and knowledge, contesting its claims through artworks that reveal shameful realities and alternate worlds.

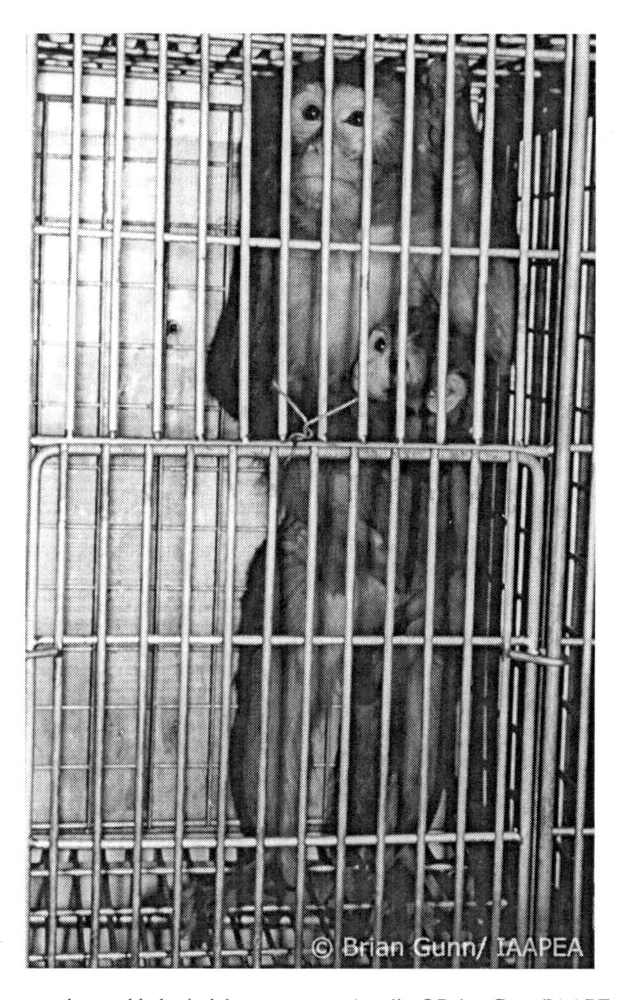

Fig. 1 Macaque mother and baby in laboratory cage (credit: ©Brian Gunn/IAAPEA)

Life as We Knew It

It is fundamental to this discussion to map out the shift in the study of organisms from that of whole, complete subjects, to what Richard Doyle (1997) has referred to as postvital bodies. Doyle has tracked the path to our current position as "Genes R Us," to a location where the body is reduced to a code. Although Doyle is more interested in the rhetoric and language of the new biology, he does establish that the move from natural history to biology produced organisms ripe to be identified through molecular biology (p. 11). This constituted a pivotal change in the study of living organisms, setting the stage for moving from a study of macro to micro, exterior to interior, into what Foucault (1994) has perceived as the "dark, concave, inner side" (p. 237). Foucault discusses natural history as the "nomination of the visible" (p. 132), the ordering and naming of organisms based on their appearance, their structure, rather than their character. The apparatuses that made visible interior space were developed in parallel with the instrument of taxonomy as proposed by Carolus Linneaus in 1735, through the publication of *Systema Naturae*. Linneaus' system of classification of plants relied solely on the characteristics of the visible reproductive organs (stamens, pistils), but when he turned his attention to animals, internal anatomical characteristics were included, implying the necessary dissections. A reliance on four variables—number, form, situation, and proportion—allowed the taxonomist the opportunity to classify any organism as part of a group in which members shared quantifiable traits. The Linnean system differentiated itself from earlier nomenclatures of animals that had included litteraria, information that pertained more to the narrative history of the organism than to science. Artists Shawna Dempsey and Lorri Millan's recent *Field Guide to North America* (2002), sponsored by the fictive Lesbian National Parks and Services, hearkens back to earlier representations of species that included subjective information. The language of this field guide, unlike the technical language of "authentic" field guides, is literary and sympathetic, the illustrations by Daniel Barrow are intentionally naïve and whimsical. Facts are presented in tandem with comments on sexual politics. Dempsey and Millan's inherent critique of Linneaus, and the acknowledgment of the naturalists' dependence on the "field guide" and its keys to classification, suggests that these artists recognize a continued dependence on the ordering of living organisms. In the section titled Mammals and You the artists inform us of our responsibility to mammals and the necessity that we recognize our inclusion in that family, adding that the Lesbian Ranger is a mammal too. This observation is followed by a key to *Homo lesbianus* subtypes: the helper-lesbian, the activist, the femme goddess and the lady golfer among others. This may seem like a gratuitous digression; however it demonstrates Dempsey and Millan's grasp of the othering process that naming, coupled with classification, facilitates.

Linneaus' project coalesced with other imperialist practices that were preoccupied with what Francis Bacon referred to as "a truly masculine birth of time" (Cited in Bordo 1987, p. 104). Evelyn Fox Keller (1985) explicates this proclamation for us. "Masculine here connotes, as it so often does, autonomy, separation, and distance. . .a radical rejection of any commingling of the subject and object" (p. 79).

Linneaus' disregard of the litteraria, the connection of organisms to their subjective histories, constitutes a rejection of such a commingling. Classification and natural history were efforts to counter the speculative science of earlier centuries. It is useful to look at other endeavors that attempted to deal with "wild" life. Ecology emerged from the seventeenth century with two very different and opposing streams. Pastoral ecology advocated a simple, humble life of peaceful coexistence with other organisms, and imperial ecology emphasized man's dominion over nature, established through reason and hard work. It was the imperialist ecologists who established a mastery of nature. Max Oeschlaeger (1991) notes, "Accordingly, all parts of nature were—like a machine—interchangeable and expendable. And the natural world was analogous to a factory to manufacture an unending stream of products for human consumption, and thus the landscape has only instrumental and not intrinsic value" (p. 105). Extrapolating from this perception of landscape, it can be assumed that the organisms it houses, flora and fauna, are equally expendable.

How does one counter these practices borne from what Charles Taylor (1989) has identified as instrumental reason? (p. 383) Taylor points to this instrumental stance as an objectification of nature, and that through the objectifying and neutralizing process we declare our moral independence from it. This may have been decidedly obvious to René Descartes (1969) and his division of the world into *res cognitans*, "a thinking and unextended thing" and *res extensa*, "an extended and unthinking thing" (p. 190), but it is less apparent with Linneaus's all encompassing taxonomic categorizations. Taylor furthers his critique, "One of the great objections against Enlightenment disengagement was that it created barriers and divisions: between humans and nature and perhaps, even more grievously, within humans themselves, and then as a further consequence, between humans and humans" (p. 383). The Romantic Movement so evident in the nineteenth century art and literature was an attempt to counter moral disengagement. For Taylor, expressivism in art during this period diversified our moral sources with nature now perceived as one of those sources. Taylor ties the desire for reparation through art directly to the rise of the instrumental stance implying that the approach of contemporary art, its assumed role as a critique of instrumental reason, is essential to the formation of the modern identity. If that is so, how do we speak to art practice that supports instrumental values? How do we consider contemporary art that abides by and submits to the contemporary equivalents of disengaged instrumentalism?

The images from Harry Harlow's lab in *Die Trauernde* imply that the trajectory was peaking in the 1950s. George Romanes, considered to be the father of comparative psychology, published a text on comparative animal intelligences in 1882. A close friend of Charles Darwin, he was born in Kingston, Ontario in 1848, subsequently spending most of his life in England. In the preface to *Animal Intelligence* we read of Romanes's struggle against instrumentalism as he tried to determine what constitutes knowledge, and whether or not facts and observations that are relayed through questionable sources will be acceptable to his project. The thesis of Romanes' text is the laying of a foundation for "Mental Evolution," but my particular interest in his work lies in Romanes' treatment of his subjects of inquiry as he moves from protozoa to primates. The notion of an anomalous subject drives

Romanes to describe a level of anthropomorphism beyond our imagination: For instance, if we find a dog or monkey exhibiting marked expressions of affection, sympathy, jealousy, rage, etc., few persons are skeptical enough to doubt that the complete analogy which these expressions accord with those which are manifested by man, sufficiently prove the existence of mental states analogous to those in man of which these expressions are the outward and visible signs. But when we find an ant or a bee apparently exhibiting by its actions these same emotions, few persons are sufficiently nonskeptical not to doubt whether the outward and visible signs here are trustworthy as evidence of analogous or corresponding inward and mental states (Romanes 1882, p. 9).

Romanes' categories of research and observation in his chapter on ants display a sympathy toward these organisms that belies future scientific research practice. Powers of memory, emotions, sympathy, powers of communication and keeping Aphides are among the initial categories. Later he moves on to even more anthropomorphized observations such as play and leisure, sleep and cleanliness, keeping pets, slave-making and wars. In spite of a racial slur in the slave-making behavior discussion, overall Romanes credits animals with full, whole lives, with emotions and decision-making capacities. Romanes was somewhat of an anomaly, torn between his interest and empathy for animals, and his allegiance to Science, and Darwin who hovers like a specter throughout *Animal Intelligence*. Whenever the psychologist veers into dangerously subjective reasoning Darwin is invoked and Romanes gets back on track distancing himself from his field of study through reason. But there is evidence in the following poem, "Scientific Research," written by Romanes (1896) that he struggled with the constraints of science:

> Why should I chafe and fret myself to find
> Some pebble still untouched upon the beach,
> Where struggling wavelets follow each on each
> Upon the tide-mark of advancing Mind. (p. 37)

In a poem titled "What is Truth?" the final lines recall his skepticism of science:

> The lock was turned; in floods of rainbow light
> I saw her pass; and then no more could see.
> Where Thought and Science access failed
> to win,
> 'Twas Art that opened, Art that entered in. (p. 32)

By studying Romanes' scientific research in conjunction with his poetry one gets the impression that he was unsettled in the necessary objectification of his subjects. His ability to reject the commingling of the subject and object was deterred by his sympathies. Margaret Bentson (1989) has argued the scientific method to be "an axiom of impoverished reality," and that if the quantifiable aspects of the phenomena are the best truth then this is compatible with male norms rather than female ones. This stripping of qualitative content, whether experiential, metaphorical, or ethical presents a reality that is diminished. Is this why Romanes felt compelled to write poetry, to fulfill this aspect of his work that was impeded by the assumptions of scientific research?

Evelyn Fox Keller (1985) draws from Barbara McClintock's work in order to think about the ways that science might be differently practiced. McClintock, considered deviant and eccentric, made her mark with her research, during the 1930s, on genetics and cytology. For McClintock, nature's complexity vastly exceeded the human imagination. "Anything you can think of, you will find" (p. 162). McClintock believed that scientists could only begin to fathom the depths of the life and order of organisms. She found organic life to be misrepresented in science, where the central dogma required an ordering that was impossible to achieve. McClintock's attitude toward 'nature' led to an emergent approach to research where it became essential to "let the experiment tell you what to do" (p. 162). Her own research on genetic transposition, studying an aberrant pattern of pigmentation on a few kernels of corn for 6 years, led her to the assertion that exceptions are not there to prove the rule, but are instead opportunities for determining relatedness. Keller points out that McClintock's respect for difference serves two functions, "a clue to new modes of connectedness in nature; and as an invitation for engagement with nature" (p. 163). Recognition of the enduring uniqueness of every organism is McClintock's contribution to the necessity of subjective and objective commingling. "I start with the seedling, and I don't want to leave it. I don't feel I really know the story if I don't watch the plant all the way along. So I know every plant in the field. I know them intimately, and I find great pleasure to know them." Keller tells us, "The crucial point for us is that McClintock can risk suspension of boundaries between subject and object without jeopardy to science precisely because, to her, science is not premised on that division" (p. 164). Lorraine Code (1988) lauds Keller's suggestion of the possibility of reciprocal subject–object relations, asserting that it is possible to weaken the boundaries between the subject and object. "The primary goal of the activity is understanding, rather than manipulation, prediction, and control" (p. 81). McClintock's interest was in how genes functioned in the 'whole' organism, rather than abiding by the divide and separate mentality of mid-twentieth century genetics, inculcated by molecular biology. She saw value in the way that a particular gene in a particular plant was not just a part of a cell and not just a part of that plant, but also part of the broader ecosystem in which it lives. One senses in Romanes's writings the desire to adhere to a similar philosophy, but the patriarchal constraints of science, coupled with his friendship with one of the preeminent scientists of that time, led to a much more traditional practice securing his position as the 'father' of comparative psychology.

At some point in our recent past, an "othering" process took place facilitating our moral and physical disengagment from animals, enabling experimentation that overtly denies the rights of the subjects. Of course ethologists would dispute this claim since the study of animal behavior suggests otherwise, but it is still incumbent on contemporary animal behaviorists to "prove" that emotions exist. The primatologist Jane Goodall has voiced her frustration at this empirical approach, and points out that in 1960, "It was not permissible to talk about an animal's mind, only humans had minds." When questioned about her insistence on allowing for each of her chimpanzees, a study of their individual personalities, "to name them and to name their emotions" (Wachtel, 2003, p. 56), Goodall gave the atypical response of having had a good teacher, her dog Rusty. "He taught me so much about animal behaviour. So even before I got to Gombe, I knew that animals had personalities, they were capable of rational thought, and that they had emotions—happiness, sadness, despair and so forth" (Goodall, 2003, p. 56).

Romanes's poems were published in 1896, an auspicious year for technological progress, with the discovery of the X-ray and following the birth of cinema. Lisa Cartwright (1995) traces the methods of medical surveillance that enabled further changes in the relationship between representations and things. Cartwright introduces her discussion of medicine and visuality by noting that the Cinématographe was an innovation of Auguste Lumière and his brother Louis for purposes of laboratory research on tuberculosis and cancer. For Cartwright, the Lumières' treatment of cinema, particularly when it is conflated with microscopic imaging processes, becomes an instrument not only of surveillance, but also of power, leading to a decomposition of the body and disengagement with its corporeality. The X-ray added to this representational deconstruction by producing an image of the internal body that hearkened back to the phantasmagoria of the early 1900s. The specter of the inner body, of the skeletal structure with its blue-gray ghostly aura, provided an image that touched on mortality and death allowing medicine and science more power to subvert the inevitable. The apparatuses of medical surveillance paved the way for the mining of knowledge and colonization of and from the body, but the ironic effect of these increasingly sophisticated tools of knowledge acquisition was the distancing of ourselves from our own bodies, the environment and the organisms it houses.

Life as We Know It

The breadth of information and knowledge that flooded the culture via medical surveillance technologies in the nineteenth and twentieth centuries enabled the processes of molecular biology and eventually genetic engineering. Richard Doyle (1997) laments the current move to sequencing and the technical, repetitive nature of genomics asking the question "is that all there is?" (p. 23). When Celera Genomics and the Human Genome Project announced a working draft of the genome in June 2000, they realized that it would take some time to rearrange the parts list that they had identified into the actual sequence of the genome itself. The artist Nell Tenhaaf offers an important counternarrative to their unfinished story, one that includes gender and mythology. Her machines and images provide access to scientific constructs that often elude us. Over the past two decades Tenhaaf has undertaken a critique of science and the accompanying illusion of objectified knowledges, through her multidisciplinary art works that incorporate photography, video, computer graphics, audio, and interaction. In the photographic work *The solitary begets herself, keeping all eight cells* (1993), Tenhaaf critiques reproductive destiny and the meiotic processes of subject determination. The artwork consists of a long thin horizontal light box, twice the length of Tenhaaf's height, containing a slice of an image of the artist lying prone. The nude body of the artist is superimposed with images of a dividing zygote, changing from two, to four, to eight cells, laboratory glassware, and two odd mythical figures. The source for the chimerical figures is Conrad de Megenberg who illustrated the *Buch de Natur* in the fourteenth century. One of the figures is a mermaid with wings replacing her arms, and the other an armless woman's torso with the lower body of a dog. This artwork supported Tenhaaf's stance at the time against the genomic project of perfecting our species. As Kim Sawchuk notes in her discussion of this art work, "Paradoxically, genetic engineering creates new hybrid species at the same time as it wants to eliminate mutations in the human gene pool" (Sawchuk, 2003, p. 18).

The call to perfection is the lynchpin of genetic manipulation as the desire for the same leads to ever more invasive reproductive technologies. Tenhaaf postulates a parthenogenetic narrative, a virgin birth, taking it out of the male–female coupling paradigm and, although it may invoke the clone, her premise is quite different. The clone is lab-manufactured, synthesized. The virgin birth remains in situ, within the uterus of the mother, subject to the random mutations that accompany human embryo development. *Homo sapiens*, and all carbon-based organisms, are emergent beings. Our new relationship to silicon-based life, since the "digital revolution," has given the impression that we have more control than we do in terms of the outcome of reproduction. Again I turn to Richard Doyle, "'Life' just isn't what it used to be" (p. 25). The tropes of mutation, as well as replication, slip back and forth, from synthesized hybrid to unintended monster. Teratological apparitions haunt our dreams of progeny, leading some of us to agree to manipulations of the proteins that determine our physical manifestation as either humans, animals, or plants. Fear and paranoia serve the scientific establishment in its quest for new sites of colonization that result in the patenting of existing and newly synthesized carbon-based life forms. As Vandana Shiva (1997) has argued, the contemporary genomic gold rush is not unlike the decree of Pope Alexander VI whose papal "Bull of Donation," of 1493, granted all islands and mainlands discovered, and not already occupied by any Christian king, to be the property of the Catholic reigning monarchs of the time (p. 1). For Shiva, patents and intellectual property rights are just a more secular version of colonization. The papal bull stems from the century following that from which Tenhaaf drew the hybrid monsters that waft over the naked splice of her body. The papal Bull of Donation represents a turning from a fear of the monster to the belief in Christianity's ability to conquer and dominate whatever might befall it. Unforeseen was the potential for profit in the pillaging of territories, peoples, animals, and plants. The resulting economic enhancement is the fallout of colonization, fortunate or unfortunate depending on one's position of subject or object. In North America in 2005, when the Christian belief in an impending apocalypse has somehow become a legitimate influence on state policy, there is evermore concern for the commodification of life that accompanies genetic engineering. Tenhaaf is present in her dream of parthenogenesis, it is her body that has been spliced for the image, that has been degraded through numerous generations of copying. Her manipulation of scientific tenets and the elements of mutation that are welcomed into this narrative compel us to look to possibilities within ourselves to combat ideals of perfection.

The conception of the hybrid, and the contemporary rhetoric that surrounds this term, is worth considering in relation to Tenhaaf's work, but also in relation to its rise in contemporary culture as a descriptor of choice. My concern lies in how non-vested agents laud hybridization as an all-encompassing trope, but fail to confront the implications of how these ideas might play out in real-time in the laboratory. The eco-feminist Ariel Salleh (1997) reiterates this concern suggesting that the anti-realist stance of postmodernism, its micro-political focus on text and its discursive pluralism all provide distractions from material reality. Contemporary theoretical stances connote paradoxical positions that problematize our relationship to hybridized life forms. Even Nell Tenhaaf, in relation to *The solitary begets herself keeping all eight cells* confesses that, "Through this fantasy narrative, I reveal my co-mingled fascination and dread in the face of the biotechnological agenda, that is, willed and controlled mutation through genetic intervention." Tenhaaf's willed and controlled mutation evinces her perception of biotechnology as the promethean force that it is, with the potential for new organisms that reside in a liminal space, between the known and unknown. Again, the potential for monsters is at play.

Rosi Braidotti (2002) has traced the monstrous m/other in the science fiction genre as laid out in literature and film arguing that our contemporary penchant for these representations acts as a "counterpoint to the emphasis that dominant post-industrial culture has placed on clean, healthy, fit, white decent, lawabiding and heterosexual young bodies. . .the fascination for the monstrous, the freaky body-double is directly proportionate to the suppression of images of both ugliness and disease in contemporary post-industrial culture" (pp. 199–200). For Braidotti, the monstrous acts as a metaphor for the anxiety of postmodernity. How do we negotiate the differences between the theoretical musings that present a seductive reading of the monstrous, of the hybrid and the material reality of genetically modified organisms? This is not a contemporary circumstance—science fiction writers have always written our future, questioning normative cultural life—but here is the slippery slope that exists between Eduardo Kac's Alba and OncoMouse™, the first animal to be patented. Donna Haraway (2000) expresses very well the position in which these animals are placed, simultaneously existing as 'real' animals but also figurations that convey new ways of living in the world. "Something has to be invented in order to be patented. It is therefore authored, is the offspring, the property of someone or some corporation, and is therefore fully alienable, fully ownable" (140). For Haraway the patented animal is a mixture of nature and labor that sits alongside the machinic-organic cyborg. Lamenting the suffering that OncoMouse™ has incurred, she identifies her debt to this invention, and the knowledge that has been gained as a result of this mouse that reliably produces breast cancer. Haraway's acknowledgment creates a portal through which we can enter and recognize our collective complicity in the project of genetic modification. We might ask at this point to whom are in we debt for Alba? For what has this rabbit suffered the degradation of its genetic makeup? Art?

N. Katherine Hayles (1999) has also documented the ushering in of the cyborg and the posthuman where leaky boundaries lead to teratologies real and imagined. Hayles's definition of the posthuman is a cautionary stance, but it is one not easily avoided in the new millennium. The posthuman "privileges informational pattern over material instantiation, so that embodiment in a biological substrate is seen as an accident of history rather than an inevitability of life"(p. 3), it regards consciousness as an epiphenomenon, and finally, "the posthuman view configures the human being so that it can be seamlessly articulated with intelligent machines" (p. 3). Hayles emphasizes the inevitability of the posthuman; however, she implores us to maintain embodiment in the face of the technological imperative. Is there a danger of compliance implied in both Haraway's and Hayles's texts? If the posthuman is inevitable, and if we are to recognize our debt to patented organisms, then are we at risk of becoming what the physicist Ursula Franklin (1990) has referred to as a culture of compliance? (p. 19). In her 1989 Massey Lecture, Franklin outlined technology as a system and a practice, broken down into holistic, bottom-up and prescriptive, top-down technologies. The genetically modified organism sits somewhere in the threshold between these two—being synthesized, but grown from an embryonic stage. Franklin cautions us in that "technology has built the house in which we all live" (p. 1) and this might be Hayles's and Haraway's point, that it is difficult to escape the inevitabilities that are the legacy of the informatics of domination. Elaine Graham (2002) postulates a more contentious reading of the posthuman seeing it as not only inevitable, but also potentially beneficial, insisting that we acknowledge the, "symbiotic relationship between humanity and its artifacts, a blurring of agent and object, external and internal, organic and artificial" (p. 33). Graham warns us against policing such boundaries, referring to our dependence on such polarities as the result of ontological hygiene. Following Bruno Latour's call for problematizing purity, Graham emphasizes that there are no pure categories of race, gender, or sexuality that can securely define the self, even questioning the evolutionary inevitability of the posthuman. So where does this leave us? If we are suspect of ontological hygiene then what is the problem with GMOs? It also bring us back to Linneaus suggesting perhaps that our desire for a lack of tampering with species hearkens back to an insistence on pure categories, all the better to classify, all the better to identify.

These questions place us in a paradoxical position that Braidotti (2002) argues is the postmodern condition, "The simultaneous occurrence of contradictory trends preoccupies most cultures in this new millennium" (p. 176). Such contradictory trends, so evident in Haraway, Hayles, and Graham, if translated onto other media of cultural expression like art practice, have sweeping ramifications. In order to stand back from policing the boundaries of the posthuman, do we become complicit in another kind of disengaged instrumentalism, the kind that, according to Charles Taylor (1989) empties life of meaning and threatens public freedom (p. 503)? Carol Gigliotti (2000), in a discussion of the ethics of porn on the internet, points to the false neutrality and objectivity of technology. She suggests that we recognize that while we collaborate with the world, we are the world and what we share with living organisms is embodiment (p. 55). Maybe that is Haraway's intent as well when she speaks of her debt to OncoMouse^{TM}, recognizing her connection to this organism, that it is part of her, as it resides in carbon-based life, with DNA as its mapmaker.

Adam Brandejs (2005) has recently produced an artwork that realizes the paradox of the monstrous and the genetically modified organism. His haunted sculptures are animatronic animals, constructed by the artist, housed in plastic packaging and hung on racks, a la retail. The products, in their mock-store setup, display streamlined, mass-produced bioengineered life, that are mammalian in their derivation as they twitch, shake, claw and head butt their packages. Tubes supply them with the electricity that acts as a stand-in for nourishment. They are convincing in their stance as a toy and collectible, for there are various states that the genpets can be purchased in, for example, the orange packaging indicates that the genpet is adventurous, confident and curious, whereas the red is athletic and energetic. The product is essentially the same and, much like the iMacs, the colors are purely a marketing ploy. With this work Brandejs takes Haraway's reading of the genetically modified animal a step further. He makes plain that these organisms are not just nature and labor, but are a conflation of nature, labor, and commerce. They are contradictory. They appear real but are animatronic; they appear to be toys but they are not for sale; they appear to have emotions and motivations but they are all the same. Are they clones? No, because they are sculptures made of latex, microchips, and other electronic mechanisms. Brandejs is a young artist, of the generation following the progenitors of genetic modification. He has inherited genetically modified organisms (GMOs) and it is incumbent on him to either adapt to them or search out alternate solutions. His decision to critique rather than embrace the possibilities in biotechnology, indicate that he has not been seduced by the practices of his seniors. Like Tenhaaf and Coburn he draws our attention to the contradictions of scientific practices and particularly genetic engineering. "Biological engineering by large companies, outside of nature, has become a terrifying reality for my generation to contend with. Today, we are well within the process of desensitizing an upcoming generation towards accepting bioengineering as 'natural.' This generation is slowly and systematically being desensitized towards owning and manipulating life through toys such as Tamagotchis, and Furbies" (paras 1–2). This comment is taken from a recent artist statement by Brandejs. By tying bioengineering to the process of desensitizing children with toys that require interaction and maintenance, he cannily denotes the continuum on which the GMO lies; certainly it is a product of the lab, but if Brandejs is correct the leap to the retail market is not far off. The environment in which any organism lies is significant in the reception of that organism by the population. As Haraway (2000) has noted, the natural habitat for OncoMouse^{TM} is the laboratory, this is where its evolutionary history lies. "And the condition of its being is not just sexual reproduction, the history of the evolution of mice (of mammals), but also the history of the development of gene transfer technology" (p. 145). For years cereal manufacturers have been inserting genetically modified soy into their product (Boyens 2000, p. 224), in fact they have more or less adopted a gene gun approach by blasting the supermarket shelves with GMOs. The benign appearance of the supermarket makes the GMO seem harmless and after all, we've been eating them for years, it's just that no one told us. They are good mimics. The invisibility of the GMO makes it impossible to detect for the lay person. The highly specialized

Fig. 2 Beagle dying from toxicity tests in laboratory (credit: ©Brain Gunn/IAAPEA)

nature of the modification, one that is only detectable at the level of the protein, leaves most of us at a loss, unable to identify, detect, resist.

It also makes the scientific process, its lexicon, and its apparatuses all the more seductive, as evidenced by the wide circulation of such within the artists' realm since the analogue to digital shift. Artworks incorporating new technologies have been marginalized within the exhibitionary complex and so appear radical and resistant. For the most part this has been the case, but perhaps not in relation to artists tampering with genes. If the driver of the technology is corporately based, as are most biotechnology initiatives—the Human Genome Project being a case in point with the United States government joining Celera Genomics for final identification of the parts list—then can artwork that engages these technologies appear radical? I think not, since the imperative of instrumental disengagement is so inherent to the biotechnological practices.

The quandary that plagues so many of the authors and artists I have cited—Goodall, Haraway, Tenhaaf, Brandejs—indicates that the consequences of tampering with the genes of living organisms has the potential for far-reaching effects. Our dependence on animals for nurturing is based on our knowledge of their communicative abilities, their ability for empathy and compassion. Jane Goodall's admission of the pedagogical relationship between her and her dog, Rusty, affirms that she questions the very science she practices. Artists are in a position to evaluate how genetic modification impacts living organisms, and cultures, but not if we embrace methods that are shared with the postindustrial corporate world. The dilemma is not new and we are well aware that many new technologies had their originary moment in the military industrial complex, but tampering with genes suggests a promethean compulsion wherein the seduction of the medium and its exclusivity implies a radical practice. The motivation for genetic manipulation has been for perfection first and profit second. How artists can insert themselves into this paradigm and maintain critical distance is yet to be seen, for until now direct involvement with genetic manipulation has been an unsuccessful proposition.

References

Bentson, M. (1989) Feminism and the critique of scientific method. In A. Miles & G. Finn (Eds), *Feminism: From pressure to politics*, Montreal: Black Rose Books.

Brandejs, A. (2005) Genpets/Why: Artist Statement (V1.3). Retrieved from http://www.brandejs.ca/portfolio/Genpets/Why

Bordo, S. (1987). *The flight to objectivity: Essays on cartesianism and culture*. Albany: State University of New York Press.

Boyens, I. (2000). *Unnatural harvest: How genetic engineering is altering our food*. Toronto: Doubleday

Braidotti, R. (2002). *Metamorphoses: Towards a materialist theory of becoming*. Cambridge: Polity Press.

Cartwright, L. (1995). *Screening the body: Tracing medicine's visual culture*. Minneapoils and London: University of Minnesota Press.

Code, L., Mullett, S. & Overall, C. (1998). *Feminist perspectives: Philosophical essays on methods and morals*. Toronto: University of Toronto Press.

Dempsey, S. & Millan, L. (2002). *Field guide to North America: Flora, fauna and survival skills*. Toronto: Pedlar Press.

Descartes, R. (1969). *Philosophical Works Vol. I and II*. In E. S. Haldane & G. R. T. Ross (Eds), Cambridge: Cambridge University Press.

Doyle, R. (1997). *On beyond living: Rhetorical transformations of the life sciences*. Palo Alto: Stanford University Press.

Foucault, M. (1994). *The order of things: An archaeology of the human sciences*. New York: Vintage.

Gigliotti, C. (2000). The ethical life of the digital aesthetic. In P. Lunenfeld (Ed). *The digital dialectic: New essays in new media*. Cambridge and London: The MIT Press.

Goodall, J. (2003). "Jane Goodall". E. Wachtel (Ed.), *Original Minds*, Toronto: Harper Collins.

Graham, E. (2002). *Representations of the post/human*. New Jersey: Rutgers University Press.

Haraway, D. (2000). *How like a leaf*. New York: Routledge.

Harlow, H. & Mears, C. (1979). *The human model: Primate perspectives*. New York: Wiley.

Hayles, N.K. (1999). *How we became posthuman: Virtual bodies in cybernetics, literature, and informatics*. Chicago: University of Chicago Press.

Keller, E. F. (1985). *Reflections on gender and science*. New Haven: Yale University Press.

Oelschlaeger, M. (1991). *The idea of wilderness: From prehistory to the age of Ecology*. New Haven: Yale University Press.

Randolph, J. (1991). *Psychoanalysis and synchronized swimming, and other writings on art*. Toronto: YYZ Books.

Romanes, G. (1882). *Animal intelligence*. London: Kegan Paul, Trench.

Romanes, G. J. (1896). A Selection from the Poems of Geroge John Romanes. London, New York: Longmans, Green and Co, 32, 37. Retrieved from http://www.archive.org/details/selectionfrompoe00romauoft

Salleh, A. (1997). *Ecofeminism as politics: Nature, Marx and the postmodern*. London: Zed Books.

Sawchuk, K. (2003). Biological, not determinist: Nell Tenhaaf's technological mutations. Fit/unfit: apte/inapte, a survey exhibition/un survol. Oshawa: Robert McLaughlin Gallery.

Shiva, V. (1997). *Biopiracy: The plunder of nature and knowledge*. Cambridge, MA: South End Press.
Taylor, C. (1989). *Sources of the self: The making of the modern identity*. Cambridge: Harvard University Press.
Wachtel, E. (2003). *Original minds*. Toronto: Harper Collins.

Meddling with Medusa: On Genetic Manipulation, Art and Animals

Lynda Birke

> *Across the clearing to the south comes a rabbit, hopping, listening, pausing to nibble at the grass with its gigantic teeth. It glows in the dusk, a greenish glow filched from the iridocytes of a deep-sea jellyfish in some long-ago experiment. In the half-light the rabbit looks soft and almost translucent, like a piece of Turkish Delight; as if you could suck off its fur like sugar. (Atwood, 2004, p.109–110)*

Abstract Turning animals into art through genetic manipulation poses many questions for how we think about our relationship with other species. Here, I explore three rather disparate sets of issues. First, I ask to what extent the production of such living "artforms" really is as transgressive as advocates claim. Whether or not it counts as radical in terms of art I cannot say: but it is not at all radical, I argue, in terms of how we think about our human place in the world. On the contrary, producing these animals only reinforces our own sense of our importance. The second theme, I explore, is the extent to which making transgenic organisms for any purposes is radical in terms of complexity. Here, I focus on the idea of complexity as a concept in developmental biology; genetic manipulation may be successful to commercial companies, but it is deeply troubling to many biologists who consider that its deeply entrenched reductionism is enormously problematic. What risks do we run by ignoring nature's own complexity—and creativity? And—in particular—what risks do we run of damaging or compromising animal welfare? The third theme turns to public perceptions of these new technologies (whether in science or art), and notes the extent of public unease. This unease is not simply a question of public ignorance about the technology, but reflects the enormously rich ways in which we make meanings about animals, and relate to them. These are, I suggest, a far more potent source of creativity than simply moving genes around to make photogenic animals.

L. Birke (✉)
Anthrozoology Unit, University of Chester, Chester, UK
e-mail: l.birke@chester.ac.uk

C. Gigliotti (ed.), *Leonardo's Choice*, DOI 10.1007/978-90-481-2479-4_7,

Keywords Art · Animals · Genetic manipulation · Complexity · Creativity · Meaning

Humankind have been interfering with non-human genetics for centuries, trying to shape animals and plants to our own ends, including aesthetic ones. Alongside that we have a long history of fear and fascination with hybrid kinds, expressed as the strange animal/human hybrids who inhabit our mythologies—as well as littering our dystopias. If new genetic technologies offer renewed means of tinkering, and if humans use them for "aesthetic" ends, then we should not be surprised.

What does surprise me, however, is the way that genetic manipulations in the name of art are so often portrayed as radical or innovative. Perhaps there is a trivial sense in which making a monkey or rabbit which glows in the dark because of genes "filched from deep-sea jellyfish" is innovative. Whether it is either aesthetic or ethical is more debatable. But it is not radical: first in the obvious sense that tinkering with other species' genetics is something we have been doing for a very long time, and second in the more restricted sense that moving genes around in this way is not the radical cutting edge of biology that so many of its advocates seem to imply.

In this chapter, I explore some of the issues that "bio-art" raises, to my eyes, for our relationship with animals, especially when the bioart entails deliberately making whole new organisms.[1] I must begin by confessing my biases. I am a biologist, but one who is not enamoured by the reductionist logic of new genetic technologies. On the contrary, I am much more excited by areas of biology that explicitly reject reductionism and seek alternative ways of understanding how life works. I am far more awed and stunned by the incredible beauty out there in nature than I am by the possibilities of fiddling around with those shapes and forms in the laboratory—or in any other space.

I explore these issues through three, rather disparate, themes—ways in which the production of these animals is, for me, problematic. The first concerns some of the claims made by artists and advocates—is the production of transgenic animals as art transgressive, for example? Does it challenge human centrality in the world? Does it involve human/animal communication in new or radical ways? The second theme is more concerned with how we think about biology, and the reductionist assumptions underpinning any work in genetic engineering, assumptions which are not at all radical. The third asks questions about public perceptions of both animals and science: does the production of these organisms promote public engagement with science, as some have claimed?

Human intervention in the processes of natural selection—by which nature changes the genetic inheritance of species—has a long history. Although humans have engaged in some selective breeding of both plants and animals for centuries, it intensified in the seventeenth and eighteenth centuries, and more and more breeds of domesticated animals and plants were produced by selective mating. The human motives were various—economic, utility, aesthetics. For several resultant breeds, there were ethical issues, as the desired characteristics created welfare problems (breathing difficulties created by the shortened nose of some pet dog breeds for instance).

The twentieth century, and the beginning of modern genetic understanding, saw further developments in breeding. Where before selection was simply a case of bringing together males and females of particular phenotypic traits, now it became possible to select for traits with known genetic linkage, and so make inheritance more predictable. In the second half of the century, the identification of DNA's structure meant that very precise genetic detail could be identified and, eventually, manipulated.

Advocates of genetic manipulation frequently argue that there are huge potential benefits to biotechnology (feeding the world, yielding new medicines, for example), and that changing the genetics of an organism through biotechnology is not fundamentally different from the breeding programmes that preceded it. Whatever one thinks about potential benefits versus possible harms, it is certainly true that human attempts to alter the genetics of other organisms are not new. We have been doing it for a long time, but we are just better now at focusing in on specific genes.

Artists, meanwhile, have long sought inspiration in the forms of nature, often seeing themselves as Leonardo's descendants, crossing the art/science border (Gigliotti, 2006). That border has undoubtedly become more permeable in recent years, partly through initiatives to promote public understanding of science. These aim not only to make complex scientific and technological ideas more accessible, but also to promote debate about ethical issues, using artworks as a means of making the science (and the art) more accessible. Thus, a recent project funded by the Wellcome Trust in London used art to explore the deeply contentious question of euthanasia and assisted suicide, for example.[2]

What makes recent developments in "bioart" new is that biological materials or organisms, and biological techniques, are now becoming part of the processes of producing art. Precisely because of this challenge to old boundaries between science and art, bioart has been called "transgressive" or "radical." And transgressive it indeed is, in the sense that it uses art to go beyond representation, to breach the boundaries of what it is to be living. Yet there are also ways in which the creation of such animals as art is not transgressive. For a start, it uses techniques that are now commonplace in the labs. Done in those places, it may be worrying for many people (wondering, what are scientists creating?), but it is hardly rare. It is transgressive only because bioart takes the techniques (partly) out of the laboratory and into the everyday world, an act that changes the relationship between laboratories and the wider society.

Not surprisingly, there have been heated ethical debates about the practice and potential of bioart, particularly when new, living, organisms are thus created. I must admit to mixed feelings, depending on what is done to living tissue in the name of art. The idea of performance art based on the extraction of tissue and its growth into "victimless meat", for instance, does appeal to me as a lifelong vegetarian (though it appeals more in the sense of it becoming commercial practice than in the sense of it as "art"). Some of it offends my sense of beauty inherent in nature—I am bewildered by the use of biotechnology to create "nonsense" patterns in living butterfly wings,

Fig. 1 Rat prepared for brain surgery (credit: uncredited)

and at the same time amused that this creation provoked much more public outcry than alterations in the name of art to a cockroach's nervous system (Catts and Zurr, 2003). The mixed message here is, of course, a speciesist one: we live in a culture that values butterfly wings and their bright colours, and associates cockroaches with dirt and disease. So, if artwork based on butterflies is more highly valued it should not come as a surprise (except perhaps to those of us who value the original butterfly more).

Breaking Down Species Barriers, or Shoring Them Up?

Crossing the borders of art and science is all very well, even desirable in some senses. But making transgenic animals is a border crossing that raises more than just eyebrows and provocative headlines. It poses a number of questions that, for me, remain unanswered. One is the issue of anthropocentrism: is crossing the boundaries of living organisms anti-anthropocentric in the way that its practitioners claim?

Anthropocentrism has indeed dominated the way we view the rest of the living world; from the Judeo-Christian traditions we in the west have inherited relationship to nature that is profoundly one of dominion, in which we arrogantly place ourselves at the centre. We undoubtedly need to challenge that in any way we can. But moving bits of DNA or pieces of animal tissue around—whether for purposes of basic science, pharmaceutical production or art—does not do so. Humans remain, as other contributors to this volume also note, right at the centre of all these endeavours: it is our interests that remain paramount, and it is human motives that shape the direction and practices of both art and science.

Moreover, for all that artists may speak of "making humans part of the continuum" and breaking down species barriers, it is not humans whose genetic integrity is thus compromised. We have not yet seen a green fluorescent protein (GFP) (fluorescent) human baby. That would, no doubt, produce a much stronger "yuk" reaction, and a sense of public revulsion, than GFP bunnies—which in itself underlines the strength of anthropocentrism. For if genetic boundary crossing is really so radical a challenge to our place at the centre of our universe, why should we baulk at making human-baby art installations?

"Manipulating life will not be as alarming as it now seems" when humans realize they are part of the continuum of life, suggest Zurr and Catts (2003) in their essay on bioart and ethics. This, notes Gigliotti, is absurd, given how much we humans have been manipulating other species for millennia (2006). It is also absurd given how much of our (Western) culture relies on maintaining the separation of we humans from "the continuum of life." None of that manipulation, including extensive breeding programmes to produce rather peculiar shapes of animal for our own purposes, has ever made a dent in our anthropocentrism. We have certainly been altering the genomes of non-human species for a very long time. But these extensive interventions do not count as an "artwork." They are usually produced by large numbers of breeders over long periods of time, which means that no one person can easily make claims that this or that is "their" creation. And breeding programmes are far from precise—so much is left to nature's complexity. What makes creatures like GFP bunny into an "artwork" is precisely the deliberate and precise introduction of a bit of known DNA into the rabbit genome for "artistic" purposes, by specific people. Whatever the origins of particular breeds of dogs or sheep, say, through old-fashioned breeding programmes, no one usually claimed artistic merit or intellectual property rights for them.

Similarly, it is only when making transgenic organisms into art that claims are made that it involves "communication between human and animal," across the species barrier. Scientists producing genetically altered organisms make no such claims. In science labs, these animals become "bioreactors"; in the context of art, they become pals. Eduardo Kac emphasises this point, presenting GFP bunny in a "social and interactive context"[3]—that is, the presentation of the animal in the social world of cultured humans. We undoubtedly need to learn more about communication between ourselves and other species: we are singularly bad at it. But why is it that Kac seeks to call his relationship between the artist and the bunny an art form? Many people strive to understand the needs of non-human animals around them, to understand what the animals are telling them, to communicate with them however limited our abilities. It is not, however, generally called art (though some of us may feel that it is more wondrous than human-made artworks). The only difference is that Alba, the GFP bunny, was made by someone who calls himself an artist and who wants to emphasise the communication.

The "social and interactive context" is, of course, a human one, into which the rabbit must fit. What humans bring to this is a multiplicity of assumptions. Kac rhapsodises on his website:

> I will never forget the moment when I first held her in my arms. My apprehensive anticipation was replaced by joy and excitement. Alba—the name given her by my wife, my daughter, and I—was lovable and affectionate and an absolute delight to play with. As I cradled her, she playfully tucked her head between my body and my left arm, finding at last a comfortable position to rest and enjoy my gentle strokes. She immediately awoke in me a strong and urgent sense of responsibility for her well being. (para 1)

In using language so redolent of the nuclear family, Kac deflects the reader from considering the more distant concept of "artwork" or "scientific artifact"—or from considering Alba in any context other than that of "cute pet". And, for all that Kac insists on understanding Alba and the project in a wider social and political context, what comes over in this narrative is a sense of Alba as his—his creation, his baby—a sense that is reinforced by the campaign (listed on his website) to "bring Alba home".

Alba's contribution to this "social and interactive context" is, not surprisingly, a largely silent one. Little is said about her specific responses or about how she has attempted to engage herself in that human social world. On the contrary, she remains apart in her alterity, her rabbitness. Family cuddles notwithstanding, it remains unclear how she (or other such creations) can effectively participate in interspecies social contexts, as active agents.

Embracing a bunny is a far cry from those researchers who try to communicate with animals on their terms, attempting to find out how animals think, what they perceive, and how humans and animals together engage in creating intersubjectivities. There is now a great deal of scholarship exploring these very issues, which seems to be omitted from the eulogies to DNA manipulation.[4] Communication between species remains a challenge to research; however, it entails subtle, bodily, responses to equally subtle cues—a kind of "becoming animal" on the part of the human and vice versa (Birke and Parisi 1999; also see Game 2001 for a discussion of the importance of bodily responses in riding horses, and Sanders' 1999 discussion of human–dog relationships). It is, moreover, very much easier to think about such communication when the participants are both mammalian than it is when one is not.

Difficult to study they may be, but what these new studies underline is the intricacies of interspecies interaction, emphasising the subjectivities and integrities of both participants. Given these burgeoning literatures on human–animal relationships and cognitive ethology, I find it hard to see what light fluorescent bunnies shed. Glow in the dark they may do, but what, precisely, does this tell us about the intricacies of communication and intersubjectivity?

Their creation might, on the other hand, promote communication about biotechnologies, as Kac suggests on his website. He clearly sees these organisms and their production as transgressive, and capable of provoking renewed discussion between various interested people (scientists, artists, legislators and so on) and a wider public about (among other things) ethics of genetic engineering, and about our relationships with the natural world. As he explains, "My transgenic artwork 'GFP Bunny' comprises the creation of a green fluorescent rabbit, the public dialogue generated by the project, and the social integration of the rabbit." The artwork, then, resides not only in the specific creation of this modified creature, but also in the social nexus in which we find her. It is a performance art, in which both humans and the GFP bunny are performers.[5]

Whether or not such an act is radical in terms of art is for other people to judge. But I do not find it radical in terms of how we think about animals. If "Art" can now include the deliberate manipulation of one's own body boundaries through cosmetic surgery (as in Orlans' bodily performance art), then it is but a small step to include that of other bodies, such as animals.[6] This does not seem to me to offer a radical rethink of anthropocentrism, as proponents claim. We have undoubtedly been manipulating animal bodies, and using animal parts, for a very long time; but then, just when we have begun to take seriously the possibility that other species have minds, we begin to devise new ways of exploiting their bodies. And, no sooner have animals been exploited in these ways in the service of art, than similar techniques are used to produce them as trivial artefacts, part of the entertainment industry—illustrated by the production of "GloFish," zebra fish manipulated to glow red in the dark, presumably to perk up jaded aquarium-keepers (Pollack, 2003).

Trivialising animals in such ways is certainly not radical; on the contrary, it draws on centuries of trivialisation, demeaning behaviour and vicious cruelty towards all kinds of others (including, of course, many other humans). It is a sad irony that, just as our arrogance begins to admit that other creatures with who we share the planet might be worthy of respect, might be subjects of a life, then we try to find renewed ways of trivialising them, of denying the complexities of their lives and experiences.

Complexity and Animals—What Kind of Biology?

If animal lives are trivialised, then, what about the complicated ethical questions surrounding genetic manipulation? I want to turn now to a different claim made by some artist advocates, which is that production of transgenic organisms for art helps to confront the complexity of the issues. Perhaps it does, in the sense that it can pose questions about the convoluted ethical questions. But here I want to use a different take on complexity, in relation to science. To begin with, it does not seem to me that the production of transgenic animals in the name of art does anything at all to help us to understand the complexity of science as a whole. On the contrary, it draws on very narrow conceptions of science—such as genetic technology. New developments in genetics certainly gain enormous numbers of column inches in newspapers, and occupy nearly all those academics who concern themselves with analysing the sciences from, say, sociological perspectives. But genetics is not all of science.

More importantly, genetic reductionism is not the only way within science to understand organisms and how they work. As Kac and others note, the new genetics is reductionist; it relies on an understanding of DNA as a master molecule, and of the organism as a kind of accident, a by-product of those selfish genes. DNA, in this story, is the blueprint for the manufacture of the huge array of proteins that act as building blocks for all living organisms. Thus, cutting up fragments of DNA and moving them into other locations within a chromosome (of the same or different organism) makes sense: what is being moved is a tiny piece of a plan, a piece which organises the production of a specific material. That is why transgenic organisms are commercially so desirable, to tap into the production of materials for particular uses. In light of the meteoric rise of biotechnology and biotechnology companies, there is clearly some truth in this: animals are indeed being produced as "bioreactors".

That does not mean, however, that we really understand what is going on, except in the relatively trivial and reductionist sense that we have altered the output of one or two particular proteins. If a gene brings about the production of a particular protein, how does that work within the enormously complex structure of the cell, in concert with other chemical constituents? How does it help to generate patterns or the form of the organism? Although they are rather drowned out by the cacophony of excitement about the wonders of DNA technology, there are a number of voices who have expressed concern and doubt about the reductionist framework and these unanswered questions. Is DNA as central as we have been led to believe, they ask? Does moving genes have implications beyond the immediate effect on a protein?[7]

What the critics draw on is a recognition of the far greater complexity of life and living organisms than the "master molecule" rhetoric implies. It is ironic that Kac and other artists producing transgenic organisms cite some of the biologists who develop complexity theory. Yet they do so while exploiting the technologies that are deeply rooted in a reductionist framework. It is also ironic that these transgenic organisms are represented as art—that is, a product of creativity—while using methods that seem to minimise nature's role in the process.[8] Biologists developing complexity theory have emphasised the creativity of the processes of life, and how these cannot easily, or only, be understood as the sum of component parts. On the contrary, they point out, complex systems create emergent order, which simply does not appear at the level of the parts. Whirling dust storms in the desert might, for example, create specific patterns of sand that are not predictable from the apparent chaos of the dust. The mammalian heart beats with regularity, yet that regularity emerges from what appears at the cellular level to be electrical chaos. In the development of living organisms, patterns emerge from the engagement of a multitude of complex processes which influence each other—the intricate whorls of a snail's shell for example: those patterns are not necessarily predictable from gene sequences alone. And even if they are predictable, in the sense that we might say that removing gene X stops pattern Y, how those patterns emerged is not really well known. Take, for example, the beautiful patterns of butterfly wings—patterns that have been altered deliberately by moving genes in yet other kinds of "artwork" installations. True, moving certain genes alters the pattern. But so what? It is a very long way from specifying a gene to the emergence of complex order, and knowing how it unfolds.[9]

Patterns emerge in nature not simply because A causes B causes C, but through the complex interactions of many processes; these act together in ways we often cannot understand through reductionist frameworks, creating new orders and patterns not predictable from the component parts. That, to me, is more creative than merely moving a gene or two.

It also underlines the reasons why critics of gene technology are so concerned. The trouble with moving a gene or two is that, while we may refer glibly to the way "this gene codes for X"—that is, that the information encoded in the DNA is crucial for the production of protein X—what we usually do not know is how that gene works in collaboration with other genes in the genome or other part of the cell. That is precisely why, say critics, we should be concerned about the production of transgenic organisms: we do not know enough about what happens to the biological processes into which the transposed gene fits, with the result that the development of the individual or of wider ecosystems might be disturbed or damaged.

Defenders of gene technologies will, at this point, throw up their hands and refer to the ways that humans have been interfering in animal and plant breeding for millennia. Transgenics is simply a more focussed way of doing the same thing, they argue. Perhaps. But, while we tend to assume that what we are selecting through breeding programmes is a "gene for X," what we usually operate with is not genotypes but phenotypes, and what we select is more likely to be a whole complex of genes and cellular components that normally operate in and around that gene.

It is precisely that complex set of unknown reactions and processes that have evolved over millennia along with the ecosystems of which they are part, and which we deliberately ignore in producing transgenic organisms. Those, by contrast, may well not inherit all the cellular paraphernalia that help to keep organisms in balance both within themselves and in their environments. Natural selection is likely to maintain those cellular and developmental checks and balances that promote survival: creating transgenics means that we risk producing organisms that do not have these balances and which would not survive at all without our intervention.

More relevant to the current theme, if we do not know enough about ecosystem consequences, neither do we know what the longer-term implications are for the welfare of the animals manufactured by gene technologies. Dazzled by the possibilities of biotechnology, it seems that artists using it take for granted that moving a particular bit of DNA has only those consequences claimed by the biotechnology lobby (and commercial companies). Animals like GFP bunny have, we are assured on websites, a wonderful life, free of the vicissitudes of pain and stress confronting many of their laboratory-based peers. This may or may not be true, but is hardly

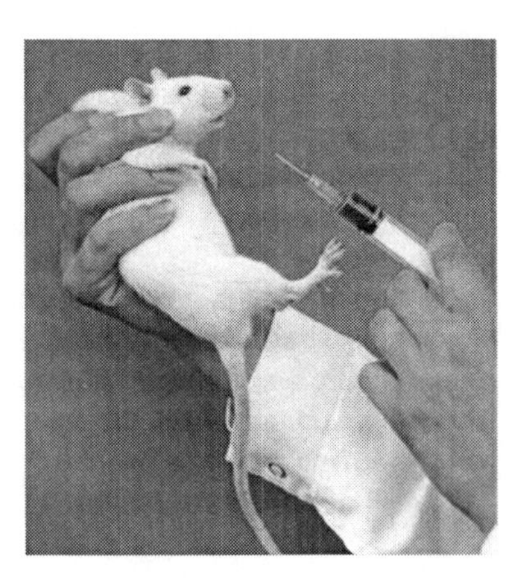

Fig. 2 Rat about to be injected (credit: uncredited)

the point. What concerns me is the possibility of creating animals whose physical integrity is clearly compromised in ways that cause them to suffer—whatever the stated purpose of producing those animals.[10]

Now, natural checks and balances are not a complete barrier to animal suffering—we have only to think of some of the breeds of dogs we have produced through breeding programmes, dogs with breathing difficulties or congenital hip dysplasia. But developmental systems do put some constraints on our interventions, constraints that might be bypassed by directly fiddling with the genes. If we can move individual genes at will, who knows how far we can go in creating appalling suffering?

I have long been hostile to reductionist biology (see, e.g., Birke 1999). Reductionism has undoubtedly been useful, allowing a great deal of predictability in experimental outcomes, a predictability which humans have made great use of in developing new medicines and all kinds of technologies. But reductionism masks so much of the complexity of biological processes and fails to explain a great deal about how, for example, organisms develop into the myriad fascinating forms that they do.

It troubles me as a biologist that the predominant framework is not encouraging us to understand these complex processes, which inhere in the whole organism. On the contrary, biology is increasingly focussing on fragmentation: whole organisms are simply not relevant. But it also troubles me for political or ethical reasons, since a way of looking at the world through fragmentation encourages acceptance of a literal fragmentation of organisms. Humans and non-humans alike thus quickly become merely sets of (and potential suppliers of) body parts. There is considerable potential here not only for human rights abuses, but also for the creation of even more animal suffering, as well as the production of environmental havoc if genetically altered animals escape (a very real risk in the case of altered zebra fish: Pollack, 2003). If we care about animal suffering, then I believe we should be arguing passionately for a more enlightened approach to understanding how animals and their bodies work, one which acknowledges and celebrates their complexity and integrity—as well as their own intrinsic aesthetic qualities—and having valued them, leaves them alone.

Widening Public Debate?

According to Machado (2000), the critics of "bioart" have a conservative bias. Bioart, advocates argue, not only challenges many of our long-held preconceptions about what it is to be human, or animal, but does so in ways that must promote wider public debate. In that I am deeply unimpressed by the kind of art under discussion, then I guess that that conservative label could well apply to me, or to the people who express disgust or dismay at the art installations. But in that I am also deeply unimpressed by the extremely unradical reliance on reductionist genetics underlying this art, then maybe not. I cannot see how using such a conservative approach to understanding living processes, thus denying their very complexity, can be called "radical."

And maybe it is as much about that denial as the "yuk" factor that makes many people wary of new developments in biotechnology. Its defenders amongst the scientific community often derogate the public, complaining that the public are often hostile out of ignorance. But ignorant of what? Recent research in the public understanding of science has found time and time again that lay people are not necessarily as "ignorant" as scientists sometimes assume; on the contrary, they may well take quite a sophisticated and nuanced approach to the difficult ethical dilemmas thrown up by new biotechnologies (Irwin and Michael, 2003). And, while we live in a culture that is clearly often cruel and abusive, it is also a society in which animals play a huge, and complex, symbolic role and in which they sometimes become "part of the family". Accordingly, some of the widespread public unease about using animals in biotechnology has to do with the way that it is seen as unnatural, and compromising animal integrity (Schroten, 1997). Not surprisingly then, surveys of public opinion indicate that, while citizens may be more willing to accept genetic alteration of plants, they are much less willing to countenance interference with animals.[11]

I suggested earlier that one way in which the production of transgenic organisms for art might be considered to be transgressive is that it takes the organism out of the laboratory. Laboratories, argues sociologist of science Latour (1983), are the key to understanding just what it is that makes the production of scientific knowledge so special, so authoritative in our culture. Laboratories, and the sometimes arcane activities taking place in them, help to create a kind of smokescreen around the production of science that renders it less accessible to wider publics and in turn helps to maintain the authority of scientific knowledge. So, breaking down the barriers between publics and laboratories might be thought of as desirable, as encouraging more access.

In writing about their use of tissue engineering to create bioart forms, Catts and Zurr (2003) note that their "semi-living" sculptures (made by growing cells) cannot be simply installed in a conventional artspace: they need laboratory apparatuses to maintain them. The tissue culture lab, in short, must be brought into the spaces where the art is installed, so transgressing boundaries of what counts as appropriate activities in particular spaces. Bringing the lab to the gallery might be seen as promoting public awareness: yet, what happens in laboratories remains something often to be abhorred. At the same time as labs in artspaces open up possibilities of public dialogue, laboratories using animals must increasingly shore up their boundaries with the public for fear of reprisals from anti-vivisectionist organisations. Not for nothing do scientists speak of feeling besieged, or "behind the barricades" (Birke and Michael, 1992). Public abhorrence is partly to do with the potential for animal suffering in lab experiments, but it is also to do with the potential of science to create new forms of animals.

Public responses to Kac's production of Alba illustrate both these reactions. His website discussion includes links to the campaign to "bring Alba home". This centred on a dispute with the French laboratory with whom Kac worked to produce the rabbit. The lab scientists claimed that Alba was only one of a number of GFP rabbits produced experimentally, while Kac claimed her special status meant that she was his, and could "come home" to Chicago. What is interesting, however, is not the dispute, but the following emails to the website. There is an overwhelming rhetoric to "free Alba," to "liberate" her from "captivity" or from her "cage." No matter that Alba herself was actually dead by the time many of these emails were written (she apparently died in 2002); what is clear from the passionate tone is that whatever else happens to her, it is preferable to being "held captive" in the laboratory. What happens behind the closed door of the labs continues to hold the public imagination, and Alba stands for (or stood for) all those rabbits "deprived of their liberty" by science. Different groups of the lay public may, of course, have different "takes" on animal biotechnology: patients with as-yet untreatable genetic diseases might feel, for example, that genetic alteration of animals is acceptable if a new cure for genetic disease could ensue. In general, however, most surveys of public opinion on new developments in science have found that people are, at best, ambivalent about animal use. If animals are to be used in experiments at all, many seem to feel, there must be a clear benefit; otherwise, the experiments, with their risk of animal suffering, cannot be justified. With genetically manipulated animals, too, people seem to seek justifications through utility. So, in studies of how people understand and react to the use of animals for xenotransplantation (as potential organ donors), lay people draw parallels with the incorporation of animal tissue into our bodies via eating meat (Michael and Brown 2004). That is, using animals as organ donors becomes less objectionable seen as a variant of a pre-existing cultural practice, eating meat.

Media representations of developments in xenotransplantation, too, reflect this need to emphasise benefits. Donor pigs, for example, are often represented in newspaper reports as heroes or saviours, dying to help save humankind (Birke and Michael, 1998).[12] Part of the public ambivalence about animal biotechnology, so regretted by scientist advocates, is based on the complex ethical dilemmas entailed; to find their way through these moral mazes, people must draw on ideas of these unfortunate animals as somehow doing good, their lives and deaths thus having a greater meaning.

Yet if the wider public struggles to find meaning for genetic manipulations that do, ostensibly, have a potential medical benefit, then what happens to that form of justification if the manipulation involves inserting a gene yielding fluorescence? And for purposes of art? If lay people use familiar examples of cultural practices (such as meat-eating) as an anchor to understand the moral issues involved in using animals as organ donors, then what kind of familiar example might be used in relation to fluorescence? The problem here is that few people are likely to see much point in using the GFP gene, even for medical research, or be able to make sense of GFP rabbits in terms of "doing good". It is not surprising that there was public furore following the announcement of Alba's production.

Writing about public anxieties about the production of animals "made-to-order" through biotechnology, Michael (2001) notes the highly complex symbolic role of animals, a symbolic potency that depends on their alterity and difference. Rabbits, for example, convey a multitude of cultural referents, in cartoons, films and children's books (and Kac himself draws on the image of the cute pet rabbit on his website). This complexity, argues Michael, is inextricably linked to a wide range of (human) identities in Western culture, such that animals become a highly significant cultural resource. This in turn is a source of public anxiety: if animals are imbued with so many meanings, then those meanings are threatened by technological interventions. Yet, at the same time, what Michael calls "technoscientific bespoking"—making animals to order—dramatically curtails their symbolic value, which can further promote unease.

A key part of public opposition to these (and other) new technologies is that people sometimes perceive science as too instrumental, too obsessed with techniques and oblivious to consequences (other than the sometimes grandiose claims used to justify the research, such as "potential cure for cancer").[13] That, I believe, is critical. Advocates of genetic manipulation, whether for purposes of science or art, often seem far too much in love with the techniques, and not the outcome, in the form of a live and sentient animal. Yet it is the outcome—what happens to the animals—with which the lay public are most concerned?[14] And this point raises further questions in relation to public perception: what will happen in the future, as new techniques are developed? What techniques will be thus appropriated as aesthetic, turned into "art-forms"? And, when these involve living creatures, what happens to the future of the relationships we have with other animals—to identities, both theirs and ours? These are the questions that we should be asking, and with considerable urgency—for these are the questions that underlie public unease.

While many geneticists might bemoan public anxieties over genetic manipulation, perceiving the public as misguided, other writers explore those anxieties. I began this chapter with a quotation from Margaret Atwood's novel, *Oryx and Crake* (2004); in this novel, she portrays a dystopian future where genetically manipulated and monstrous animals have escaped and cause havoc in a world laid waste. Among these are the green glow-in-the-dark rabbits "with gigantic teeth". These strange (and usually dangerous in the story) figures draw upon our profound anxieties about crossing species boundaries. But the story goes further, for the other side of the genetic experiments that led to disaster was the creation of tailor-made humans made, among other things, with built-in obsolescence—a terrible vision for a species so obsessed with dreams of immortality.

And we should heed where our dreams and symbols can take us. Perhaps we might pay attention to the way that those glow-in-the-dark genes came from jellyfish, whose free-swimming adult form is called a Medusa. Remember Medusa? Her vanity about her own beauty got her into a great deal of trouble with the goddess Athena, and she was made into a horrible monster, with her hair becoming a halo of hissing serpents. She was a monster so terrible that no one could look on her without being turned to stone.

Notes

1. Some forms of bioart involve tissue engineering, using cells derived from (dead) animals to grow into artificial media. Here, I focus more on the production of live animals through genetic manipulation, although all forms of bioart raise a number of troubling questions.
2. "Euthanasia and assisted suicide," Sciart project developed by Tracy Mackenna and Edwin Janssen, from Wellcome Trust website, Jan 2005: www.wellcome.ac.uk/node2530.html .
3. See Kac's website: www.ekac.org/gfpbunny .
4. See, for example, papers contributed to the journal *Society and Animals*.
5. Although Alba is a performer only in a limited sense. Birke et al. (2004) have used the idea of performativity (developed in feminist theory, for example) to apply it to the human–animal relationship, asking how we might understand the animal's part in terms of its own agency and engagement. Given Alba's creation by very explicit human intervention and the role she plays on a human stage, the potential for her own engagement is limited indeed.
6. Indeed, the production of artificial ears from human tissue, a production by Zurr and Catts, has been developed as an artwork for potential transplantation onto the head of another performance artist, Stelarc.
7. See for example various critical articles published on the web pages of ISIS—the Institute for Science in Society (http://www.i-sis.org.uk/). Many of these articles point to evidence that manipulated genes can have multiple and unpredictable effects, and can cause problems if organisms escape into other environments. They also suggest that genetically manipulated genomes are not as stable as advocates claim.
8. Arguably, there is more of nature's creativity at work in the use of tissue engineering to create artworks, such as that by Zurr and Catts. They use tissue from dead animals (pigs, for example) and allow the cells to grow along artificial structures. The "art" is thus partly created by whatever forms the cells take as they move along these structures.
9. There are several theorists developing complexity theory, such as Kauffman (2000), and Sole and Goodwin (2000). Also see discussions by contributors to Oyama et al. (2001).
10. There are, furthermore, regulatory frameworks that help to control the scientific production of transgenic organisms. These may be partial, even inadequate (and are criticised by many), but they can help to set limits. There may be no such limits if transgenic organisms are made for art.
11. Survey conducted by the Pew Initiative on Food and Biotechnology. See http://pewagbiotech.org/research/2003update/4.php .
12. The trope of the animal as dying for our sins—our saviour—is a recurrent motif in science, as Haraway (1997) has argued.
13. Jasper and Nelkin (1992) point out how the rhetoric of animal rights and environmental campaigners draws on an explicitly anti-instrumentalist stance, opposed to what they see as the excessive instrumentalism of modern science and technology.
14. I am grateful to Mike Michael for drawing my attention to this point. In noting public concern about the outcome, I mean to include both concern over animal integrities and public concerns

over potential medical outcomes. These are far more important to lay observers, I would argue, than the techniques.

References

Atwood, M. (2004). *Oryx and crake*. London: Virago.

Birke, L. (1999). *Feminism and the biological body*. Edinburgh: Edinburgh University Press.

Birke, L., Bryld, M., & Lykke, N. (2004). Animal performances: An exploration of intersections between feminist science studies and studies of human/animal relationships. *Feminist Theory* 5, 167–183.

Birke, L. & Michael, M. (1992). Views from behind the barricade. *New Scientist* 4, 29–32.

Birke, L. & Michael, M. (1998). The heart of the matter: Animal bodies, ethics, and species boundaries. *Society and Animals* 6, 245–262.

Birke, L. & Parisi, L. (1999). Animals, becoming. In H. P. Steeves (Ed.) *Animal others: On ethics, ontology and animal life*. New York: State University of New York Press.

Catts, O. & Zurr, I. (2003). The art of the semi-living and partial life: Extra ear. *Tissue Culture and Art Project*. Retrieved Dec, 2004 from http://www.tca.uwa.edu.au

Game, A. (2001). Riding: Embodying the centaur. *Body & Society*, 7(1), 1–12.

Gigliotti, C. (2006). Leonardo's choice: the ethics of artists working with genetic technologies. *AI & Society*, 20(1), 22–34.

Haraway, D. (1997). *Modest_Witness@Second_Millennium. FemaleMan_Meets_Oncomouse: feminism and technoscience*. London: Routledge.

Irwin, A. & Michael, M. (2003). *Science, social theory and public knowledge*. Maidenhead, Berkshire: Open University Press,

Jasper, J. & Nelkin, D. (1992). *The animal rights crusade*. New York: The Free Press.

Kauffman, S. (2000). *Investigations*. Oxford: Oxford University Press.

Latour, B. (1983). Give me a laboratory and I will raise the world. In K.D. Knorr-Cetina and M. Mulkay (Eds.), *Science observed: Perspectives on the social studies of science*. London: Sage.

Machado, A. (2000). Towards a transgenic art. In S. Britton and D. Collins (Eds.), *The eighth day: The transgenic art of Eduardo Kac*. Tempe, AZ: Institute for Study in the Arts.

Michael, M. (2001). Technoscientific bespoking: Animals, publics and the new genetics. *New Genetics & Society*, 20, 205–224.

Michael, M. & Brown, N. (2004). The meat of the matter: Grasping and judging xenotransplantation. *Public Understanding and Science*, 13, 379–397.

Oyama, S., Griffiths, P.E. & Gray, R.D. (Eds.) (2001). *Cycles of contingency: Developmental systems and evolution*. Cambridge, MA: MIT Press.

Pollack, A. (2003, November 22). Gene-altering revolution nears the pet store: Glow-in-the-dark fish. *The New York Times*. Retrieved from http://www.mindfully.org

Sanders, C. (1999). *Understanding dogs: Living and working with canine companions*. Philadelphia, PA: Temple University Press.

Schroten, E. (1997). Animal biotechnology, public perception and public policy from a moral point of view. In A. Nilsson (Ed.), *Transgenic animals and food production: Proceedings from an international workshop in Stockholm, May 1997*. Sweden: KSLA. Retrieved from http://www.kslab.ksla.se/tranpdt.html

Sole, R. & Goodwin, B. (2000). *Signs of life: How complexity invades biology*. New York: Basic Books.

Zurr, I. & Catts, O. (2003). The ethical claims of bio art: Killing the other or self-cannibalism? *The Australian and New Zealand Journal of Art*, Art Ethics 4, 167–188

Transgenic Bioart, Animals, and the Law

Taimie L. Bryant

Abstract This chapter considers aspects of the complex relationship between law and transgenic bioart, which is the genetic alteration of living beings as a type of performance art. Transgenic bioart is ambiguously situated in relation to bioengineering and law. As a type of "scientific" endeavor, bioengineering receives preferential treatment under several laws, including anticruelty statutes, but art does not consistently receive similarly favorable treatment. Thus, deciding whether a particular transgenic bioart project is "art" or "science" becomes a necessary, but surprisingly difficult, interpretive exercise if one is seeking to protect animals. That difficulty arises not only because of the nature of transgenic bioart but also because of variance in the type of collaborative relationships that exist between bioartists and scientists. There is no case law on the subject of whether or when transgenic bioart in general or some specific projects are considered "scientific," and it is difficult to predict whether or how specific anticruelty statutes would be applied. However, one thing is clear: the law has not thus far impeded any bioartistic endeavor even when such endeavors result in the suffering of sentient animals. To date, bioartists have faced only one legal challenge, which concerned the acquisition of biological materials and not the nature of the project itself. Laws could be enacted or interpreted with the result of curtailing bioartistic manipulation of nature, despite First Amendment protections of expressive speech, if doing so is considered necessary to protect "compelling governmental interests." However, there would have to be sufficient social and political will to do so. That social and political will to protect nature, including animals and the environment, cannot originate in law because law, first and foremost, protects those interested in exploiting nature. Exploitation of nature is so deeply entrenched in law that legal reform can, at this point, only follow and complement societal change in the direction of respecting and protecting nature.[1]

T.L. Bryant (✉)
University of California, Los Angeles, School of Law, Los Angeles, CA, USA
e-mail: bryant@law.ucla.edu

C. Gigliotti (ed.), *Leonardo's Choice*, DOI 10.1007/978-90-481-2479-4_8,

Keywords Bioart · Transgenic art · Animal law · Animal rights

Introduction

Once understood as whole biological systems, plant and animal species are increasingly understood to be inventories of transferable genes available to serve human purposes. As chemical ecologist Thomas Eisner has phrased it,

> [a] species is not merely a hard-bound volume of the library of nature. It is also a loose-leaf book, whose individual pages, the genes, might be available for transfer and modification of other species (as cited in Midgley, 2000, p. 6).

Eisner suggests that species can be at one and the same time hard-bound volumes *and* loose-leaf books. That may be true of humans, but, as far as nonhuman species are concerned, we seem well on the way to a conception of species *only* as loose-leaf books. Such a conception enhances preexisting distinctions drawn between human and nonhuman species. We cause pigs, monkeys, fish, mice, rabbits, even the testicles of mosquitoes, to glow fluorescent green—with little more justification than that we have the technological capacity to do so and the possibility of some future utility to humans. The same cannot be done to humans.

Among those who engage in bioengineering are "transgenic artists," who genetically alter living beings, such as bacteria, plants, and animals, as a type of performance-based art.[2] It is because such artists perform actual genetic alteration of living beings that the genre is particularly controversial and feels so "dangerous" or "subversive" (Kennedy, 2005, para 7). Critics distinguish "proper" and "improper" purposes for the genetic manipulation of life, but, to transgenic bioartists, bioart fulfills important societal functions and constitutes as valid a reason as any to manipulate the genetic basis of life (Lynch, 2008, p. 180; Sholette, 2005; Kac, 2003). Indeed, to some bioartists whose goal is demystification of science, anyone should be able to engage in the genetic manipulation of living beings (Sholette, 2005).

Because of the close relationship between biotechnology as a scientific endeavor and transgenic bioart, there is a complex relationship between transgenic bioart and the law. The law generally favors biotechnological development, but there are legal restrictions regarding such matters as the use of human cells, access to nonhuman life forms for the purpose of conducting bioengineering, and the circumstances of release of altered life forms into the environment. Those restraints exist because of concerns about risks to humans—risks of health, safety, and self-definition. Thus far, concerns about risks to nature in general or animals in particular have not been expressed in the law.

The presence of legal restrictions to protect humans and the apparent absence of legal restrictions to protect nonhuman species impacts transgenic bioartistic expression in various ways. Transgenic bioartists claim to challenge the idea of species boundaries through their art (Gigliotti, 2009 and Birke, 2009). However, because there are legal limits on creating transgenic human beings, such bioartists end up reinforcing values that underlie legal rules that limit their transgenic art to the genetic manipulation of nonhuman plant and animal species only. Transgenic bioartists cannot help but deliver the message of acceptability of traditional views of a sharp distinction between human and nonhuman species and human subordination of nonhuman species; if they did not accept those views, they could not ethically engage in the act of genetically manipulating life under legal restraints that restrict them to use of nonhuman species only. As it is, transgenic bioart and the law act in concert to maintain a traditional conception of the organization of life even as transgenic bioartists proclaim that their work does otherwise.

Similarly, transgenic bioartists may attempt to critique the idea that only scientists can or should engage in transgenic alteration of life, to educate the public about risks associated with bioengineered organisms, to call for public scrutiny of scientific research, or to illustrate the need for more regulation of bioengineering (Kennedy, 2005). However, trangenic bioartists' collaboration with scientists in order to lawfully procure life forms, supplies, facilities, or performance of some techniques reinforces the importance of scientists as gatekeepers of the technology.[3] Further, to the extent that the law privileges scientists' genetic manipulation of life but not artists' genetic manipulation of life, such collaboration provides legal cover to transgenic bioartists even as they proclaim that "anyone" can manipulate life forms.

As compared to transgenic bioart, which involves the actual genetic alteration of living matter, representational art can be more conceptually radical because there are relatively fewer legal and ethical limitations on the ideas one can express through means that do not themselves involve genetic alteration of life forms. One can legally include humans in *representations* of genetic alteration of life; one cannot legally include humans in *performances* of genetic alteration of life. Moreover, representational art does not require collaboration with scientists whose status or activities can be criticized only obliquely through transgenic bioart due to transgenic bioartists' dependence on scientists. It would seem, therefore, that transgenic bioartists' work contradicts and undermines their stated aims. At the very least, transgenic bioartists sacrifice expression of many of the ideas they claim they want to express—ideas that can be expressed only through means that do not involve the actual genetic alteration of life—when they reject representational artistic expression in favor of transgenic bioartistic expression.

Some transgenic bioartists may not particularly care that their art entrenches concepts of human dominance of nature. Those who do accept the criticism as valid may, nonetheless, emphasize what they contend are positive features of transgenic bioart as a distinctive form of performance art. They may also overestimate their ability as artists to protect the lives they create or alter. For instance, Eduardo Kac (2003) has represented his endeavor to create Alba, a rabbit genetically modified by the insertion of jellyfish genes, as performance art "comprise[d of] the creation of a green fluorescent rabbit. . ., the public dialogue generated by the project, and the integration of the rabbit into a social environment" (p. 97). By Kac's own definition of his project, it failed. Alba was not integrated into a social environment, at least not the type of social environment Kac found acceptable. Kac waged a very public campaign to secure her release to his care, but Alba lived out her life in a research laboratory under the control of research scientists. Kac and other transgenic bioartists seem unpersuaded by concerns about harms that could result from release of genetically altered organisms into the environment. As we shall see, the law operates ambiguously in that regard, although public health and safety rationales for restricting ownership and sale of genetically altered life forms abound.

What of the public's interest in the health, safety, and well-being of Alba herself? Even if Alba was not subjected to other experiments, it is unlikely that she was treated with the "commitment to respect, nurture and love" that Kac (2003, p. 97) identified as a requirement of producing transgenic art. Moreover, having stated categorically that *all* life generated by transgenic art be created "above all, with a commitment to respect, nurture, and love the life created," Kac (2003, p. 97) developed an installation, entitled "The Eighth Day," in which sentient transgenic creatures would live in a four-foot Plexiglas dome, for purposes of "making visible what it would be like if these creatures would, in fact, co-exist in the world at large" Kac (2001, para 1) described the installation as follows:

> As a self-contained ecological system it resonates with the words in the title, which add one day to the period of creation of the world as narrated in the Judeo-Christian Scriptures. All of the transgenic creatures in "The Eighth Day" were created through the cloning of a gene that codes for the production of green fluorescent protein (GFP). As a result, all creatures express the gene through bioluminescence visible with the naked eye. The GFP creatures in The Eighth Day are GFP plants, GFP amoeba, GFP fish, and GFP mice... By enabling [human] participants to experience the environment inside the dome from the point of view of the biobot, "The Eighth Day" creates a context in which participants can reflect on the meaning of a transgenic ecology from a first-person perspective. (2001, paras 2–4)

Is this installation indicative of "respect, nurturance, and love of the lives" created for artistic purposes? Or are those values applicable only if and after the artist has satisfied his own expressive needs? As discussed below, the law is ambiguous as to whether performance art such as "The Eighth Day" constitutes the infliction of suffering on sentient beings and, if it does, whether such suffering is acceptable as "necessary" to the attainment of valued human objectives.

Legal Inadequacy, Ambivalence, or Intentional Under-Regulation?: The Coordinated Framework for the Regulation of Biotechnology

Just as debate about bioart ranges widely, so, too, do perspectives about the role of law in bioengineering generally and transgenic bioart specifically. There are signs of legal resistance to some types of bioengineering that threaten humans' definitions of themselves,[4] but, thus far, federal law has facilitated humans' exploitation of nonhuman nature through bioengineering.

The Coordinated Framework for the Regulation of Biotechnology was created in 1986 to encourage cooperation and coordination among federal agencies charged with regulatory oversight of processes and products that might contain bioengineered components: the Environmental Protection Agency, the Food and Drug Administration, and the Department of Agriculture.[5] Under that framework, each new product derived from genetic alteration of a single organism is subjected to a level of review that corresponds to its use as a food or drug.

Proposed use in drugs is given more scrutiny than proposed use in food products, but, as a general matter, if members of a genetically altered species are considered "substantially equivalent" to nongenetically modified members of the species, then they and products derived from them are legally presumed to be safe. Determinations of "substantial equivalency" rest on comparison of isolated features of the modified and unmodified species that relate to the purposes for which the modified species will be used. For example, in 2003 the United States Department of Food & Drug Administration (FDA) issued a draft risk assessment of cloned livestock recommending the presumption that "food products from healthy animal clones and their progeny that are not materially different from corresponding products from conventional animals are as safe to consume as their conventional counterparts" (p. 5). The basis of risk assessment is compositional analysis of the food products derived from animal clones in comparison to food products from animals that are not clones. The assessment is so myopic that it has led an observer to conclude that, "[a]lthough [genetically modified] products are likely to have been altered to an extent that is sufficiently "novel" to qualify for patent protection for the developer, the vast majority of food products submitted for commercial marketing thus far have continued to meet the United States' definition of substantial equivalence, thus incurring no special regulatory scrutiny" (Lawrence, 2007, p. 238). According to Lawrence (2007), the Pew Initiative on Food and Biotechnology reported poll results indicating that only 34% of those polled thought that genetically altered foods were safe and that Americans generally favor regulation of such foods (p. 249).

The Coordinated Framework is vulnerable to multiple specific criticisms besides being generally unsuited to perform the tasks that would be responsive to public concerns. In addition to the breadth of the definition of "substantial equivalence," criticisms include the extent of judicial deference to agency determinations, especially since agencies may be subject to special interest group capture; the presumption of safety until a product of genetic modification is proven dangerous; and gaps in oversight, such as failure to review genetically altered life forms developed for purposes other than food and drugs (Lawrence, 2007, pp. 238–247). Yet, despite criticisms and proposed reform, American law continues to be fundamentally receptive to claims that biotechnology is beneficial to humans overall and that future benefits to humans will offset any harms that may result. Needless to say, evaluation of products from the standpoint of the extent of suffering inflicted on animals is not now and never has been considered a significant consideration.

One reason for that apparent receptivity to biotechnology may be simply the speed with which technological development has outstripped society's capacity for appropriately reactive legal change. Another reason may be a deep-seated belief that development should not be impeded legally unless there is extremely clear evidence that there is a need to do so. Yet another may be humans' longstanding self-granted entitlement to appropriate and exploit all things "natural." Having invested labor in appropriation and alteration of something "natural," humans conceptualize the result as no longer natural but as "manmade" and, therefore, appropriately subject to human ownership and further modification.

The Case of *Diamond v. Chakrabarty*

The ideology of human prerogative to use, to own, and to exploit nature is evident in the 1980 United States Supreme Court decision of *Diamond v. Chakrabarty*, which concerned the patentability of bacteria genetically altered in ways that enabled the bacteria to break down crude oil. Perhaps signaling the ultimate basis for its decision, the court notes early in its opinion that the bacteria "is believed to have significant value for the treatment of oil spills" (p. 305). Chakrabarty submitted patent claims of three types: the processes employed to alter the bacteria, the material that sustained and supported the bacteria while it was floating on water, and the bacteria themselves. Although the patent examiner allowed patents for the first two types of claims, he rejected Chakrabarty's claim of ownership in the bacteria themselves. The examiner based his rejection on grounds that micro-organisms are "products of nature" and that living beings cannot be the subjects of patents. The United States Supreme Court disagreed, stating that the "new bacterium [has] markedly different characteristics from any found in nature and one having the potential for significant utility. [The patentee's] discovery is not nature's handiwork, but his own; accordingly it is patentable subject matter" (p. 310).

The court rejected the argument that Congress and only Congress should decide the specific question of whether genetically altered organisms are patentable because "the judiciary 'must proceed cautiously when... asked to extend patent rights into areas wholly unforeseen by Congress'" (p. 315). The court decided that validating Chakrabarty's patent did not "extend patent rights"; in the court's view the facts of the case fit unambiguously with the language, intent, and legislative history of the Patent Act. Quoting the Constitution, the court noted that Congress has "broad power to legislate to 'promote the Progress of Science and useful Arts, by securing for limited Times to Authors and Inventors the exclusive Right to their respective Writings and Discoveries'" (p. 307).[6]

Arguments premised on predicted harms to nature and humans were not absent from the litigation. Indeed, the court makes explicit reference to those claims.

> [The Patent Office]... points to grave risks that may be generated by research endeavors such as [Chakrabarty's]. The briefs [submitted to the Court] present a gruesome parade of horribles. Scientists, among them Nobel laureates, are quoted suggesting that genetic research may pose a serious threat to the human race, or, at the very least, that the dangers are far too substantial to permit such research to proceed apace at this time. We are told that genetic research and related technological developments may spread pollution and disease, that it may result in a loss of genetic diversity, and that its practice may tend to depreciate the value of life. These arguments are forcefully, even passionately, presented; they remind

> us that, at times, human ingenuity seems unable to control fully the forces it creates—that with Hamlet, it is sometimes better 'to bear those ills we have than fly to others that we know not of.'
> It is argued that this Court should weigh these potential hazards in considering whether [Chakrabarty's] invention is patentable subject matter... We disagree. The grant or denial of patents on micro-organisms is not likely to put an end to genetic research or to its attendant risks. The large amount of research that has already occurred when no researcher had sure knowledge that patent protection would be available suggest that legislative or judicial fiat as to patentability will not deter the scientific mind from probing into the unknown... Whether [Chakrabarty's] claims are patentable may determine whether research efforts are accelerated by the hope of reward or slowed by want of incentives, but that is all. (pp. 316–317)

The Supreme Court's validation of the patentability of genetically altered bacteria affirms that living beings can be the property of humans, which may be unremarkable given the fact that humans have been the legal owners of domesticated animals for as long as there have been laws to protect such claims. The court distinguished "Chakrabarty's handiwork" from "manifestations of nature," which are not patentable. But, the court also noted that even if they are not patentable, "manifestations of nature" are "freely" exploitable by "all men."

> Thus, a new mineral discovered in the earth or a new plant found in the wild is not patentable subject matter... Such discoveries are "manifestations of... nature, free to all men and reserved to none." (p. 309)[7]

The Patent Office acted on that premise when it readily issued patents for the *processes* by which Chakrabarty altered the bacteria; there was no controversy about Chakrabarty's claims that he could patent the processes by which he altered bacteria. Ownership of those processes insured commercial advantage associated with the exploitation of "manifestations of nature." The only question before the court was whether the genetically altered bacteria themselves could be patented. There are clear advantages to owning a patent in the beings that result from genetic manipulation, but ownership of patents related to processes of exploitation would drive exploitation of living matter, even without ownership of patents in the beings that result. Moreover, the particular physical beings that result from genetic modification research would be owned, and their subsequent exploitation could be controlled by contract. Considering existing legal means of commercially exploiting the processes by which life is genetically altered, it was only a relatively small step to validate the patentability of genetically altered beings themselves.

Concepts of Nature and Ambivalent Legal Responses to Biotechnology

Part and parcel of this point of view is the idea that nature and "manifestations of nature" have no identifiable or legally assertable interests of their own. As a conceptual matter, nature and its constituent parts *could* have legally cognizable interests separable from those of humans. Indeed, various legal approaches to recognizing the interests of nature have been proposed. For instance, lawyer Cormac Cullinan (2002) argues that *all* laws should be structured in light of our interdependency with nature, rather than relying exclusively on piecemeal legal recognition of the interests of one or more aspects of nature. Cullinan's approach would require legal recognition that entities other than humans have legal interests, but his focus is on reorganizing society rather than simply privileging or restraining the conduct of any particular species. Legal scholar Christopher Stone (1996) similarly supports laws that would curtail human privilege in order to protect various entities in nature, such as turtles, trees, and rivers. He asserts that it is not necessary to go so far as to establish legal rights for such entities so long as their "legally considerable" interests are recognized (Stone, 1996, p. 51).

Both of these proposals could be understood to operate from an instrumental, positive law perspective that legal entitlements can be established for *any* entity if there is a socially recognized benefit in doing so; establishing such entitlements need not turn first and foremost on characteristics of the potential holder of the entitlements (Posner, 2004). From that standpoint, the Cullinan and Stone proposals are not particularly radical. Moreover, depending on the definition of "rights," it is possible to argue, as has legal scholar Cass Sunstein (2003), that animal rights is not a controversial idea and that some animals already have "rights" to the extent that laws exist to protect them from suffering. The reason that Cullinan's and Stone's approaches can seem radical is not that they speak of "rights" or "legal interests" for entities that exist in and of nature; it is that they advocate reformulating the whole of our relationship with the natural world by incorporating in law an ethic of environmental protection and reduction in human privilege.

Instead of moving in a direction through which humans resituate themselves in relation to the rest of nature, humans use biotechnology to further situate nature in relation to human preferences. When human conduct so poisons the earth that we *must* resituate ourselves in order to save ourselves, the law might serve as a mechanism through which we do so. In the meantime, legal responses to bioengineering are likely to be ambivalent and piecemeal. Such is evident in the fact that the primary regulatory mechanism available in the United States is the Coordinated Framework, which only cobbles together preexisting federal agencies and laws without overhauling the legal means by which products are evaluated.

A central problem with the Coordinated Framework is its focus on the intended use of the bioengineered products and the gaps in oversight that result. This problem is illustrated by the example of GloFish, fluorescent zebra fish that glow red or green due to the insertion of sea anemone genes or jellyfish genes, respectively (The Associated Press, 2003, para 4). GloFish were originally developed with the expectation that they could be used to detect water pollution (Lamb, 2004, Ban on GloFish, para 5). When that turned out not to be feasible, GloFish were produced for the amusement of people who keep fish in aquariums. As such, they were given only cursory review by the United States Food and Drug Administration (FDA), which focuses on evaluation of the safety of food products and drugs. GloFish were deemed "substantially equivalent" to unmodified zebra fish, and there was no clear evidence at the time they were evaluated that they would cause harm if they ended up in the wild (Lawrence, 2007, p. 257). Accordingly, the FDA declined to regulate the sale of GloFish (U.S. Department of Drug & Administration [FDA], 2003). Although the FDA's decision was challenged as an abuse of authority, the court in *International Center for Technology Assessment v. Thompson* (2006) decided that it was within the FDA's discretion to reach the decision it made in the case of GloFish. Thus, with minimal review, GloFish became the first transgenic creature available for commercial marketing in the United States (Lawrence, 2007, p. 255).[8]

Because GloFish were not going to be marketed for food or drug purposes, the federal agency charged with evaluation of GloFish, the FDA, left legal regulation up to the states. While other states debated whether to allow the sale of GloFish, by a vote of three to one, California's Fish and Game Commission decided not to exempt GloFish from the state's ban on selling genetically engineered fish (The Associated Press, 2003, para 3). The decision is not explained by hostility to the scientific method by which the fish were genetically altered; the Commission had already approved the state's license to conduct research on genetically modified fish. Rather, for at least two of the Commissioners the problem was the purpose for which the GloFish would be sold. Commissioner Sam Schumchat was quoted as saying, "For me it's a question of values, it's not a question of science. I think selling genetically modified fish as pets is wrong... To me, this seems like an abuse of the power we have over life, and I'm not prepared to go there today" (The Associated Press, 2003, paras 2, 15). Similarly, Commissioner Jim Kellogg was quoted as saying that the question before the Commission was one "of values, not science" (Bacher, 2003, para 21). In other words, the FDA did *not* prohibit the sale of GloFish because they would be pets, but California *did* prohibit the sale of GloFish because they would be pets. That a majority of the Commissioners voted against the sale of GloFish on

Fig. 1 Danio GloFish (credit: www.glofish.com)

grounds of values indicates that people can decide in favor of limiting human prerogative to genetically alter living beings under some circumstances. It is not yet a completely foregone conclusion that nature will continue to be genetically altered for all purposes or at the whim of those with the technological ability to do so, including transgenic bioartists.

The Problem of Legal Categories

This discussion of the effect of different legal rules affecting the availability of fluorescent fish depending on the purposes for which they are produced illustrates the importance of legal categories for understanding how the law currently functions and for predicting future developments. Categorical uncertainty as to specific "facts" results in uncertainty about legal rule application. Because it is difficult to categorize "transgenic bioart," the law replicates the complicated social discourse of transgenic bioart.

A feature of the story of Alba, the fluorescent green rabbit, that is particularly useful for considering the problem of legal categorization is the lack of clarity about whether Alba was purposely "made to order" for Kac or whether Kac simply made use of the existence of a particular rabbit who was produced as part of an existing scientific project to produce fluorescing rabbits (Lynch, 2008, p. 193). Did Kac only generate an aesthetic discourse about an "artifact" of scientific research or did Kac participate in generating both the artifact and the aesthetic discourse? Stated differently, is Alba a creation of science or bioart? In an article about bioart and bioterrorism, George J. Annas raises this issue from the perspective of ambiguity in *all* bioart and biotechnology:

> Like defensive and offensive bioweapons research, bioart and biotechnology may be impossible to distinguish by anything other than the researcher's or creator's intent. Thus, Alba, the bunny with the inserted jellyfish gene, is considered to be and is accepted as a creation of bioart, at least in the contemporary art community; whereas ANDi, the monkey with the inserted jellyfish gene, is considered to be a creation of science, at least in the biotechnology community. (Annas, 2006, p. 2718)

To Annas, a professor of health law, bioethics, and human rights, the lack of clear distinction is problematic. He criticizes bioart for provoking fears about biotechnology and bioterrorism such that the public is easily confused, potentially leading to overreaction to biosafety rule violations. To Lisa Lynch, a professor of media studies, such ambiguity is relatively unimportant, at least as to the matter of particular artists' purposes:

> Kac's interest was to bring the product of genetic research closer to the public, but the product he delivered was stripped of its production context; this is why, in the case of "GFP Bunny," the media could imagine Kac as the actual creator of Alba and why scientists could accuse him of being a sacrilegious tinkerer. In fact, Kac was not tinkering at all; he was not practicing amateur science, but amateur semiotics, taking a preexisting object and trying to get it to signify differently. (Lynch, 2008, p. 198)

From a legal perspective, distinguishing artistic from scientific endeavors is important because intent of the creator, where the creation took place, and who was involved in the creation may all be relevant factors in whether particular laws apply. These factors cannot always be easily sorted. Not only are transgenic bioartists using scientific techniques, many transgenic bioartists have collaborative relationships with scientists, even if only for limited purposes. Certainly if a bioartist does not do the genetic alteration himself but instead relies on scientists to produce the genetically altered life form to the bioartist's specification, there will a significant collaborative component. That collaborative aspect means that some transgenic bioart projects could be framed *legally* as scientific endeavors even if the life form was altered for no other purpose than artistic expression and the altered life form is subsequently popularly characterized as an artistic creation. Categorical uncertainty about whether transgenic bioart is art or science complicates application of the few laws that could be deployed in an effort to protect sentient animals from harm. It affects the operation of state anticruelty laws as well as the federal Animal Welfare Act.

State Anticruelty Laws

American state anticruelty laws prohibit only intentional or grossly negligent infliction of suffering on "animals" as defined by the statute. The use of words such as "torment" and "mutilation" indicate that the suffering inflicted on an animal must be substantial and severe. However, even as to severe, intentionally inflicted suffering, anticruelty statutes balance animals' interest in not suffering against humans' interests in such activities as consumption of animal-based products, hunting, pest control, and scientific research. Accordingly, only severe suffering that is inflicted "unnecessarily" on a statutorily protected animal is prohibited by anticruelty statutes. Because the law—especially criminal laws—operate only on specific sets of facts, as to each instance of transgenic bioartistic performance the relevant questions would be (1) whether the animal harmed is covered by the statute, (2) whether the suffering inflicted on the animal is of the type covered by the statute; and, (3) if so, whether it was inflicted for no other reason than to cause suffering or as incident to the pursuit of valued human goals. It is difficult to predict outcomes because the application of law to particular facts is susceptible to interpretation; prosecutors, defense attorneys, jurors, and judges often interpret the same fact-law situation quite differently.

State Anticruelty Laws: What Is an "animal"?

The difficulty associated with the first question—whether transgenic animals are "animals" for purposes of state anticruelty statutes—preexisted the development of bioengineering. The broadest state statutory definitions of "animal" include "any dumb creature," "any sentient being," and "vertebrate animal," but it is not only the characteristics of the animal him- or herself that matter. Context also matters. A rabbit is certainly a "dumb creature," a "sentient being," and a "vertebrate animal," but if she is a garden "pest," a "laboratory animal" in a "bona fide" research experiment, or raised for human or animal food consumption, she is not covered by anticruelty statutes at all regardless of the extent of suffering that may be inflicted on her. Very few animals are actually protected by anticruelty statutes, despite the fact that such statutes would seem to include most sentient animals, because there are so many exemptions for activities humans value more than they value sparing animals any amount of suffering.

A problem with the application of anticruelty statutes to bioengineering in general and to bioartistic genetic manipulation of animals in particular is that blurring of animal species—a stated purpose of bioart—could result in further complication. For example, consider Alba's situation in light of Alaska's anticruelty statute, which defines an "animal" as "a vertebrate living creature not a human being, but does not include fish" (Alaska Stat. §11.81.900) or Georgia's statute, which states that "'animal' shall not include any fish" (Ga. Code. Ann. § 16-24-4). If Alba has become a rabbit-jellyfish, would the jellyfish part of her result in exclusion from the statutory definition of an "animal?"

Most genetically modified animals produced to date can be identified as primarily one type of animal. Whether legal outcomes would correspond or not, most people would identify a GloFish as a fish and Alba as a rabbit, for instance. However, even if there is some degree of clarity (or presumptive clarity) about some transgenic animals, the fact remains that the more an animal becomes an "artifact"—the more distant it is from one or more of its originating forms—then the more distant the end-product animal is from the contemplation of and application of the anticruelty statutes. Most importantly, the *will* to apply concepts of anticruelty to "monstrous" or "manmade" creatures invented by humans may diminish as a result of such blurring.

State Anticruelty Laws: What Constitutes "Suffering"?

The second relevant legal question—that of the type of suffering covered by general anticruelty statutes—is equally difficult. Bioengineering results in the birth of many deformed and disabled animals, in addition to the only relatively lucky few who become "poster children" for support of transgenic alteration. The fact that many animals suffer and die for every one that survives the process of transgenic alteration may be considered legally analogous to the breeding of companion animals and livestock, which is not actionable as "cruelty." That transgenically altered animals suffer serious disabilities during their lives may be considered similarly analogous to the fact that many companion animals and livestock also suffer disabilities resulting from purposeful breeding. Medical ethicist Art Caplan alluded to this problem when he was interviewed about GloFish. Caplan noted that it would be ethically troubling if there were "changes that cause an animal to suffer, such as the hip problems in German Shepherds caused by inbreeding or breathing problems that bulldog breeds develop" (Lamb, 2004, Ethics of Bioengineering section, para 3). It might be ethically troubling, but it has not been legally problematic in the case of traditional breeding practices. Traditional methods of altering the nature of animals through breeding, even when such breeding has resulted in severe disabilities and considerable suffering, have not been actionable as cruelty. The legal question would be whether genetic modification of animals for purposes of bioart *is* analogous to other types of breeding and genetic alteration of animals that similarly result in the suffering of many animals who die or are killed as part of the enterprise. Do humans "need" art in the same way that they "need" other commodities that cause animal suffering?

Although most genetically modified animals displayed to the public, such as fluorescent rabbits, pigs, monkeys, and fish, do not appear to be suffering from physical ailments, it is difficult to know if such animals experience unexpressed pain or if they experience the alteration as degrading or humiliating. This is a problem even outside the context of transgenic alteration of animals. For instance, when young mice were surgically attached to older mice to determine whether the physical joinder would confer youth-restorative advantages to the older mice, there may have been no indication of suffering once the wound was healed (Brownlee, 2005).[9] Nevertheless, forcible physical joinder significantly altered the life circumstances of both mice, and we cannot know how those mice experienced life thereafter.

Bioengineered animals create even more difficult assessment problems than we have experienced before. Given legal definitions of "cruelty" and "suffering," is it legally "cruel" to a mouse to genetically modify him so that he no longer experiences fear or anxiety in the presence of cats? (The Associated Press, 2007). Since anticruelty statutes deal with physical manifestations of suffering, they are inadequate for reaching the deeper question of human prerogative to alter the mice at all. That is true whether the mice were surgically or genetically altered or whether they were altered for reasons of artistic expression or science.

The statutory infrastructure for recognizing kinds of suffering other than severe "unnecessary" physical suffering does not exist in general anticruelty statutes. However, specifically enacted anticruelty statutes can prohibit a wider range of acts than those that cause direct suffering to animals. For example, many states have anticruelty statutes that specifically prohibit the sale of artificially colored animals such as chickens, rabbits, or ducks, even if the process of artificial coloring itself does not cause the animal to suffer. While it would still be legal to artificially color an animal, not being able to sell or give away such animals could limit the extent to which people artificially color animals at all. Connecticut is one such state. It prohibits, as a matter of cruelty to animals, "any person" from selling or giving away, "living chickens, ducklings, other fowl or rabbits, which have been dyed, colored, or otherwise treated so as to import to them an artificial color..." (Conn. Gen. Stat. Ann. tit. 53, ch. 945 § 53–249a). GloFish would not be covered by Connecticut's prohibition because they are fish and, therefore, not "animals" covered by the statute. The fluorescent green mice in Kac's "The Eighth Day" would not be covered, either.

A facial reading of the statute would suggest that it would be illegal in Connecticut to sell or give away a rabbit like Alba because the rabbit has been "treated so as to import to [the rabbit] an artificial color." However, counter-arguments exist that (a) the words "treated" and "rabbit" do not apply when the process of importing an artificial color is genetic alteration and the result is a rabbit-jellyfish, and (b) that the legislative history does not support the application of the law in a context the legislature could not have anticipated. If the legislature intended to prevent the production of artificially colored animals typically sold or given away at Easter or fundraising events and did not anticipate use of the statute to prohibit the sale or giving away of a genetically altered rabbit, there is an argument that it would be improper to assume that legislators would have addressed artificial coloration of animals through genetic alteration in the same way that they addressed the artificial coloration of animals through other means. Yet, if it is cruel to "import" an artificial color to "a chicken, duckling, other fowl, or rabbit" through non-genetic means how could it not be cruel to alter the very genetic make-up of an animal with the same result?

State Anticruelty Laws: Animal Suffering as "Necessary" to Achieve Valued Human Goals

Ultimately, the third relevant legal question—the reason for altering the color of the rabbit—would play an important role in deciding questions of statutory application. That would entail a consideration of who altered the color of the animal—scientist, commercial vendor of animals, or bioartist—and the circumstances under which the animal was genetically altered. The most legally protected categories are those of "scientist" and "scientific research." Even the most egregious suffering inflicted on an animal will not be actionable as "cruel" if that suffering occurs as a result of "scientific research." However, statutory ambiguity lies in the definition of "scientific research." All states have exemptions for scientific research, but what constitutes "scientific research" varies.

In apparent blanket approval for the scientific enterprise and scientists' entitlement to define what that means without interference, some statutes define "scientific research" as activities conducted in licensed research facilities or by qualified scientists.[10] Nevertheless, it is clear that not all acts are "research" simply because of where and who engages in those acts. Beating an animal to death as part of "scientific research" on humans' reactions to the suffering of animals might well not be prosecuted as "cruelty," but beating an animal to death because the scientist was frustrated with the animal's lack of cooperation with a research protocol could be. Obviously, many other activities, such as boarding companion animals or providing veterinary surgical operations at the request of people who are not part of the research establishment would not constitute "research" just because those activities are conducted by a properly licensed scientist in a properly licensed research facility. Therefore, a scientist's performing the genetic alteration of an animal at the request of a bioartist would not necessarily immunize the act as "scientific."

Some statutes capture those and other distinctions by exempting "bona fide research"[11] or "medical research"[12] or "research governed by accepted standards."[13] Since bioengineering for artistic purposes is distinguishable from bioengineering to serve other human purposes, it is possible that bioartistic creations, even if they take place in licensed research facilities by licensed scientists, fall outside the purview of the scientific research exemption in a state anticruelty statute. There are no explicit exemptions for bioart, and, without protection of the scientific research exemption, the bioartist and scientist would be vulnerable to prosecution for cruelty if animals covered by the statute suffer harm that could be characterized as "torment" or "mutilation" as a result of transgenic bioartistic activities.

The weight of prosecution would then rest on whether "artistic expression" is sufficiently "necessary" to justify the incidental infliction of suffering on animals. Arguably, *any* use of animals and *any* amount of suffering inflicted on them should pass legal muster as long as humans may legally subject animals to horrific suffering in factory farms and slaughterhouses for no better reason than that humans like the taste of animal flesh. Similarly, as long as scientists can lawfully inflict any amount of suffering on animals as part of scientific research, which is entirely legal under state anticruelty laws and the federal Animal Welfare Act, why should artists not have similar latitude?

Given the history of enactment and application of anticruelty statutes, bioartists might well claim that *all* justifiable uses of animals should be protected and that the operation of the anticruelty statutes should address purely gratuitous, senselessly inflicted suffering on animals. Of course, bioartists would not identify transgenic bioart (or other types of bioart) as gratuitously inflicted suffering, even though alternative means of expressing the same ideas exist. For those of us who care about animals' interests in non-interference, let alone freedom from human-inflicted suffering, there are *no* justifiable uses of animals, and the operation of the anticruelty statutes should protect animals in many more ways than they currently do. However, the law tends to reflect the values of consumers, legislators, jurors, and others who make distinctions among the purposes for which suffering is inflicted on animals. The distinctions drawn in the case of Glofish suggest bioengineering for purposes of advances in human health may be considered a "necessary evil" while bioengineering for such purposes as artistic expression or critiquing science, scientists, or bioengineering may not be—or at least not when it comes to the genetic manipulation of or infliction of suffering on sentient creatures for such purposes.

In fact, "necessity" is a highly freighted concept in the context of anticruelty statutes. When it comes to factory farming practices, the "necessity" to eat animal-derived products has not been successfully challenged. "Necessity" has bearing only on whether a particular practice is necessary to the production of those products, which are themselves presumed to be necessary (Francione, 2007, p. 10). Hence, it would be "cruel" to deny food to a cow because doing so serves no purpose, but it would not be "cruel" to deny food to a laying hen if doing so results in her laying more eggs when access to food is restored. Similarly, it is not possible to challenge the "need" to hunt or to use animals in experimentation. Only acts that cause terrible suffering without any socially justified purpose are deemed "cruel" under state anticruelty statutes.

It is not clear whether any type of bioart, including transgenic bioart, would receive similar presumptive approval as "necessary" to fulfill a socially justified purpose. The defense of "artistic expression" did not exculpate a vegetarian Canadian who produced a film of his torture and killing of a stray cat ostensibly as a social commentary on meat-eating (*The Queen v. Power*, 2003). To the extent that transgenic bioart results in the same type of visible suffering, it might well not pass legal muster. However, bioart that does not result in that level of visible suffering might not be actionable because it is not the type of suffering general anticruelty statutes are intended to prohibit or because the production of art is considered to be as necessary as the production of meat. The suffering of all the animals used in the process and the disabilities that plague even the "success" stories would have to become sufficiently visible to offset the images of "happily green" rabbits and mice.

Certainly it is not strictly necessary to actually genetically alter or harm living animals in order to express points of view that involve the genetic alteration of animals. Artist Adam Brandjes's use of animatronic animals suggests that such artificial animals can be effective means of communicating various perspectives, including disturbing or otherwise complex relationships between apparently real and altered beings (Langill, 2006, p. 59). Similarly, sophisticated methods of interactive juxtaposition of images from existing video footage can accomplish many of the same expressive goals. Some artists may bridle at the thought of suggested substitution and argue that restrictions of the means by which ideas are communicated is antithetical to the very nature of art. However, Caroline Langill (2006), speaking from the perspective of an artist, argues persuasively that "[a]rtists are in a position to evaluate how genetic modification impacts living organisms and culture, but not if we embrace methods that are shared with the post-industrial corporate world" (p. 61). The fact is that transgenic bioart is already limited; humans are not subjects of transgenic bioart. That nonhuman animals *can* be used is not recognition of a "need" for artistic license; it is uncritical, unself-conscious replication of the lesser status of animals as mere resources. The problem for animals and the people who care about them is that the legal definition of "necessity" in general anticruelty statutes has not historically protected most animals from most forms of human-inflicted suffering.

The Federal Animal Welfare Act

All of the problems associated with the applicability of state anticruelty statutes to bioengineering, transgenic bioart, and transgenic animals are present at the federal level. For instance, the problem for animals of noninclusive statutory definitions of "animals" is nowhere more serious than in the federal Animal Welfare Act (AWA) of 1970, as amended in 1985. The AWA defines an "animal" as "any live or dead dog, cat, monkey (nonhuman primate mammal), guinea pig, hamster, rabbit, or such other warm-blooded animal, as the Secretary [of Agriculture] may determine is being used, or is intended for use, for research, testing, experimentation, or exhibition purposes" (§ 2132 [g]). Alba, the fluorescent green rabbit, and ANDi, the fluorescent monkey, would be covered by the AWA, if they were determined to be a rabbit and monkey, respectively. Transgenic fish, birds, and mice would not be covered at all. Indeed, it is estimated that 90% of animals used in scientific research are birds, mice, and rats—animals that are not considered "animals" under the AWA and, therefore, not protected at all (Cornutt, 2001, p. 57, 69, 448). For decades, legal advocates for animals petitioned the Secretary to use his discretion to designate them as animals. When it appeared that they would finally be successful, Congress amended the AWA explicitly to exclude them (Francione, 2007, p. 31).

The loss to mice, rats, birds, and other non-enumerated animals that they are not included as "animals" is not as great as first appears. That is because the AWA is extremely limited in scope; it is grossly inadequate for protecting even the most minimal interests of animals who *are* covered by the AWA. First, although one of the AWA's purported purposes is "to insure that animals intended for use in research facilities or for exhibition purposes... are provided humane care and treatment" (§ 2131 [2]), for the AWA to apply at all there must be use of a federally inspected or funded research facility or the transport in interstate commerce of animals who are the subject of the bioart/biotechnology. Second, the definition of a "research facility" is linked to the definition of "animal" in that only entities that use live "animals" as defined by the AWA are "research facilities" covered by the AWA (§ 2131 [e]). Moreover, the Secretary of Agriculture has authority to exempt by regulation any entity that would be covered by the AWA, if that entity does not use live dogs or cats or "substantial numbers" of other live animals for biomedical research or testing (§ 2131 [e]).

The AWA states that the "Secretary [of Agriculture] shall promulgate standards to govern the humane handling, care, treatment, and transportation of animals by dealers, research facilities, and exhibitors" (§ 2143 [a][1]), but the AWA also states explicitly that the Secretary does not have the authority to act in ways that affect the design or performance of research (§ 2143 [a][6][A][i]–[iii]). Moreover, in cases of "scientific necessity" it is possible to subject animals to painful experimentation without the use of any form of pain relief at all (§ 2143 [3][C][v]). In other words, the AWA does not constrain any means by which or purposes for which animals are subjected to experimentation. Finally, even if meaningful protection of some kind were conferred by the AWA, it is enforceable only by the United States Department of Agriculture, which, for a variety of reasons, fails to enforce the law.

The federal AWA burdens scientists only with paperwork and not with restrictions on how they can exploit animals or how much suffering they can inflict on animals. As such, it is a real boon to scientists and research facilities because its existence suggests to the public that there are safeguards in place to protect animals. In response to animal advocate protests, the heads of research facilities can and do state truthfully that their facilities are in full compliance with federal laws that regulate the humane treatment of animals. They are not asked or required to describe those laws or to recount the limitations of those laws for protecting animals even from intense and prolonged suffering. So great is the protection afforded scientists (and not animals) that, if the Animal Welfare Act did not exist in its current form, scientists and research institutions would surely seek its enactment.

Of Bioart and the First Amendment

Ultimately, only legislation that cuts deeply at the root of human prerogative to exploit nature—not just sentient animals—will have much effect on bioengineering, bioart, animals, and nature. Existing law is not of much use in resolving the difficult problems that bioart exacerbates through its replication of existing patterns of disregard for nature; it does not even assuredly guarantee the health and safety of sentient beings used in bioart. Yet any deep-cutting proposals that would have the effect of curtailing bioartistic endeavors would undoubtedly meet with objections based on parity of treatment with other exploitative activities and claims of constitutional First Amendment protection.

It is striking that even in a case of alleged improper procurement of biological materials, bioartists claim that the primary objective of prosecution was persecution of bioartists and erosion of First Amendment rights of free speech.[14] This much debated case began after Steven Kurtz, a founder of the Critical Art Ensemble (CAE), called for emergency assistance when he discovered that his wife had stopped breathing.[15] Emergency responders found Hope Kurtz dead in a residence where there was also "a biological lab, with an incubator, centrifuge and bacterial cultures growing in Petri dishes. Windows nearby were covered with foil, and on the shelves sat books like 'The Biology of Doom' and 'Spores, Plagues and History: The Story of Anthrax'" (Kennedy, 2005, para 2). Kurtz was initially suspected of bioterrorism, but, when the cultures were found to contain only benign bacteria, legal proceedings related to bioterrorism ended. Nevertheless, prosecutors went forward with legal charges related to the means by which Kurtz procured the bacteria samples.

Prosecutors alleged that Kurtz and geneticist Robert Ferrell committed mail and wire fraud when Ferrell ordered samples of bacteria on behalf of Kurtz, allegedly in violation of the contractual terms by which American Type Culture Collection (ATCC) supplies biological materials to scientists and research facilities.[16] Ultimately, Ferrell pleaded guilty to lesser charges, and a judge dismissed the case against Kurtz (*United States v. Kurtz*, 2008). The court dismissed the case on the basis of the indictment as written but explicitly left open the question of whether the prosecution could go forward if the indictment were written differently.[17] In other words, the court made quite clear that it was making no determination about whether Kurtz and Ferrell committed mail and wire fraud or any other unlawful act. Regardless, defenders of Kurtz and Ferrell contend that there should have been no legal proceedings against either of them once it was established that the bacterial cultures were harmless. They argue that prosecution for "technical" violations constituted persecution of bioartists and an overreaction to the events of September 11, 2001.[18]

Bioethics professor George Annas seems to support this view but blames Kurtz and CAE for eliciting that overreaction. In his critical review of the relationship between bioart and bioterrorism, Annas notes that as a founder of CAE, Steven Kurtz positioned his art and commentary so as to provoke fears about biotechnology, going so far as to encourage critics of biotechnology to engage in "fuzzy biological sabotage," such as releasing visibly genetically altered flies at restaurants "to stir up paranoia" (Annas, 2006, p. 2717). Annas takes the position that governmental reactions—including, potentially, overreactions—based on fear should be expected, if bioart "aims at the heart of our fears" and intentionally "disturbs" (2006, p. 2717).

There may well be some truth to the argument that a particular group that promotes "fuzzy biological sabotage" might be subject to more suspicion than other groups, but Annas's general argument about transgenic bioart provoking such suspicion seems overly simplified both as to transgenic bioartists' intentions and as to public responses. As to the first—transgenic bioartists' intentions—surely it can be said that there is an intent to "provoke" and to engage the public in a discourse. For instance, Kac used Alba to express different ideas, including that "molecular biology is not a rarefied language spoken by experts beyond the reach of ordinary citizens" (as cited in Lynch, 2008, p. 194) and that people should "overcome their fear of transgenic creatures" (Lynch, 2008, p. 193). Similarly, CAE used different kinds of projects to elicit a varied discourse on biotechnology, some of which included prompts related to bioterrorism. However, any given prompt can signal a variety of intentions on the part of the bioartist.

As to the second—public responses—there are many possibilities besides simply fear or dislike of transgenic bioart and overreaction to the potential for harm to human health or the environment. For instance, people responded to Alba in many different ways, including praise for her appearance, "anxieties about mutant monstrosities," attention to the plight of animals in "laboratory prison(s)," and concerns about "the irresponsibility of contemporary conceptual art" (Lynch, 2008, p. 192, 196). Similarly, there were probably a number of different reactions to the circumstances that surrounded the filing of charges against Robert Ferrell and Steven Kurtz. Therefore, fears provoked by paraphernalia associated with a project to simulate bioterrorist attacks may or may not have played a significant role in the prosecutions for mail and wire fraud. As Annas argues, it is possible that members of the public (including the federal grand jury) and law enforcement do not like the idea of transgenic bioart, even if—or perhaps especially when—it is practiced by a group that purports to criticize biotechnology. On the other hand, that charges related to the acquisition of bacteria samples remained after bioterrorism charges were dropped may be indicative only of what can happen when some but not all charges are determined clearly to be without legal basis. As noted earlier, there was no clear legal resolution of the matter. The mere fact that a judge decided ultimately to dismiss the case because the indictment was written in a particular way does not mean that Kurtz and Ferrell's conduct was lawful or that the prosecution was trivial. Certainly there is room for Annas's point that '[b]iosafety regulations are not merely legal technicalities. They constitute some of the terms of the pact between science and the public that established the public trust" (2006, p. 2718). And, as transgenic bioartist Joe Davis has pointed out, there is no reason to expect less of bioartists than scientists when it comes to biosafety rules (Kennedy, 2005, para 23). It is not persecution of transgenic bioartists if they are held to a standard of care and rule compliance that is no higher than that applied to others who work with biological material.[19]

Since law is as much an interpretative process as any other interpretative endeavor, it is not possible to rely solely on one explanation for the legal events that ensued after emergency responders came to Kurtz's home any more than it is possible to rely solely on one interpretation of CAE's art or Kac's production of Alba. Just as it is possible to claim that public lack of understanding of bioart led to overreaction to the circumstances that surrounded the death of Hope Kurtz, it is possible to claim that some artists' lack of understanding of legal process and constitutional free speech protections led to overreactions to the prosecutions of Kurtz and Ferrell.

Of course, there is no mistake that free speech is considered a fundamental right. Free speech protection is supported by numerous rationales, including respect for the individual, concerns about governmental suppression of political speech, and the belief that protection of speech results in a "marketplace of ideas" through which society explores and selects those ideas that best serve society's interests. Bioart serves purposes for which free speech is protected, and bioart is presumptively protected because it is artistic expression. Nevertheless, application of the Constitution's right to free speech in the context of bioart is no more straightforward than is the application of state anticruelty statutes or the federal Animal Welfare Act. That is because fundamental rights are not absolute rights. There are circumstances under which rights guaranteed under the Constitution can be permissibly negatively impacted by governmental action. For instance, government can, under some circumstances, regulate some types of conduct that have an impact on speech. There is a free speech right to discuss the results of scientific research, but courts have not found an independent free speech right to conduct the underlying research itself, which may be regulated in ways that affect free speech (Adams, 2003; Andrews, 1998). Indeed, government already regulates scientific research in ways that have an impact on speech, including transgenic bioart. Laws that pertain to experimentation on human subjects already seriously constrain transgenic bioartistic expression of a number of ideas. Government could further regulate all uses of living matter for reasons of public health, environmental protection, or prevention of cruelty to animals. As long as bioart is not specifically targeted because of the ideas it seeks to explore or disproportionately negatively impacted by an overly broad law that sweeps in bioart without needing to do so to meet the objectives of the law, there would not be a constitutional violation.[20]

First Amendment law also allows government to directly limit free speech on the basis of content when there are compelling governmental interests in doing so. For instance, government can regulate commercial speech and obscene speech. Framed as a type of prohibition on obscene speech, Congress, in 1999, passed a law to criminalize the production and sale of depictions of animal cruelty for which animals were intentionally subjected to cruelty for the sexual gratification of the viewer (U.S. Code Annotated 18, § 48). The law was enacted with the goal of assisting local law enforcement in prosecuting those in the "crush video" business. "Crush videos" are videos in which women taunt, torture, and gradually crush an animal to death with their high heels. Various types of animals—mice, rats, worms, kittens, lizards, guinea pigs, and insects—are used. They are immobilized, such as being taped to the floor, and capturing their experience of torture and death is the object of crush video production. Such videos appeal to those who are sexually stimulated by identification with the victim.

Many of the acts depicted in crush videos violate state anticruelty laws, and local law enforcement agencies persuaded Congress that there were enforcement problems that could be resolved by enacting a federal law. (U.S. Code, Title 18, § 48) seems sweeping in that it covers the creation, sale, and possession (with intent to sell) of depictions of animal cruelty of all kinds in order to "dry up the market" for depictions of cruelty to animals, which drives the production of such depictions. In fact, overbreadth was one reason given by the appellate court that overturned the first conviction under the law in a case involving sales of depictions of dog-fighting (*United States v. Stevens*, 2008). However, three dissenting judges in the *Stevens* case argued that the law is not impermissibly overly broad because the law as enacted is so limited in scope. For instance, depicted acts of cruelty must be illegal in the state where the depiction is created, sold, or possessed with intent to sell; one does not have a right to commit a crime for the sake of art. Also, there are many explicit exemptions in the law, including depictions that have serious religious, political, scientific, educational, journalistic, historical or artistic value.

At the heart of the disagreement between the majority and dissenting judges in the *Stevens* case is whether protecting animals is a sufficiently important justification for restricting free speech. Unlike the majority, the dissenting judges argue that there is minimal, if any, value in depictions of acts of cruelty to animals harmed only for the purpose of making such depictions. They argue further that protecting animals from such gratuitous acts of violence is a compelling governmental interest sufficient to restrict even otherwise protected speech, as evinced by anticruelty statutes enacted in all 50 states, and that U.S. Code, Title 18, § 48 as enacted is sufficiently narrowly tailored to accomplish the goal of protecting animals from the gratuitous infliction of suffering.

It remains to be seen whether appellate courts in other jurisdictions will adopt a different interpretation of U.S. Code, Title 18, § 48. However, even before those appeals materialize, the *Stevens* case reveals that, although government can, with narrowly tailored laws, restrict free speech interests in favor of other compelling governmental interests, challenges to free speech-restrictive laws enacted to protect animals are inevitable because there is ambivalence about whether preventing animal suffering is important enough to warrant restricting speech.

Conclusion: The Role of Law

Much of the foregoing has concerned sentient beings. Yet, at its heart, the problem of humans' ruthless exploitation of sentient beings is how we treat nature. As a theoretical matter, laws could be enacted to prevent harm to nature and animals that result from biotechnology. As a pragmatic matter, laws are reactive, piecemeal, and controlled by special interest groups that would consider a respectful relationship to nature to be a major hindrance to fulfilling those groups' interests in exploiting nature. The shallow reach of existing laws attests to the existence of only minimal regard for animals and nature, and it is difficult to imagine a reliable legal pathway to a more respectful relationship. Many advocates for such a relationship propose legal avenues for disclosing to the public how animals are treated. However, as Steven Best (2009) persuasively points out, mere factual disclosure of what happens to animals and nature may not have transformative effect. Alternative, differently contextualized presentations of the facts—contextualized in various ways to illuminate the extent of human disrespect of nature and sentient beings—can provide opportunities through which differently situated people become aware of nature and sentient beings independently of their utility in satisfying human desires. That is the power of language of all sorts, including art. However, that transformative potential will be hindered to the extent that a linguistic or artistic form of expression reinforces a primary structure by which humans' oppression of nature is sustained. That is why transgenic bioart by itself cannot have the transformative effect that its advocates promise.

Law cannot now serve as a primary means to protect nature because law is currently primarily protective of human interests that are realized through exploitation of nature. Other, extra-legal voices will have to be the primary voices of opposition to the course humans have taken to degrade nature. Nevertheless, there will also and always be people grounded in law who will use their training to seek to limit the damage and direction of biotechnology, and who will thereby join with others in turning human minds and hearts in the direction of hearing those of animals and nature.

Notes

1. The author thanks Vicki Steiner, Library Services Analyst at UCLA Law Library, for her invaluable research assistance and Katherine Sharif for assistance with checking citations and formatting the manuscript. Research for this chapter was funded by Bob Barker's generous endowment gift to UCLA Law School for purposes of teaching and scholarship in the field of animal law.
2. Technically, "transgenic art" is a sub-category of "bioart," which includes other types of art that do not involve genetic manipulation of living beings but do involve manipulation or killing animals for artistic expressive purposes. For instance, bioartist Garnet Hertz (2007) killed frogs, replaced their internal organs with web server components, and suspended them in a clear glass container of mineral oil, and bioartist Hubert Duprat removed caddis fly larvae from their natural habitat and placed them in environments in which the only materials available with which to build external skeletal material were beads, pearls, and 18-karat gold

(Duprat and Besson, 1997). "The activities of the caddis worm, as manipulated by Hubert Duprat, are prompted by the "noise"—beads, pearls and 18-karat gold pieces—introduced by the artist into the insect's environment." (Duprat and Besson, 1997, para 1). Nevertheless, the term "bioart" is frequently used to refer specifically to "transgenic art" (e.g., Kennedy, 2005; Annas, 2006) as well as nontransgenic bioart forms.

3. Randy Kennedy (2005) describes collaborative relationships that range from the bioartist depending on a scientist to acquire bacteria to working as an unpaid research assistant to a scientist in order to have full access to a laboratory. Also, as suggested by Lisa Lynch's (2008) discussion of Alba, the genetically altered rabbit, bioartists may make use of living beings who have been altered by scientists.
4. See, for example, D. Scott Bennett's (2006, pp. 357–360) discussion of the U.S. Patent Office's reluctance to issue patents to Jeremy Rifkin and Stuart Newman for the processes by which to produce a human–animal chimera and for the resulting chimeras themselves.
5. See Sheryl Lawrence (2007, pp. 211–219) for a discussion of the history preceding the creation on June 26, 1986, of the Coordinated Framework for the Regulation of Biotechnology.
6. The court is quoting Article I, § 8, cl. 8 of the Constitution, which authorized Congress to enact the law to be applied in this case, the Patent Act (35 U.S.C. § 101).
7. The court is citing the previously decided U.S. Supreme Court case of *Funk Brothers Seed Company v. Kalo Inoculant Company*, 333 U.S. 127 (1948).
8. In the fall of 2008, the FDA proposed guidelines for review of certain genetically engineered animals before those animals, their parts, or products made from them could be sold. The FDA proposes treating those animals like drugs in terms of the standards of safety review that would be required. Critics claim that there are significant differences between the two such that wholly different review procedures should be used for genetically altered animals. They also criticize the proposal on grounds that the approval process would be secretive in order to protect producers' commercial interests and that there is insufficient attention to the environmental impact of inadvertent release. Moreover, the proposed regulations do not cover cloned animals, research animals, and most pets (Maugh and Kaplan 2008, para. 1, 4, 5, 10).
9. Christen Brownlee (2005) reported for *Science News* that "scientists removed skin along the flanks of young mice, aged 2–3 months, and old mice, aged 19–26 months. They then sutured pairs of the mice together. The mice quickly adjusted to life as partners, cooperatively eating and roaming their cage. Within several weeks, each pair's blood systems merged. After giving each mouse a leg injury, the researchers found that among the young–old pairs, young mice healed more slowly than did young mice that hadn't been joined and old mice showed significantly faster healing rates than old, solo mice did" (para 3).
10. For example, Idaho's anticruelty law does not apply to "research carried out by professionally recognized private or public research facilities or institutions" Idaho Code § 25-3514 (3). Maryland's anticruelty statute does not apply to "animals that are used in privately, locally, State, or federally funded scientific or medical activities" Md. Crim. Law § 10-602 (7).
11. For example, Kentucky exempts "bona fide animal research activities of institutions of higher education" in addition to "business entities registered with the U.S. Department of Agriculture or subject to the other federal laws governing animal research" Ky. Revised Stat. § 525.130 (2) (f). Hawaii exempts "activities carried on for scientific research governed by standards of accepted educational or medicinal practices" Haw. Revised Stat. § 711-1108.5 (2) (b).
12. For example, Connecticut exempts suffering inflicted on animals "while performing medical research as an employee of, student in or person associated with any hospital, educational institution or laboratory" Conn. Gen. Stat. Ann. § 53–247 (b).
13. Alaska exempts the infliction of suffering on animals that is "part of scientific research governed by accepted standards" Alaska Stat. § 11.61.140 (c).
14. *United States v. Kurtz*, No. 04-CR-0155A, 2008 WL 1820903 (W.D.N.Y. April 21, 2008). Bioartists' reactions to the prosecution of Steven Kurtz and Robert Ferrell are described by Kennedy (2005, para 15–25), Lok See Fu (2005–2006, pp. 102–104), and Annas (2006, p. 2718).

15. There appears to be considerable consistency among the various accounts of the background facts of the case. A good general introduction is that of Randy Kennedy, reporting for *The New York Times* on July 3, 2005 in an article entitled "The Artists in the Hazmat Suits." George J. Annas (2006) discusses the prosecution of Dr. Thomas Butler and the prosecutions of Kurtz and Ferrell and raises the possibility that blurring of lines between bioart and bioterrorism has the potential to lead to overreaction.
16. The progress of the case is reported at http://www.caedefensefund.org.
17. The court stated: "The Court passes no judgment on whether the indictment could have been drafted in such a way, based on the facts and circumstances as they have been presented here, to allege sufficiently a "no-sale" theory of fraud. It simply finds that the indictment, *as currently written*, [italics added] fails to allege such a theory" (*United States v. Kurtz*, 2008, p. 5).
18. See, for example, the critical reviews of Gregory Sholette (2005) Disciplining the avant-garde: The United States versus The Critical Art Ensemble; Joyce Lok See Fu (2005–2006) The Potential decline of artistic creativity in the wake of the Patriot Act: The Case surrounding Steven Kurtz and the Critical Art Ensemble; George J. Annas, (2006) Bioterror and bioart—A Plague o' both your houses.
19. George Annas (2006) reports extensively on the case of Dr. Thomas Butler who served a two-year sentence for failing to account for some samples of plague cultures, mislabeling shipments that contained such samples, and transporting such samples without proper permits. There was no evidence of harm as a result of these lapses, and it was possible to characterize the violations of the shipment rules as "technical" or "trivial." Nevertheless, such rules are hardly trivial to the public concerned about accidental tragedies, not just intentional terrorist attacks. That Kurtz and Ferrell were involved in the transport of benign bacterial cultures does not change the fact that rules exist partially to avoid case-by-case line-drawing about when rule compliance is actually required.
20. It is beyond the scope of this chapter to fully examine First Amendment jurisprudence as applied to bioart. However, it is discussed at length in an article by Nahmod (2000–2001) in which he concludes that there is no easy solution to problems posed at the intersection of the First Amendment and bioart (p. 484). The devil is in the details of specific proposed laws and the application of specific laws to particular bioartistic projects.

References

Adams, N. A. (2003). Creating clones, kids & chimera: Liberal democratic compromise at the crossroads. *Notre Dame Journal of Law, Ethics & Public Policy*, *17*, 71, 110.

Alaska Stat. Ann. § 11.81.900 (b) (3).

Andrews, L. B. (1998). Constitutional challenges to bans on human cloning. *Harvard Journal of Law & Technology*, *11*, 643, 661–664.

Animal Welfare Act, 7 U.S.C. §2131 et seq. (1970).

Annas, G. J. (2006, June 22). Bioterror and bioart—A plague o' both your houses. *New England Journal of Medicine: Legal Issues in Medicine*, *354*, 2715, 2718

Bacher, D. (2003, Dec 29). *California ban on glofish ignites debate over 'frankenfish.'* Retrieved June 5, 2008 from the Dissident Voice Web site: http://www.dissidentvoice.org/Articles9/Bacher_Frankenfish-CA.html

Bennett, D. (2006). Chimera and the continuum of humanity: Erasing the line of constitutional personhood. *55 Emory Law Journal 347*, 357–360.

Best, S. (2009). Genetic science, animal exploitation, and the challenge for democracy. In C. Gigliotti (Ed.), *Leonardo's choice: Genetic technologies and animals*. Dorchedt: Springer.

Birke, L. (2009). Meddling with medusa: On genetic manipulation, art and animals. In C. Gigliotti (Ed.), *Leonardo's choice: Genetic technologies and animals*. Dorchedt: Springer.

Brownlee, C. (2005, Mar 5). Healing secret lies in blood. *Science News, Vol. 167(10)* Retrieved June 5, 2008 from Science News Web site: http://sciencenews.org/view/ generic/id/5943/title/Healing_secret_lies_in_blood

Connecticut General Statutes Annotated, Chapter 945, § 53–249a.
Coordinated Framework for Regulation of Biotechnology, 51 Fed. Reg. 23302-01 (1986).
Cornutt, J. (2001). *Animals and the law: A sourcebook.* Santa Barbara, California: ABC-CLIO, Inc.
Cullinan, C. (2002). *Wild law: Governing people for earth.* Claremont, South Africa: Siber Ink.
Diamond v. Chakrabarty, 447 U.S. 303 (1980).
Duprat, H. & Besson, C. (1997). *The Wonderful Caddis Worm.* Retrieved March 14, 2008, from http://www.leonardo.info/isast/articles/duprat/duprat.html
Francione, G. L. (2007). Reflections on animals, property, and the law, and Rain without thunder. *Law & Contemporary Problems*, 70, 9, 10, 31.
Georgia. Code Annotated § 16-12-4 (a) (1).
Gigliotti, C. (2009). Leonardo's choice: The ethics of artists working with genetic technologies. In C. Gigliotti (Ed.) Leonardo's choice: Genetic technologies and animals. Dorchedt: Springer.
Hertz, G. (2007). *Experiments in Galvanism: Frog with Implanted Webserver.* Most recently displayed at the Dutch Electronic Art Festival in Rotterdam, Netherlands from April 2007. Retrieved March 14, 2008, from the World Wide Web: http://www.conceptlab.com/frog/
International Center for Technology Assessment v. Thomson, 421 F. Supp.2d 1 (2006).
Kac, E. (2001). *The Eighth Day: A Transgenic Net Installation.* Displayed at Arizona State University from October 25 to November, 2001. Retrieved June 5, 2008 from the World Wide Web: http://www.ekac.org/8thday.html
Kac, E. (2003). GFP Bunny. *Leonardo special project Global Crossings: The Cultural Roots of Globalization. Vol. 36, No.2,* p. 97–102. Retrieved June 5, 2008 from the Project Muse Web site: http://muse.jhu.edu
Kennedy, R. (2005, July 3). The Artists in the Hazmat Suits. *The New York Times*, Section 2; Column3; Arts and Leisure Desk; Art; p. 1.
Lamb, G. M. (2004, Jan. 22). *Glofish zoom to market.* Retrieved June 5, 2008 from the Christian Monitor website: http://www.csmonitor.com/2004/0122/p14s02-sten.html
Langill, C. (2006). Negotiating the hybrid: Art, theory and genetic technologies. *AI & Society,* 20:49–62, 59, 61.
Lawrence, S. (2007). What would you do with a fluorescent green pig?: How novel transgenic products reveal flaws in the foundational assumptions for the regulation of biotechnology. *Ecology Law Quarterly* 34, 201, 238–247, 255, 257.
Lok See Fu, J. (2005–2006). The Potential decline of artistic creativity in the wake of the Patriot Act: The Case surrounding Steven Kurtz and the Critical Art Ensemble. *Columbia Journal of Law & the Arts,* 29, 83, 93–97.
Lynch, L. (2008). Culturing the pleebland: The idea of the public in genetic art. *Literature and Medicine* 26(1), 180, 193, 198.
Maugh, T. H. and K. Kaplan. (2008, Sept 19). FDA proposes approval process for genetically engineered animals. *The Los Angeles Times.* Retrieved December 3, 2008 from The L.A. Times website: http://articles.latimes.com/2008/sep/19/science/sci-genetic19
Midgley, M. (2000). Biotechnology and monstrosity: Why we should pay attention to the 'yuk factor'. *The Hastings Center Report,* 30(5), 6–15.
Nahmod, S. (2000–2001). The GFP (green) bunny: Reflections on the intersection of art, science, and the First Amendment. *Suffolk University Law Review,* 34, 473, 478–481.
Posner, R. A. (2004). Animal Rights: Legal, Philosophical, and Pragmatic Perspectives. In Sunstein, C.R. & Nussbaum, M.C. (Eds.), *Animal Rights: Current debates and new directions.* (pp. 51–77). New York: Oxford University Press.
Sholette, G. (2005). *Disciplining the avant-garde: The United States versus the Critical Art Ensemble.* Retrieved June 5, 2008 from The Critical Art Ensemble Defense Fund website: http:www.caedefensefund.org
Stone, C. D. (1996). *Should trees have standing? And other essays on laws, morals and the environment.* New York: Oceana Publications.
Sunstein, C.R. (2003). The Rights of Animals. *University of Chicago Law Review* 70, 387.

The Associated Press. (2007, Dec 15). Modified mouse stands up to cat. *The Los Angeles Times*, p. A18.

The Associated Press. (2003, Dec 4). *California blocks sales of 'Glofish Pets'*. Retrieved June 5, 2008 from CNN.com website: http://cnn.com/2003/TECH/science/12/04/fluorescent.fish.ap/index.html

The Queen v. Power, 176 C.C.C (3d) 209, 174 O.A.C. 222 (Ontario Ct. App. 2003).

U.S. Code Annotated, Title 18, §48.

U.S. Department of Food and Drug Administration. (2003). *Animal Cloning: A Risk Assessment: Draft Executive Summary.* Retrieved June 5, 2008 from the U.S. FDA Web site: http://www.fda.gov/cvm/Documents/CLRAES.doc

U.S. Department of Food and Drug Administration. (2003, Dec 9). *FDA Statement Regarding Glofish*. Retrieved June 5, 2008 from U.S. FDA Web site: http://www.fda.gov/bbs/topics/NEWS/2003

Part III

Dis/Integrating Animals: Ethical Dimensions of the Genetic Engineering of Animals for Human Consumption

Traci Warkentin

Abstract Research at the intersections of feminism, biology and philosophy provides dynamic starting grounds for this discussion of genetic technologies and animals. With a focus on animal bodies, I examine moral implications of the genetic engineering of "domesticated" animals—primarily pigs and chickens—for the purposes of human consumption. Concepts of natural and artificial, contamination and purity, integrity and fragmentation and mind and body feature in the discussion. In this respect, Margaret Atwood's novel, *Oryx and Crake*, serves as a cogent medium for exploring these highly contentious practices and ideas as it provides hypothetical narratives of possibility. Moreover, it is used to highlight contemporary hegemonic assumptions and values in ways that make them visible. Particular attention is paid to issues of growing human organs in pigs for xenotransplantation (resulting, for Atwood, in "pigoons") and the ultimate end of the intensive factory farming of chickens through the genetic engineering of "mindless" chicken tumours (or, as Atwood calls them, "ChickieNobs"). Integral to these philosophical considerations is the provocative question of the genetic modification of animal bodies as a means to end the suffering of domestic food animals. The ultimate implications of this question include an ongoing sensory and moral deprivation of human experience, potentially resulting in a future mechanomorphosis, the extreme manifestation of an existing mechanomorphism.

Keywords Animal ethics · Genetic engineering · Mechanomorphosis · Oryx and Crake · Speculative/science fiction · Transgenic organisms

> Like The Handmaid's Tale, Oryx and Crake is a speculative fiction, not a science fiction proper. It contains no intergalactic space travel, no teleportation, no Martians. As with The Handmaid's Tale, it invents nothing that we haven't already invented or started to invent. Every novel begins with a what if, and then sets forth its axioms. The what if of Oryx and Crake is simply, What if we continue down the road we're already on? How slippery is the slope? What are our saving graces? Who's got the will to stop us?—Margaret Atwood[1]

T. Warkentin (✉)
Department of Geography, Hunter College, The City University of New York, USA
e-mail: twarkentin@hunter.cuny.edu

C. Gigliotti (ed.), *Leonardo's Choice*, DOI 10.1007/978-90-481-2479-4_9,

A biotechnological age is here and contemporary Western society is steeped in its controversies of possibilities and problems. Debates on its potential for inspiring hope, as a tool for environmental salvation, and for rousing fear, as in opening Pandora's Box and creating Frankensteins, have raged for long enough now to even become clichè. News of a new medical finding or agricultural epidemic makes headlines daily. Indeed, as I write, this week's leading headline on the ScientificAmerican.com website reads: "Mouse Research Bolsters Controversial Theory of Aging," and is accompanied by an all-too-familiar image of a white mouse, an icon of laboratory research.

The first paragraph reads,

> Aging is a process we humans tend to fight every step of the way. The results of a mouse study underscore the potential of antioxidants as a tool in that battle: animals genetically modified to produce more antioxidant enzymes lived longer than control animals did. (Graham, 2005)

While this news article may seem commonplace now, it is headlines such as this one that inspired Margaret Atwood to write a book of speculative fiction on how genetic engineering may continue to shape life on Earth in the future. Her novel, *Oryx and Crake*, published in 2003, opens with a description of an ominous and barren landscape and it becomes immediately apparent that Atwood's vision of the biotechnological future is dystopian to say the least. The take-home messages are deeply humbling as Atwood presents her readers with provocative and disturbing possibilities. As such, *Oryx and Crake* provides a transitional narrative space for the discussion of current biotechnological philosophies and practices in Western society and where they might lead to in the not-so-distant future. While the book covers many aspects of society and technology worthy of discussion, this chapter focuses on issues of genetically modified organisms (GMOs), particularly "transgenic organisms."[2] A transgenic organism is one that has been microgenetically engineered so that its genome contains genetic material derived from a different species (Wheale and McNally, 1990, p. 285). For example, a "geep" is a sheep and goat hybrid, containing genetic material from both species (Wheale and McNally, 1990, p. 276).

Focussing on transgenic animals, I explore the complex concepts and assumptions of value embedded within such practices of genetically engineering animal bodies. For instance, Leesa Fawcett[3] calls the use of animals as medical models of human disease, the ultimate practical expression of anthropomorphism (personal communication, 5 April 2005). The mouse model in the aging study quite literally becomes a metaphor for human physiology. This is particularly important in terms of challenging dominant Western understanding of humans and animals, of nature and culture, and related patriarchal dualisms,[4] which tend to define who and what we are. Thus, as Lynda Birke asserts, these Western scientific ideas and practices have integral connections with social and political, not to mention economic, issues and deserve attention from an ecofeminist perspective (1994, pp. 10–11). This treatment of animal bodies as biofactories is a clear expression of the strong reductionist trend in Western sciences in general, and biotechnologies in particular, which has resulted in a predominant view of organisms as machines. The body-as-machine metaphor is prolific and powerful, and further exemplified in an application—which enjoys the most attention in this chapter—in which the bodies of animals are modified to grow in ways that result in commercially valuable and consumable "parts" and processes for medicinal or agricultural purposes. The production of these biofactories, of commodified transgenic flesh and viscera, is now a chilling reality, and may have even more startling implications with regard to both animal welfare and human moral sensibilities in its future development.

As such, research at the intersections of feminism, biology and philosophy (like Lynda Birke's and Leesa Fawcett's) provides dynamic starting grounds for this discussion of genetic technologies and animals. With a focus on animal bodies, I examine moral implications of the genetic engineering of "domesticated" animals—primarily pigs and chickens—for purposes of human consumption. Concepts of natural and artificial, contamination and purity, integrity and fragmentation and mind and body feature in the discussion. In this respect, Atwood's *Oryx and Crake* serves as a cogent medium for exploring these highly contentious practices and ideas as it provides hypothetical narratives of possibility. Moreover, it is used to highlight contemporary hegemonic assumptions and values in ways that make them visible, for example, by taking a current biotechnological process/practice to a seemingly absurd end. Particular attention is paid to issues of growing human organs in pigs for xenotransplantation (resulting, for Atwood, in "pigoons") and the ultimate end of the intensive factory farming of chickens through the genetic engineering of "mindless" chicken tumours (or, as Atwood calls them, "ChickieNobs"). Integral to these philosophical considerations is the provocative question of employing the genetic engineering of animal bodies as a means to end the suffering of domestic food animals. This discussion ultimately leads to its implications for the future of human experience and morality.

Dis/Integrity Is a Virtue?

In a brief survey of literature that discusses genetic engineering and animals, two prevailing concerns emerge, those that focus on issues of animal welfare and those that focus on issues of animal rights. Advocates of animal rights claim that animals, particularly vertebrates, should not be used at all,[5] while those taking up animal welfare accept the uses of animals to varying extents and direct their concerns to how animals are treated and whether harm can be justified by the benefits to humankind (Becker and Buchanan, 1996, p. 8). The moral equation of utilitarianism is commonly employed to calculate the overall "greater good" by subtracting the costs from the benefits, which is always done from an anthropocentric, or human-centred, perspective. That is to say that the benefits always relate to human needs and desires because only humans are morally responsible to other human beings. This is the ethical position that typically forms the basis of policy (Bowring, 2003, p. 127).[6]

However, the idea that there is indeed something inherently wrong with the genetic manipulation of animals is taken up in a different, yet equally popular, philosophical approach known as deontological ethics. This ethical approach, most commonly developed from the philosophy of Immanuel Kant and applied to animals rather than just human beings, is drawn upon to argue that living beings should never be treated as merely a means to an end (Bowring, 2003, p. 134; Holland, 1990, p. 170). In a similar line of argument, others claim that it is wrong to treat animals as instruments solely for human purposes because they have value unto themselves, an intrinsic value, regardless of any instrumental value they may hold for human beings. But, it is not just that animals are considered instruments, the bigger problem is that they are viewed as mechanical instruments (Holland, 1990, p. 170). Stressing the full implications of this view, Val Plumwood states that "a nature represented in mechanistic terms as inferior, passive and mindless, whose only value and meaning is derived from the imposition of human ends is simply replaceable by anything else which can serve those ends equally well" (2002, p. 49).

This is an important contention because it exposes the extremely reductionist, value-laden nature of Western technoscience, with its fundamental ideology and language of mechanism which essentially makes genetic engineering possible. As Bowring succinctly sums up,

> the idea that the functioning of organisms can be distilled to discreet and transferable units of information is the dominant fiction which underpins and legitimizes the practice of genetic engineering. (2003, p. 1)

Rather than attributing or recognizing human characteristics in animals, known as "anthropomorphism," technoscience favours "mechanomorphism," labelling animal bodies, and describing behaviour, in mechanical terms (Cenami Spada, 1997, pp. 43–44). This mechanistic way of thinking and talking about organic life results ultimately in a dramatic reduction of actual bodies in biotechnological practices. Bowring further adds,

> the disturbing image engendered by these developments is thus of a scientific community indifferent to the natural patterns, features and divisions of organic life, and which is content to address the alter instead as an assemblage of inherently disposable artefacts to be manipulated and reconstructed according to human whim. (2003, p. 118)

It may be that what makes this mechanical reduction so disturbing is that it represents a deliberate corruption of an integrity of being, or telos.[7] This notion of integrity makes the fragmentation, the mutilation of bodies into machine-like components, unsettling in subtle and sometimes dramatic ways.[8] It appears to be an organic quality arising in an organism through its own bio-physical processes of development and through interaction with the environment within which it is immersed. It is also aligned with a notion of intrinsic value, which, as opposed to instrumental value, is independent of any human judgment of utility. Of course, the word "integrity" itself is loaded with meaning and particular moral value in English language and in Western culture, which likely heightens the discomfort felt by many when confronted with the threats to organic integrity that biotechnology is said to pose. This way of thinking is grounded in the pervasive Western assumption of a radical discontinuity between human beings and nature. By this logic, anything that human beings do is unnatural, and anything "manmade" [sic]—such as a transgenic animal—is therefore also unnatural or artificial. The assumption that "natural is good" is remarkably prevalent in the literature on transgenics even though it is contested and problematized across cultures, ethnicities and religions. While many stick to this designation in their arguments, some make a distinction between the process and the animal itself. So, where the process of genetic engineering, the way in which a being is modified, is understood as unnatural because it is biologically impossible outside of a laboratory, the organism itself still possesses integrity of its own, and is neither natural nor unnatural in definitive terms (Holland, 1990, p. 169).[9]

Taking this understanding further, in her essay "The Promises of Monsters: A Regenerative Politics for Inappropriate/d Others," Donna Haraway disrupts any and all divisions between natural and artificial when she states that "if organisms are natural objects, it is crucial to remember that organisms are not born; they are made in world-changing technoscientific practices by particular collective actors in particular times and places" (2004, p. 65).

Moreover, Haraway insists that although life may seem to be overwhelmingly "denatured" by these human practices, "it is not a denaturing so much as a particular production of nature" (2004, p. 66). Haraway proposes that nature is continually co-constructed by various organic and technological actors, some human and some non-human (2004, p. 66). In spite of this, notions of natural goodness maintain a tenacious hold on Western imaginations and the debate often turns to the integrity of species, rather than the individual.

Pigoons, Purity and Perversions of Boundaries

Interestingly, the possibilities of actually transgressing the "natural" boundaries between species via genetic engineering effectively call into question the very definition and concept of species (Becker and Buchanan, 1996, p. 7). It points to a certain cultural arbitrariness both in scientific definitions of species as well as in the moral stances that species boundaries support. As Gerhold Becker puts it,

> to bridge the species barrier brings to the fore not only the very concept of species itself but also its significance and function in the natural order of beings. It is noteworthy that in most cultures the crossing of species lines used to be the subject of taboos for humans. (1996, p. 7)

If I may be allowed some cultural and historical generalizations, in the West there has been an acceptance of the idea that all organisms can be categorized into different species based upon biological characteristics, particularly in terms of mating. So defined, species have become relatively fixed and bounded phenomena and there is a long history regarding the ideological and practical maintenance of such boundaries in Western society.[10] Much has been written on this history of ideas already, of great chains of being and of cultural taboos, too much to summarize here; rather, I take up the underlying notions of purity and contamination in the discussion as they relate to attitudes and values regarding transgenic animals and xenotransplantation.

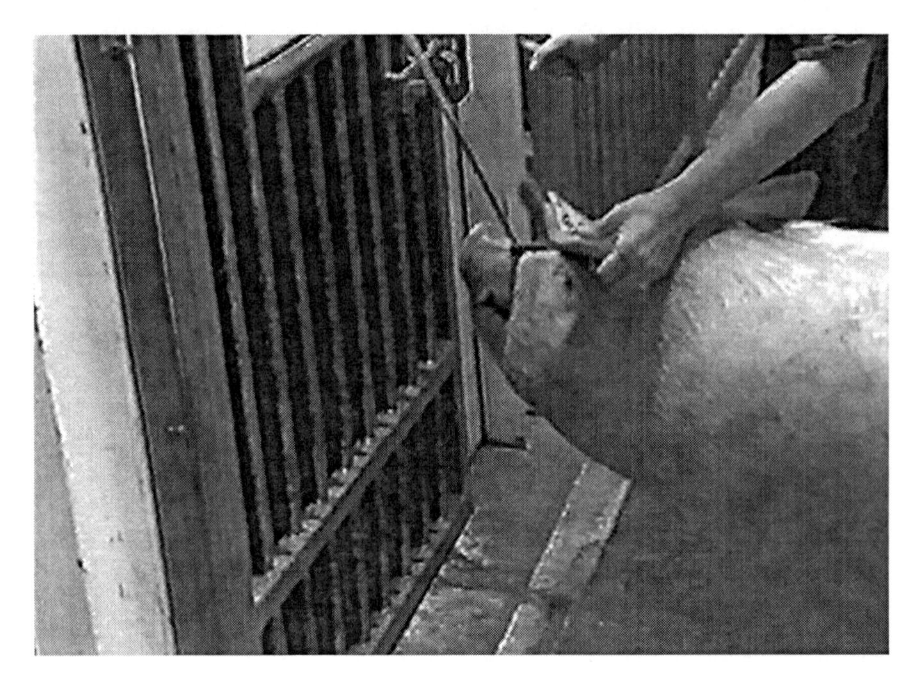

Fig. 1 Pig used for organ xenotransplantation (credit: Courtesy ITV/Carlton)

Both involve the deliberate crossing of the so-called species barriers and both result in the actual combination of genetic material from two or more different species.

In the fictional world created by Atwood in *Oryx and Crake*, such transgenic animals, called "pigoons" (or by their Latin title "*sus multiorganifer*"), play a pivotal role in human medical and commercial enterprise. At first they are used to provide various means of human physical enhancement and prolonging life, yet, ironically, in the end they become a tangible threat to human survival. For Jimmy, the principle character in the story, pigoons represent his family's livelihood.

His father is employed as a genetic engineer at OrganInc Farms, a huge biotechno-industrial compound, which also serves as a gated-community for all of its employees and their families. As Atwood explains in the narrative,

> The goal of the pigoon project was to grow an assortment of foolproof human-tissue organs in a transgenic knockout pig host—organs that would transplant smoothly and avoid rejection, but would also be able to fend off attacks by opportunistic microbes and viruses, of which there were new strains every year. A rapid maturity gene was spliced in so the pigoon kidneys and livers and hearts would be ready sooner, and now they were perfecting a pigoon that could grow five or six kidneys at a time. Such a host animal could be reaped of its extra kidneys; then, rather than being destroyed, it could keep on living and grow more organs, much as a lobster could grow another claw to replace a missing one. (2003, pp. 22–33)

It is interesting to note that in naming them "pigoons," Atwood hints at the idea that since they have undergone dramatic genetic modifications they are no longer "pigs," that they have become a new animal altogether. This is contrary to the language currently employed in biotech rhetoric, where genetically modified pigs are still called pigs, and yet it also points to another controversial aspect in biotechnology: the patenting of GMOs. A patent is only granted if an organism is considered to be completely novel and artefactual, which is what the renaming actually represents.[11] To be sure, the boundary between reality and fiction is porous. As Atwood stated in the passage at the beginning of this chapter, this is not merely the stuff of speculation, or science, or fiction and there are presently many examples in which human genetic material is being inserted into the genomes of other animals. For example, transgenic pigs, known as "Beltsville"[12] pigs, "were produced from single-cell embryos that had been injected with a human growth hormone gene" (Bowring, 2003, p. 124). The corporeal results of this experiment failed dramatically in terms of animal welfare, as Bowring explains:

> The pigs showed improved weight gain, greater feed efficiency, and reduced subcutaneous fat, but at the cost of a wide range of pathological side-effects, including 'gastric ulceration, severe synovitis [joint inflammation], degenerative joint disease, pericarditis and endocarditis [inflammation of the outer and inner lining of the heart], cardiomegaly [enlargement of the heart], parakeratosis [cracking of the skin], nephritis [inflammation of the kidney] and pneumonia. 'In addition,' the researchers disclose, 'gilts were anoestrus [infertility due to quiescence or involution of the reproductive tract] and boars lacked libido'. (Bowring, 2003, p. 125)

Clearly, there is no moral consideration for the individual pigs involved, only the potential human benefits of increased agricultural efficiency are of value. Strangely enough, however, these transgenic pigs present a challenge to such blatant anthropocentrism, and may serve to undermine species-based morality (Ryder, 1990, p. 190). It comes back to ideas of purity and species boundaries and we are forced to ask at what point does a pig with human genes stop being a pig and become something more human, or a porcine–human hybrid? Richard Ryder (1990) raises this provocative idea in his essay, "Pigs Will Fly," when he poses questions like

> How many human genes make a sufficiently human creature to have human rights in the eyes of the law? How many human genes can you give a humanized pig before you feel obliged to send it to school rather than to the slaughterhouse? (p. 190)

Ryder is hopeful that the perforation of species boundaries "spotlights the absurdity of our species-based morality", particularly by drawing attention to practices in which human growth hormone genes

> have already been injected into the embryos of pigs. The aim was to produce bigger and juicier pork chops. But wait a minute. This would mean eating human genetic material! It might only be a minute proportion of the chop, but all the same, would it not be a partial cannibalism? (Ryder, 1990, p. 190)

Playing up the revulsion of this very real possibility in her fiction, Atwood has Jimmy reflect on the same questions while dining at the OrganInc Farms cafeteria where, "to set the queasy at ease, it was claimed that none of the defunct pigoons ended up as bacon or sausages: no one would want to eat an animal whose cells might be identical with at least some of their own" (2003, pp. 23–44). Beyond their role as living pork factories, and like the pigoons, present day pigs are also being genetically modified to provide a "source of replacement organs (the so-called "spare parts factories") for humans" (Bowring, 2003, p. 121). To this end, the use of biotechnology is vital. Bowring explains that

> Though pig heart valves have been used in human cardiac surgery for some years, the future prospects for xenotransplantation using living organs like hearts and kidneys are naturally limited by the aggressive rejection and often destruction of foreign matter by the human immune system (even in successful cases the recipients must take immunosuppressant drugs for the rest of their lives). For this reason progress is being made in adding human DNA sequences to pig embryos in order to produce animals which express human proteins that prevent or reduce hyperacute rejection of the organ by the recipient. (2003, p. 121)

The purpose of this process is to make a pig's body less pig-like[13] so that it can become more compatible with human bodies, and in essence then, more human (at least on a physiological level). The impetus for ridding the pig's organ of all traces of its "pigness" perhaps also arises in part out of sentiments of a general distaste, a "feeling that there is something undignified for the recipient in receiving a pig's heart" (Aldridge, 1996, p. 132). Atwood mirrors this attitude of aversion through a sardonically humorous exchange between Jimmy's parents, in which his father enthusiastically announces a recent breakthrough at work to his mother:

> 'We've done it,' said Jimmy's father's voice. 'I think a little celebration
> is in order.' A scuffle: maybe he'd tried to kiss her.
> 'Done what?'
> Pop of the champagne cork. 'Come on, it won't bite you.' A pause: he
> must be pouring it out. Yes: the clink of glasses. 'Here's to us.'
> 'Done what? I need to know what I'm drinking to.'
> Another pause: Jimmy pictured his father swallowing, his Adam's
> apple going up and down, bobbity-bobble. 'It's the neuro-regeneration
> project. We now have genuine human neocortex tissue growing
> in a pigoon. Finally, after all those duds! Think of the possibilities, for
> stroke victims, and...'
> 'That's all we need,' said Jimmy's mother. 'More people with the
> brains of pigs. Don't we have enough of those already?' (p 56)

While fears of contamination appear justified[14] and provide reason for caution, general concepts of blood purity and contamination may be more dangerous in terms of racism and classism. Notions of pedigree have been around for centuries, revered in royal family lineage and designations of "pure blood" and "blue blood", and have extended to breeding practices of "domesticated" animals, such as dogs, cats and horses.[15] This is not an unfamiliar idea, and so does not require elucidation. A more novel and complex implication of "blood purity" however, is its exposure of how certain beliefs arise out of bodily experiences which may not be reconcilable with what are considered to be rational truths and facts (Kirmayer, 1992, p. 329). Such beliefs and fears can play out even when the species barrier is not transgressed, for instance, in cases of human-to-human blood transfusion. For example, Laurence Kirmayer illustrates what he calls "the body's insistence on meaning" through a challenging case study of a man, called "Mr. Y", who refuses to undergo a life-saving blood transfusion because of his belief that the "foreign" blood will contaminate him with the donor's, "genetic material that carries personality traits" (1992, p. 325). Exasperated by the patient's irrationality, his doctor explains that it is impossible and refuses to investigate the experiential basis of the patient's belief which, upon psychiatric consultation, is found to stem from traumatic childhood experiences in hospitals with the doctors treating his kidney disease (Kirmayer, 1992, p. 326). During the interview, the psychiatrist notices that Mr. Y makes repeated references to feeling vulnerable in terms of his body boundaries and to his impressions of the physicians' insensitivity and repeated transgressions of those boundaries (Kirmayer, 1992, p. 326). It was also apparent that Mr. Y is "fastidious" about caring for his body, and he admits to eating only organic foods and drinking purified water (Kirmayer, 1992, p. 327). But, as Kirmayer surmises, his body issues go much deeper and have wider social dimensions:

> Mr. Y is a foreigner, a businessman who has come from Europe to Canada. He views the locals as less cultured and sophisticated than himself. He feels their coarseness is intrinsic, part of their bodily constitution, and fears receiving blood from them since this would taint his body and with it his mind. This incipiently racist theme of "blood purity," although it may receive impetus from Mr. Y's childhood traumas, finds support in collective representations and discriminatory practices. (1992, p. 329)

Promises of Failure and Risks of Success (or, It's All About Perspective!)

With this kind of visceral discrimination, I can only imagine the horror of Mr. Y's reaction to the idea of receiving blood or organs from an animal of another species. Nevertheless, biotechnological research in this area gallops on with great fervour and promise for medical progress. Meanwhile, there is a greater uncertainty to this work that I have yet to see specifically addressed in the literature on transgenics. While there is much written about the fears of failure in which the GMO is unable to survive out in the world due to unforeseen problems (such as increased susceptibility to disease that was coded for on a gene that was removed for other reasons (Rollin, 1995, p. 110), and about welfare arguments concerned with failure (as in the Beltsville pigs' case) there is very little about the risks of success. Admittedly, concerns of ecological risk have received some attention, but they all tend to focus on microscopic organisms, invertebrates (mainly insects) and transgenic plants, that have been modified for agricultural purposes. The conventional warning is that, "unlike chemical substances, genetically engineered organisms have the capacity to mutate, migrate and multiply" (Holland, 1990, p. 173) and that "a genetically engineered organism once free in the environment is impossible to recall" (Bereano, 1996, p. 30).

As with the fear of blood contamination due to xenotransplantation, this fear of "polluting" an ecological system with a "foreign" organism has an experiential basis; experiments in biological pest control have provided all too many lessons learned, the hard way, of what can happen. So the logic follows that any novel (i.e. transgenic) organism has no natural habitat, no place of origin in this world, and therefore always presents an ecological risk if released from the laboratory. But, what of those transgenic animals never intended for release? Although human beings are capable of altering the genetic makeup and phenotypic expression of that modified genotype, we are not actually in control of evolution, and even transgenic animals kept in a laboratory or otherwise confined may change and adapt over time, through mutation and reproductive processes. What, indeed, might be the potential long-term implications of splicing human genetic material into pig embryos? Has the scientific mind become so accustomed to a severely reductionist view of the world that it can only "see" animals as passive machine-like objects, an assemblage of inert parts to be tinkered with? Some humility and wonder is in order; a respect for all that is still not known or understood about bodies and whole biological systems, and the findings of biological sciences which suggest their dynamic complexity and chaotic nature.[16] For all of its sophistication, the science of genetic engineering is founded upon an overzealous faith in the technology itself, and upon an over-simplified idea of living processes and bodies. The former, known as technological determinism, creates a hold upon scientific thought and effectively eliminates worries of risk. The latter, known as biological determinism, defines animals as only the physical products of their DNA.[17]

In this regard, Atwood's pigoon becomes not only a kind of allegory for the ecological risks of transgenic organisms, but also a warning for what may come to pass, quite literally. In Jimmy's own lifetime and as a result of deliberate human action, a microorganism wipes out all human life on Earth. In this post-apocalyptic world, pigoons have escaped from the laboratories and now roam freely. Possibly emphasizing the agency of all transgenic animals, Atwood writes, "They were always escape artists, the pigoons: if they'd had fingers they'd have ruled the world" (p. 267). In this scenario, from the pigoon's perspective of course, they are very successful in terms of ecological adaptation and ability to thrive. On the other hand, from a human perspective, pigoons become an experimental failure, or monstrous mistake, as they now express both the desire and talents to hunt and eat human flesh. Apparently, the practice of mixing human and pig genetic material for numerous generations has endowed pigoons with a certain amount of human similarity. Jimmy contemplates this phenomenon when he finds himself in a vulnerable position. He ruefully reflects

> There are too many pigoon tracks around here. Those beasts are clever enough to fake a retreat, then lurk around the next corner. They'd bowl him over, trample him, then rip him open, munch up the organs first. He knows their tastes. A brainy and omnivorous animal, the pigoon. Some of them may even have human neocortex tissue growing in their crafty, wicked heads. (Atwood, 2003, p. 235)

The pigoons have clearly evolved human-like traits as a result of their genetic alteration, and Atwood illustrates the irony through their apparent enjoyment of the tables being turned. For instance, at one point Jimmy narrowly escapes his predators by fleeing up a steep flight of stairs, knowing that pigoons are not physically built to climb a staircase. In his panicked haste, he drops his plastic bag carrying what amounts to his worldly belongings:

> He starts cautiously downward. As he's stretching out his hand, something lunges. He jumps up out of reach, watches while the pigoon slithers back down, then launches itself again. Its eyes gleam in the half-light; he has the impression it's grinning. They were waiting for him, using the garbage bag as bait. They must have been able to tell there was something in it he'd want, that he'd come down to get. Cunning, so cunning. (Atwood, 2003, 271)

The idea of pig–human hybrids running rampant in the streets may be taking current GMOs to an absurdly extreme end of the range of possibility. However, experience does suggest that organisms can and will respond to biological and ecological changes in unpredictable ways. Atwood's pigoons remind us of this agency of animals, which tends to be ignored or denied all too often. Therefore, although "mechanical models express the denial to nature of any uniqueness, agency and power" (Plumwood, 2002, p. 49), warnings of ecological risk speak loudly and clearly of the agency and power of transgenic organisms. Such agency is expressed when genetically modified wheat suddenly appears in an organic farmer's field several kilometres away, and when the Beltsville pig's body develops severe arthritis. Unfortunately, the denial of animal agency is underpinned by a reification of the "gene" itself, the ultimate reduction of life into its smallest component part, which guides genetic engineering. Birke cautions that "reifying them allows us to see only the genes, devoid of their physiological context, the organism itself and its environment" (1994, p. 83). Once removed from their original context, how can scientists be sure of how the gene will function in an essentially alien one? Expressing this concern, Birke questions

> what happens to the inserted gene apart from its attributed role of molecule factory. How does it interact with the sheep's genes? Or with its cells? Or with its wider environment? And what are the consequences of those actions? (1994, p. 83)

Thinking of a gene as just interchangeable DNA matter brings the discussion back to the question of species hierarchy, presenting a challenge to the superiority dominantly attributed to the human species over all others. Success in genetic engineering depends upon similarity between and among different individuals and species, otherwise it could not work. As a result, "the boundaries start to dissolve", and Birke recognizes that "to avoid that worrying prospect, we can label the genes as embodying essence. Human genes must have something special about them, an essence of humanness, that gives them a territorial boundary" (1994, pp. 83–84). This presents us with an apparent paradox in which "genetics as a system of representation both challenges concepts of species as fixed (in the practice of, for example, transgenics) and reinforces them (by incorporating notions of essence)" (Birke, 1994, p. 84).

Rather than struggling with the contradiction, it appears that biotechnoscience is content to ignore it and to maintain one set of rules for defining human beings and another set for all other animals. To continue this hierarchy of species by defining animals as only products of their biology poses perhaps the greatest threat to animal agency and welfare, resulting in the actualization of the "animal is machine" metaphor.

Designed for Deprivation

Factory farming is already based upon thinking of animals as machines to be tinkered with so they can be made more efficient. For instance, cows are thought of as milk machines and given genetically engineered bovine growth hormones, known as Bovine Somatotropin (BST), to increase their production of milk (Wenz, 2001, p. 222; Wheale and McNally, 1990).[18] As the conditions within industrial animal agriculture intensify, the occurrences of disease and disharmony among the animals themselves increase dramatically (Bowring, 2003, p. 131). Unfortunately, rather than changing the conditions of production, many look to genetic engineering for quick-fix solutions. As Caroline Murphy points out, "the 'Trojan horse' of disease resistance may provide a means whereby genetic engineers can design animals to cope with conditions that no animal, genetically engineered or not, should be expected to endure" (1990, p. 16). Modifications to animal bodies via genetic engineering thus effectively enable continued, and possibly even increased, deprivation as animals are treated as nothing more than meat factories, confined to "cramped and unsanitary conditions" (Bowring, 2003, p. 132). This in itself is cause enough for alarm in terms of animal welfare, yet becomes even more horrifying as proponents attempt to alleviate all causes of suffering through changing the very nature of the animals.

Those in favour of redesigning animals to be even better adapted to factory environments call it simply the "latest stage in the historical process of 'domesticating' animals" (Bowring, 2003, p. 132), insisting that it is no different, in theory, from pastoral practices of domestication through selective breeding. But the stakes are much higher, as Bowring explains, this means that now "one could, therefore, produce chickens which lack a desire to nest" (2003, p. 136). What's more, advocates insist that since transgenic animals could, in principle, be created without causing suffering, and for that matter engineered "... to suffer less, there is in the eyes of most scientists nothing inherent in genetic engineering which makes it an unethical tool" (Bowring, 2003, p. 132).

By contrast, opponents assert that this would be morally reprehensible, and turn again to the argument that the telos of beings should not be violated under any circumstances (Bowring, 2003, pp. 133–134; Holland, 1990, pp. 170–171; Fox, 1990, pp. 32–33). Expanding upon his earlier deontologic argument, Holland claims that

> there is a distinction between using another creature's ends as your own—which is acceptable—and disregarding that other creature's ends entirely—which is not. A problem, however, which Kant's notion does not seem to address... comes when the genetic engineer starts to redesign those ends. (1990, p. 170; emphasis added)

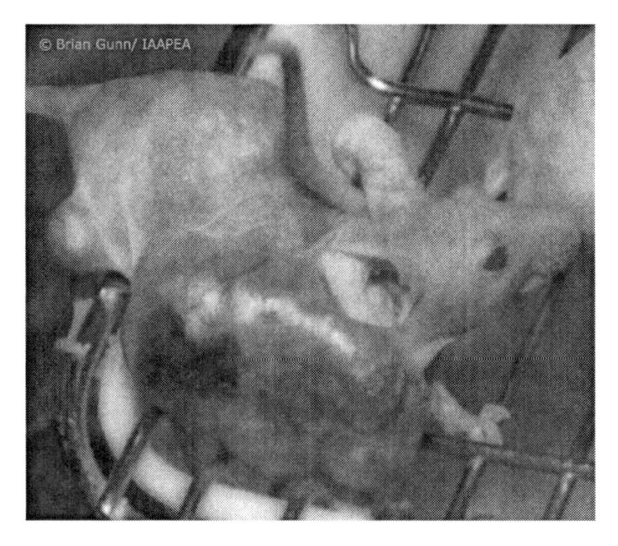

Fig. 2 An oncomouse—bred for cancer (credit: Brian Gunn/IAAPEA)

The End of Animal Suffering = The End of Animals?

Incidentally, this is precisely what Bernard Rollin advocates in his "new social ethic for animals" (1995). Rollin argues that if animals can be modified to be "happier" in the confinement conditions of factory farms, there should be no moral opposition to it. It is interesting that he does not argue that it should be a moral imperative, only that it should not be immoral to alleviate suffering using genetic modification to better fit animals to their conditions. In response to claims of the inviolability of telos, Rollin addresses his critics by emphasizing that he believes there is nothing wrong with changing the nature, or telos, of an animal as long as the animal's interests—which emerge from that telos—are not violated (1995, p. 171).[19] For example, he supports changing the nature of a chicken, when normally confined in a battery cage, which suffers frustration at not being able to nest to lay her eggs so that the "new kind of chicken" no longer expresses the urge to nest and instead experiences satisfaction at laying eggs in a cage (Rollin, 1995, p. 172). Rollin admits that public apprehension to his argument may emerge from "a queasiness that is at root aesthetic. The chicken sitting in a nest is a powerful aesthetic image.... A chicken without that urge jars us"(1995, p. 175).

However, he is quick to point out that this pastoral image is already forsaken to an immense degree in current practices of factory farming, which should cause as much if not more concern than hypothetical applications of genetic engineering. Indeed, his argument becomes harder to dispute as he recounts that Western culture has "a historical tradition as old as domestication for changing (primarily agricultural) animal telos (through artificial selection) to fit animals into human society to serve human needs. We selected for non-aggressive animals, animals disinclined or unable to leave our protection and so on" (Rollin, 1995, p. 174). Here, Rollin's argument begins to appear very pragmatic in the face of the West's "current exploitative business context" (1995, p. 182) of intensive industrial animal agriculture.

He is not without sympathy, nor ignorant of the full ramifications of what he suggests; rather, he presents himself as a realist—justifiably cynical of society's lack of will to change such strongly rooted economic, political and social practices—admitting that to change the nature of an animal to fit a poor environment rather than changing the environment itself is merely the "lesser of two evils" (Rollin, 1995, p. 192), but a necessary one when the likelihood of the system changing to fit animals' needs is slim.

Up to here, Rollin presents a fairly persuasive case. However, he takes it too far when he suggests

> ...if we could genetically engineer essentially decerebrate food animals, animals that have merely a vegetative life but no experiences, I believe it would be better to do this than to put conscious beings into environments in which they are miserable. (1995, p. 193)

To suggest genetically modifying a chicken which lacks the urge to nest is provocative enough, but to suggest a chicken which lacks a brain is downright shocking. And it is here where we come to the most extreme manifestation of the mechanical metaphor, and the division of mind and body, with the proposed production of actual meat machines.

This prospect is celebrated among some scientists in the field and has apparently been around for quite some time. In fact, Rollin's comment on "decerebrate food animals" was inspired by a conversation with a genetic engineer who claims that

> ...in the long run, biotechnology will make the whole debate about agricultural animal welfare moot. Eventually, he told [Rollin], we will be able to create the relevant animal proteins in fermentation vats produced by bacteria genetically coded to do so. Thus we will be able to have animal products without animals. (1995, p. 193)

Furthermore, Peter Roberts recalls that

> ...those attending a cattle breeders conference in Cambridge England, in 1988 were somewhat stunned when they were told that within 15–20 years the production of meat need no longer involve the rearing and slaughter of animals (Ridley, 1988). In this case they were not being told about textured soya proteins, but about genetically engineered meat, not grown in the stockyard but cultivated in vitro (in glass dishes) in commercial laboratories, from cells selected from rump steak or chicken breasts. (1990, p. 201)[20]

Beyond Suffering: "ChickieNobs Bucket O'Nubbins"

Bringing this terrifying prospect to life in the world of Oryx and Crake, Atwood confronts her readers with an evocative image that they will surely find unforgettable. Jimmy is visiting his best friend, Crake, at his new high school called the Watson–Crick Institute. As its name implies, it's an institute of genetic technology, with students specializing in areas such as "Botanical Transgenics," "NeoGeologicals," and "Décor Botanicals." Crake takes Jimmy on a tour of these departments, winding up in "Neo Agriculturals" or "AgriCouture" as it is nicknamed. Here, Jimmy is presented with "a large bulblike object that seemed to be covered with stippled whitish-yellow skin. Out of it came twenty thick fleshy tubes, and at the end of each tube another bulb was growing" (Atwood, 2003, p. 202). He is told that they are essentially chicken parts, some just breasts, others just drumsticks. They have been modified to have no head, just a mouth into which nutrients are dumped:

> This is horrible,' said Jimmy. The thing was a nightmare. It was like an animal–protein tuber. 'Picture the sea-anemone body plan,' said Crake. 'That helps.' 'But what's it thinking?' said Jimmy. (Atwood, 2003, p. 202)

In response, Jimmy is told by one of the student scientists involved that it's not supposed to think that "they'd removed all the brain functions that had nothing to do with digestion, assimilation and growth" (Atwood, 2003, p. 203). "'No need for added growth hormones,' said the woman, 'the high growth rate's built in.... And the animal-welfare freaks won't be able to say a word, because this thing feels no pain'" (Atwood, 2003, p. 203).

Jimmy's immediate response of horror and his question of "what's it thinking" speak to an agency in chickens that has been dramatically violated and distorted. Atwood then makes allusions to the already tight relationship between present-day research in biotechnology and big business through its commercial applications:

> 'Those kids are going to clean up,' said Crake after they'd left. The students at Watson–Crick got half the royalties from anything they invented there. Crake said it was a fierce incentive. 'ChickieNobs, they're thinking of calling the stuff.' 'Are they on the market yet?' asked Jimmy weakly. He couldn't see eating a ChickieNob. It would be like eating a large wart. (Atwood, 2003, p. 203)

As unpalatable as they seem to Jimmy at first, eventually they become familiar, cheap and convenient enough for him to overcome his revulsion. With his acquiescence to ChickieNobs Bucket O'Nubbins, Atwood is definitely commenting upon the power of marketing and on contemporary society's cultural amnesia regarding many once controversial environmental issues. As warnings of global warming and species extinction become increasingly monotonous and commonplace in the news and everyday life, it seems that the public tends to either tire of them or finds them so completely overwhelming that it becomes preferable to ignore them and let the politicians work it out. Ironically, this seemingly pervasive apathy in dominant Western society is exactly what Rollin had been referring to for supporting the non-fictional production of ChickieNobs.

To get beyond these circular arguments of suffering, Holland stresses freedom as grounds for objection, which he defines as "the capacity to exercise options;" he abhors the idea that an animal might be modified to have a diminished capacity in this respect (1990, 171; also see Fox, 1990, p. 34). With sober wit he acknowledges that if capacity is to be modified, a kind of genetic lobotomy is the inevitable result since

> ...the genetic engineer is unlikely to deal with the problem of the unhappy pig by increasing the animal's sophistication to the level where it is capable of being philosophical about its

> condition and learns not to mind. Rather, the capacities of the animal would be reduced to the state where it was, from a sentient point of view, more vegetable than animal. (Holland, 1990, p. 172)

Depriving Ourselves: Mechanomorphosis and Other Ethical Consequences

Alternatively, it may be more generative to argue that genetic engineering presents a violation of "being" in terms of living processes rather than seemingly static qualities such as capacities. To this end, Bowring cautions that the continued treatment of animals as malleable artefacts is as much a threat to "human's own vital sensibilities" as it is to animal welfare, and that it will "progressively erode the scope for and substance of human's moral existence" (Bowring, 2003, p. 3). In other words, through the philosophy and practices embedded within genetic engineering that ultimately reduce all animal life into biological machines, human beings are distorting their own experience of the world, and thus their values and belief systems along with them. In phenomenological terms, human beings are situated within our own unique bodies, which connect us with the world, enabling sensory experiences of it, while simultaneously limiting what and how we know to our own embodied experiences (Merleau-Ponty, 1962, p. 82). There can be no absolute objectivity, nor isolated subjectivity, as we are always in relation to other beings, materials and processes. To borrow Maurice Merleau-Ponty's fabric metaphor of existence, I imagine numerous threads stretching from my body out to everything and everyone around me (Gill, 1991, p. 4). And just as each thread stretches in both directions, so too is perception reciprocal, facilitating my experiences as both an observer and as one who is observed by others. Extending this reciprocity of perception to animals, David Abram provides an elegant example of an ant walking upon his arm. He senses the ant visually and through the touch of the ant's feet upon his skin, and realizes that the ant also senses him as he notices the ant respond to his arm movements (Abram, 1996, p. 67). Counter to a mechanical view of nature, a phenomenological perspective enables such experiences of intentionality and agency in others, particularly other animals. Emphasizing the importance of human–animal interaction through direct embodied experiences, Bowring stresses that

> While there are... obvious dangers involved in romanticising traditional agricultural practices for being 'closer to nature', it may still be argued that the genetic approach to farming, and the reductionist science which underpins it, marks a watershed in the development of humans' relationship to nature in so far as it involves a decisive degradation in their subjective qualities of feeling and perception. (Bowring, 2003, p. 138)

In doing so, we are losing the capacity to relate to other animals, our own bodies and other human beings. Such an impoverishment of experience is, as evident in the discussion throughout this chapter, intimately connected with an inanimate language, reinforcing and reinforced by a dominant mechanical worldview. To speak of organisms as machines legitimizes our treatment of them as artefacts, as completely knowable and transparent objects and of their lives as having no ethical significance. As such, Bowring warns that

> "...genetic engineers' favoured language of 'bioreactors,' 'spare-parts factories,' 'nutraceutical fruits,' 'live-stock pharming,' 'biofacture' and so on, are not just commercially sanitised descriptions of what remain stubbornly recalcitrant natural phenomena...: this mechanistic language and philosophy is also disturbing for the way they seem to express the degradation of scientists' own physical and moral sensibilities, and a diminishing care for the integrity of life which has troubling ramifications for human relations themselves. (2003, p. 142)

Behind commercial profits from the commodification of bodies and lingering enlightenment fantasies of revealing all the secrets of nature, there lurks the ominous promise of our own sensory and moral deprivation. Ultimately, by transforming nature into an insensate, decerebrate, wholly objectified product, devoid of independent well-being, the biotech programme may thus deliver its golden promise by relieving us of care. The mechanization of nature will lead to the mechanization of ourselves, our sentiments, judgments, fears and dreams (Bowring, 2003, p. 143).

In this respect, Bowring worries that "this process is already foreshadowed by the cultural fetishization of the cyborg, and of course in the assertion that human nature itself should now become the object of the biotech enterprise" (2003, p. 143). While I can appreciate Bowring's fear of the cyborg, I do not find it to be problematic in itself. Truth be told, I find the reality of the cyborg metaphor much more appealing than the current mechanical one, particularly in terms of Haraway's understanding of the cyborg's potential to expose and dissolve hierarchical divisions between mind and body, nature and culture and so on that are the direct result of a mechanical (and patriarchal) worldview (1991). For, as Haraway reminds us in her infamous "cyborg manifesto," a cyborg is a "hybrid of machine and organism, a creature of social reality as well as a creature of fiction" (1991, p. 149). If we are already cyborgs, as Haraway contends and Bowring fears, then we are least still partly human animals. As cyborgs, we retain some form of organic sensory perception while also enjoying technological augmentations. Thinking of ourselves in this way may enable the humility so desperately needed in our social relations with other animals, particularly as it encourages us to engage our imaginations, to recognize both the generative and destructive fictions in our worldviews, as we struggle to comprehend our ontological predicament (the meaning of life, the universe and everything).[21]

If we continue along the biotechnological path without questioning its ideological basis, we risk much more than becoming actual cyborgs, we risk manifesting our mechanomorphism via mechanomorphosis. That is, we gamble with becoming machines ourselves, in the most reductionist sense, inheriting only the mechanistic part of our cyborg heritage and leaving all traces of humanimality behind. With the loss of embodied sensibility, of our modes of social relatedness, we run the risk of eliminating our ability to ponder metaphysics, to question our own actions and fundamental beliefs, and with it the desire or need for ethics at all. In due course, we need to resist our conversion into insensate automatons and imagine what is at stake. Will the dis/integration of our corporeal-selves and other animals, and re/integration through mechanomorphosis lead to amoral relationships? Do we really want to achieve a future without suffering by way of disembodiment?

Notes

1. Quoted from an essay by Margaret Atwood about the writing process of the book on the Oryx and Crake website at http://www.randomhouse.com/features/ atwood/essay.html
2. For an accessible scientific explanation of genetic engineering and the processes of producing transgenic organisms, please see Aldridge, S. (1996). *The Thread of Life: The Story of Genes and Genetic Engineering*. Cambridge: Cambridge University Press.
3. In her foundation course on Environment and Culture: Nature, Technology and Society, Leesa Fawcett, an associate professor in the Faculty of Environmental Studies at York University, has lectured extensively on complex social and biological issues raised by genetic technologies, including: the dangers of genetic discrimination and the oppression of differently abled people arising from "designer babies"; the diverse cultural and religious attitudes towards xenotransplantation; and, the implications of phenomena such as zoonoses (transfer of disease across species) which challenge "pure" boundaries assumed by hierarchical dualisms and their corresponding ethical perspectives, like anthropocentrism.
4. For cogent explanations of the patriarchal (male hierarchy) basis of modern, Western sciences, please see Plumwood, V. (1993). *Feminism and the Mastery of Nature.* London: Routledge and Plumwood, V. (2002). *Environmental Culture: The Ecological Crisis of Reason*, London: Routledge; also see Haraway, D. (1989). *Primate Visions*. London: Routledge.
5. For a thorough argument against the use of animals in experimentation and in favour of animal rights using dominant Western philosophical perspectives, such as utilitarianism and deontological ethics, see Wacks, R. (1996). Sacrificed for Science: Are Animal Experiments Morally Defensible? In J. K. Becker and J. P. Buchanan (Eds). *Changing Nature's Course: The Ethical Challenge of Biotechnology*. Hong Kong: Hong Kong University Press.
6. "Most advisory reports in the UK and the European Union on the 'ethical implications' of biotechnology have thus taken the utilitarian view that there is nothing inherently wrong with the genetic manipulation of animals, but scientists involved in such practices must demonstrate that there are tangible human benefits to be gained from their work" (Bowring, 2003, p. 127).
7. Michael Fox provides a dramatic view of this when he states: "Genetic engineering makes it possible to breach the genetic boundaries that normally separate the genetic material of totally unrelated species. This means that the telos, or inherent nature, of animals can be so drastically modified (for example by inserting elephant growth hormone genes into cattle) as to radically change the entire direction of evolution, and primarily towards human ends at that" (1990, p. 32).
8. In a paper she recently presented at the American Association of Geographers conference, Emma Roe, of Cardiff University, explores similar notions of integrity and cow flesh in the beef industry stating that: initially, the living cow has integrity, wholeness; then, it is dis-integrated through the butchering process into fleshy parts; and, finally, the flesh is virtually re-integrated through the marketing of the meat product, through the product history information and common image of a cow on the package label (*Growing beef: The 'branded' non-human(s) and the retail distribution of the bovine body part(s)* presented at the American Association of Geographers (AAG) Annual Conference in Denver, Colorado, 5–9 April 2005).
9. In support of this claim, Holland cites Charles Darwin's work regarding "naturally occurring" animals and those that have been modified through selective breeding, stressing that "nearly all of the distinctions which Darwin observes between natural and artificial forms of life. . .reduce to a difference of degree rather than of kind" (1990, p. 169).

10. Employing cannibalism as a theme, Stuart Newman (1995) discusses the historical and particularly religious bases of boundaries between species in Carnal Boundaries: The Commingling of Flesh in Theory and Practice, in L. Birke. & R. Hubbard (Eds), *Reinventing Biology: Respect for Life and the Creation of Knowledge*. Bloomington and Indianapolis: Indiana University Press.
11. Take OncoMouse for example. Designed to develop various forms of cancer under the total control of the genetic technician, OncoMouse is the "first patented animal in the world" (Haraway, 1997, p. 79). According to an advertisement by Du Pont, the corporation which actually holds the patent, "Each OncoMouse carries the ras oncogene in all germ and somatic cells. This transgenic model, available commercially for the first time, predictably undergoes carcinogenesis" (Haraway, 1997, p. 81). So named, OncoMouse is an artefact, a trade-marked commodity, and no longer considered a mouse.
12. The pigs are named after the town of Beltsville in Maryland, USA, where they were "produced" in a US Department of Agriculture laboratory. The failure was very widely publicized and considered a "public relations disaster" for genetic engineering (Murphy, 1990, p. 15; for more information also see http://www.bbc.co.uk/science/genes/gene_safari/wild_west/bigger_and_better.shtml
13. It is worth noting here that genetic scientists are also developing a strategy for "deleting the porcine gene that codes for an enzyme (called *a*-1,3-galactosyltransferase) which makes the sugar that, recognized as a foreign antigen by the primate immune system, causes hyperacute rejection in humans" (Bowring, 2003, p. 121).
14. Beyond such aesthetic revulsion, there is a definite and well-grounded fear that disease and its vectors will also transgress the species barrier in dangerous and unpredictable ways. Bowring warns of the "widespread concerns amongst ecologists and medical researchers that xenotransplantation will allow new and unknown microorganisms, harmless to their natural hosts, to cross the species barrier, causing infectious disease, spreading cancer-causing retroviruses, and potentially creating mutant viruses as deadly as HIV, Ebola, or BSE. Pigs are already known to harbour endogenous retroviruses which have been found to infect human cells in vitro" (2003, pp. 303–304). Bowring further notes that "the Ebola and Marburg monkey viruses have caused large disease outbreaks in humans, and HIV is widely believed to have derived from a monkey retrovirus, and millions of people in the 1950s were infected with the non-virulent monkey virus SV40 after vaccines were contaminated by the monkey kidney cells in which they were produced" (2003, pp. 303–304).
15. The practical extensions of such human ideals into animal breeding are cited by Lynda Birke: "Even the breeding of pet animals have moved from Victorian 'fancies' into the technological society of the twentieth century. At the time of writing, a doggie sperm bank had recently started in California. Reminiscent of the human sperm bank set up in the 1970s to provide sperm from men with high IQs, the canine version (the Canine Cryobank and Animal Fertility Clinic) offers sperm from 'blue blood dogs and cats' (such as particular sled dogs) and abortions offered (at $350 each) in the event of fertilization from 'substandard sperm'" (1994).
16. In this respect, Bernard Rollin points out the similar warning made by Michael Crichton in his speculative/science fiction novel *Jurassic Park*: "Crichton's fictional vehicle for making his point is the prospect of re-creating dinosaurs by genetic engineering for a dinosaur wildlife park. Although scientists build into both the animals and the park various clever mechanisms to prevent the animals from reproducing or ever leaving the confines of the preserve, things do not go as planned. The conceptual point underlying Crichton's story is drawn from the relatively new branch of mathematics known as chaos theory, which postulates that the sorts of intrinsically predictable systems beloved by Newtonians, determinists and introductory philosophy professors... are few and far between, and, in any case, are essentially irrelevant to complex new technologies like genetic engineering" (1995, p. 72). The bottom line is that biological systems are too complex and chaotic to make thorough predictions about, something unanticipated can and likely will occur.

17. Arguing against biological determinism, both Birke (1994, p. 84) and Haraway (2004, p. 65) refer to Simone de Beauvoir's well-known assertion that "women are made, not born," an assertion often repeated by feminists in opposition to the biological determinism of gender.
18. Wheale and McNally (1990) devote an entire section to the discussion of "Genetically Engineered Bovine Somatotropin (BST)," with essay contributions from four additional authors in their book *The Bio-Revolution: Cornucopia or Pandora's Box?* London: Pluto Press. The chapters by Bulfield and Webster are particularly relevant. Also see my discussion of the ethical and practical implications of the cow as milk machine metaphor in Warkentin (2002). It is not just what you say, but how you say it: an exploration of the moral dimensions of metaphor and the phenomenology of narrative, *Canadian Journal of Environmental Education* 7:241–255.
19. Also see Rollin (1998). On telos and genetic engineering. In A.Holland & A. Johnson (Eds.). *Animal Biotechnology and Ethics*. London: Chapman and Hall.
20. The reference to Ridley that Roberts makes in this quotation is curiously impossible to find. All of my attempts so far have been unsuccessful, but I encourage anyone interested to try (Ridley, 1988).
21. Unless, of course, we accept the answer of "42," given in Douglas Adams' very popular work of science fiction, *The Hitchhiker's Guide to the Galaxy*.

Acknowledgements I thank Leesa Fawcett for her boundless intellectual, editorial and personal generosity towards revising this chapter and well beyond. I thank Carol Gigliotti for her tireless efforts and good humour in putting together this special issue.

References

Abram, D. (1996). *The spell of the sensuous*. New York: Random House.

Aldridge, S. (1996). *The thread of life the story of genes and genetic engineering*. Cambridge: Cambridge University Press.

Atwood, M. (2003). *Oryx and crake*. Toronto: McClelland & Stewart Ltd.

Becker, G. & Buchanan, J. P. (Eds). (1996). *Changing nature's course: The ethical challenge of biotechnology*. Hong Kong: Hong Kong University Press.

Bereano, P. (1996). Some environmental and ethical considerations of genetically engineered plants and foods. In: G. Becker & J. P. Buchanan (Eds), *Changing nature's course: The ethical challenge of biotechnology*. Hong Kong: Hong Kong University Press.

Birke, L. (1994). *Feminism, animals and science: The naming of the shrew*. Buckingham: Open University Press.

Bowring, F. (2003). *Science, seeds and cyborgs: Biotechnology and the appropriation of life*. London: Verso.

Bulfield, G. (1990). Genetic manipulation of farm and laboratory animals. In P. Wheale & R. McNally (Eds.), *The bio-revolution: Cornucopia or Pandora's Box?* London: Pluto Press.

Cenami Spada, E. (1997). Amorphism, mechanomorphism, and anthropomorphism. In R. W. Mitchell, T. S. Thompson & H. L. Miles (Eds.). *Anthropomorphism, anecdotes, and animals*. Albany: State University of New York Press.

Fox, M. (1990). Transgenic animals: Ethical and animal welfare concerns. In P. Wheale & R. McNally (Eds.), *The bio-revolution: Cornucopia or Pandora's Box?* London: Pluto Press.

Gill, J. (1991). *Merleau-Ponty and metaphor*. New Jersey: Humanities Press International Inc.

Graham, S. (2005) Mouse research bolsters controversial theory of aging. Scientific-American.com Retrieved May 6, 2005 from http://www.sciam.com/article.cfm?id=mouse-research-bolsters-c

Haraway, D. (1989). *Primate visions*. London: Routledge.

Haraway, D. (1991). *Simians, cyborgs, and women: The reinvention of nature*. New York: Routledge.

Haraway, D. (1997). Modest_Witness@Second_Millennium. FemaleMan _Meets_OncoMouse TM. New York: Routledge.
Haraway, D. (2004). *The Haraway Reader*. New York: Routledge.
Holland, A. (1990). The biotic community: A philosophical critique of genetic engineering. In P. Wheale & R. McNally (Eds.), *The bio-revolution: Cornucopia or Pandora's Box?* London: Pluto Press.
Kirmayer, L. (1992). The body's insistence on meaning: metaphor as presentation and representation in illness experience. *Medical Anthropology Quarteryly,* 6, 323–346.
Merleau-Ponty, M. (1962). *Phenomenology of perception.* C. Smith (Trans.). London: Routledge & Kegan Paul Ltd.
Murphy, C. (1990). Genetically engineered animals. In P. Wheale & R. McNally (Eds.), *The bio-revolution: Cornucopia or Pandora's Box?* London: Pluto Press.
Newman, S. A. (1995). Carnal boundaries: The commingling of flesh in theory and practice. In L. Birke & R. Hubbard (Eds), *Reinventing biology: Respect for life and the creation of knowledge.* Bloomington, Indiana: Indiana University Press.
Plumwood, V. (1993). *Feminism and the mastery of nature.* London: Routledge.
Plumwood, V. (2002). *Environmental culture: The ecological crisis of reason.* London: Routledge
Ridley, M. (1988). 'another slice from the tumour, dear?' Meat Industry, March, 10.
Roberts, P. (1990). Blueprint for a humane agriculture. In P. Wheale & R. McNally (Eds.), *The bio-revolution: Cornucopia or Pandora's Box?* London: Pluto Press.
Rollin, B. (1995). *The Frankenstein syndrome: Ethical and social issues in the genetic engineering of animals.* Cambridge: Cambridge University Press.
Rollin, B. (1998). On Telos and genetic engineering. In A. Holland & A. Johnson (Eds.). *Animal biotechnology and ethics.* London: Chapman and Hall.
Ryder, R. (1990). Pigs Will Fly. In P. Wheale & R. McNally (Eds.), *The bio-revolution: Cornucopia or Pandora's Box?* London: Pluto Press.
Wacks, R. (1996). Sacrificed for science: are animal experiments morally defensible? In J.K. Becker & J.P. Buchanan (Eds.), *Changing nature's course: The ethical challenge of biotechnology.* Hong Kong: Hong Kong University Press.
Warkentin, T. (2002). It is not just what you say, but how you say it: An exploration of the moral dimensions of metaphor and the phenomenology of narrative, *Canadian Journal of Environmental Education* 7: 241–255.
Webster, J. (1990). Animal welfare and genetic engineering. In P. Wheale & R. McNally (Eds.), *The bio-revolution: Cornucopia or Pandora's Box?* London: Pluto Press.
Wenz, P.S. (2001). *Environmental ethics today.* Oxford: Oxford University Press.
Wheale, P. & McNally, R. (Eds.), (1990). *The bio-revolution: Cornucopia or Pandora's Box?* London: Pluto Press.

The Call of the Other 0.1%: Genetic Aesthetics and the New Moreaus

Susan McHugh

Abstract Remakes of popular novels and films indicate how animals have become a primary means of representing not only the rapid development and proliferation of genetically modified organisms in plant form but also the interplay of aesthetics and scientific technologies in the post-Darwinian emphasis on species as a social form. Where vivisection worked in the 1896 novel *The Island of Dr. Moreau* as a scientific mechanism for social dominance, ensuing versions of the story over the past 100 years have come to position eugenic breeding and, most recently, transgenic splicing as the trope for playing out the central cultural work of ordering species in the distinction of human species being. But the contradictions of representing animals also open up other possibilities for genetic aesthetics. Especially in the 1996 film remake, the isolation of the successful transgenic animal (akin to what artist Eduardo Kac calls "the beautiful chimera") and her alignment with the human become the conditions of possibility for a theriocentric (nonhuman animal-centered) community of animal transgenics that emerges in opposition to the human genetic aesthetic. Animals remain the medium in which these struggles are (re)enacted, but their transgenic forms enable an investigation of how and why the human is increasingly defined as genetically 0.1% removed from the animal.

Keywords Genetic aesthetics · Cognition · Film · The Island of Dr. Moreau · Transgenic animals · Genetically modified organisms (GMOs)

Animals may not be big in the biotech industry at large, but they sure are big in its public perception. Fast becoming figures of evil in popular science fictions, genetically modified organisms (GMOs) are anything but the poster children of their industries. Frequently imagined as super-powered animals run amok, fictional genetic chimera instead seem to take the rap for another kind of emergent creature, the multinational agribusiness now producing and distributing patented transgenic organisms[1] on a global scale. Even narratives critical of these new economic structures of globalization focus on the rare GM animal in a world in which every year

S. McHugh (✉)
University of New England, Biddeford, ME, USA
e-mail: smchugh@une.edu

C. Gigliotti (ed.), *Leonardo's Choice*, DOI 10.1007/978-90-481-2479-4_10,

markets for GM plants and plant products grow astronomically, a trend that suggests a broader representational problem: why do animals (and not plants) loom large in the transgenic imaginary while plants (and not animals) become the medium of daily encounters with transgenic organisms?

While the problems of transgenic plants are vast and pressing, this representational link suggests that their transformation from test subject to supermarket product (in the US at least)[2] depends on the ways in which animal bodies contain ideas of the transgenic—the ways in which these technologies are both located in and confined to animal forms in the popular imagination. One effect of the spectacle of animal transgenics becoming a mainstay of science fiction is that GMOs in plant form fade into the background of both the fictions and consumer concern. For instance, the bioengineered fauna, not flora, of the blockbuster *Jurassic Park* novels and films are variously read as figures of old and new capital in critical discussions that hinge on whether and how these transgenic creatures are seen as genetically manipulated, particularly their function as naturalistic animals or cloned dinosaurs.[3] And in the process what drops out of the picture is any recognition of the specific problems posed by artificial genetic alteration.[4] The focus on the transgenic as animal here makes it difficult if not impossible to see how the proliferation of Jurassic stories and creatures alike depends on feral GMOs breeding beyond human control (arguably the situation already unfolding with agricultural deployment of these technologies).[5]

Their representational utility then derives from a central paradox, one that illuminates the broader problems of imagining the world shaped at the genetic level through human-centered models of social agency. As animal subjects gone wild, so to speak, they challenge human dominion yet the lab or factory setting of their production structurally ties them to the human, mitigating their social agency. What I want to explore here is how, in this conflicted process, the transgenic animal has come to operate as the psychic subject of genetics and, moreover, how these ideas of genetics, animals, and transgenics only begin to suggest the potential social and scientific reformations already under way. In what follows, I will examine how another narrative of animal transgenics that takes shape in the development of the explicitly transgenic 1996 film adaptation of H.G. Wells's (1896/1988) novel *The Island of Dr. Moreau* more explicitly connects these problems to models of human identity and cognition, suggesting that the interplay of genetic technologies and aesthetics is what is at stake in these transgenic animal fictions.

Neither a commercial nor a critical success, the film can be read like many other current transgenic films as a genetic translation of Wells's story of a scientific attempt to create humans from animals, in which the transgenic technology spectacularly backfires and becomes a means of separating or marking the beasts as different from the humans. However, starting from the premise that Moreau's story is never that simple, that the fluctuating status of the human in relation to other beings undercuts scientific hubris at every turn,[6] I argue that the film adaptations or "new Moreaus" imagine the transgenic animal in terms of not simply a posthuman but rather a theriomorphic aesthetic of genetics, one that is centered on animal forms. Taking at face value Wells's Moreau's claim that his choice of the human form as his model is purely aesthetic—it "appeals to [his] artistic turn of mind" (p. 112), which was "never troubled about the ethics" of his own experiments (p. 115)—these narratives trace an aesthetic critique through increasingly genetic contexts. From this approach, the 1996 film's conclusion, positing Moreau's mixed-species creatures as a self-determining community that turns away from human life, does not just digress from the original storyline but, more importantly, offers a rare glimpse of how the specter of mixed species communities eclipses anthropocentric approaches to genetics.

GMOs presented in the familiar forms of nonhuman animal species appeal to and at the same time expose the inner workings of the biotech paradigm of control, in which animals are seen at best as "quasi-human"[7] and more commonly as test models, licensed products, even production facilities or "walking drugstores."[8] In these conditions, isolation becomes a central feature of both the everyday practice of animal husbandry as well as a pivotal feature of biotech ideology. Articulating the concerns of the industrially implemented animal more generally within a Marxist framework, Noske (1997, p. 20) elaborates how the "alienation from species life," the systematic deprivation or gross distortion of social contact with members of the same and other animal species, becomes not just a human problem but more broadly a key indicator of how the ideology of domestication changes under capitalism.[9] If animals more generally appear to have become (like humans) not natural alternatives to but rather central components of the economic regimes of late capitalism, then the management of GM proximities in these stories of animal transgenics involves deferring not just animal subjectivity but more profoundly the possibilities of collaborative and interdependent social forms.

As the new Moreaus suggest, divisions of and from species being in practice prove difficult to maintain. Transgenic animals produced for biomedical applications especially complicate this aesthetic of isolation; when human genetic materials are incorporated into the patented genomes of other species, these creatures lay bare the contradictions of human subjectivity in genetic fact and fiction.[10] This is how I begin to account for the allure of animal transgenics, for why genetic neomorphs represented as fictional animals threaten humans (often with direct violence) and more importantly portend radical restructurings of economic, ecological, and other social systems. Fear of contact with the transgenic animal has become the stuff of science fiction, reflecting and fueling a more broadly disproportionate concentration on animal–human border crossings in scientific and social critiques of recombinant DNA research[11] and, less clearly, a disavowal of the profound alienation presumed by these technologies. Bodying forth the aspirations for, as well as apprehensions of, how gene manipulation alters the future of life as we know it, this apparition of the transgenic animal becomes incorporated into the new genetic and especially genomic qualifications of human species being.[12]

As the increasingly familiar statistics go, only 5% of DNA separates all the known genomes, the uniquely human part of which is limited to 0.1%. Yet the social implications of these genetic overlaps remain far from clear. For some, this evidence that the total biodiversity of our planet can be accounted for in terms of a very small portion of any given genome seems cause for celebration, legitimating a human sense of interconnectedness across species boundaries that Wilson terms "biophilia"(1984). For others, the uncertainties of the so-called language of DNA offer new means of using statistics like these in order to reinforce old hierarchies even among humans, most recently through the racially specific pharmacogenomic product BiDil (Kahn, 2004). The logic of inclusion and exclusion triggered at both extremes suggests that genetic technologies do not determine so much as destabilize a (knowledge of) species, above all human being as the basis of politico-ethical life. In this context, humanist ethics even more clearly appear to be based not on an inherent quality (an ontological difference) so much as an ongoing production of human being (a difference that makes a difference only in relation to other species).[13]

To reckon with human distinctiveness in terms of genetics—to hear the call of the other 0.1%, so to speak—means reconceptualizing the human as not a stable form but as one that, in lockstep with the transgenic animal, is ceaselessly rearticulated, not through humanist ethics but rather scientific, especially genetic, aesthetics. Moreover, this genetic reinscription of human being as endlessly contested involves situating it amid an ongoing struggle for distinction amid the diverse histories of what have always been profoundly mixed communities of species, a process elaborated by the narratives that spin out of Wells's novel *The Island of Dr. Moreau*, including the 1996 film as well as the 1977 film of the same title and the 1933 version that Wells disavowed, released as *The Island of Lost Souls*. Starting each time with the brilliant scientist Moreau's attempts to isolate something intrinsic to human beings through animal experimentation, the stories ultimately destabilize the human center of such distinctions.

As the creatures more clearly stage a conscious and collaborative overthrow of his increasingly eugenic aspirations, they raise questions about the relationship of genetics to species as a social and scientific identity form. Although there are important differences among these texts, together the novel and its film versions tell a story not only of why anxieties about species differences are increasingly articulated through transgenics but also of how it is that through the past century the genetic aesthetics of the new Moreaus proliferate through the animal form, alienated from species life and threatening humans.

Experimental Aesthetics of the Old Moreau

Long after its initial publication, the author of *The Island of Dr. Moreau* took a backward glance at its conception, suggesting that this novel was a critical response to the immediate social conditions of science as well as more broadly humanist aesthetics:

> There was a scandalous trial about that time [1895], the graceless and Pitiless downfall of a man of genius, and this story was the response of an imaginative mind to the reminder that humanity is but an animal, rough-hewn to a reasonable shape and in perpetual internal conflict between instinct and injunction. The story embodies this ideal, but apart from this embodiment it has no allegorical quality. It is written just to give the utmost possible vividness to that conception of men as hewn and confused and tormented beasts. (Wells, 1931, ix)

The "man of genius" at the center of a notorious British sodomy trial in that year was Oscar Wilde, so HG Wells seems to invite a literary–historic connection that illuminates the homosocial context of Moreau's scientific pursuit of male parthenogenesis.[14] Furthermore, the refusal of allegory in this passage casts the novel's trouble with human–animal distinctions not as a metaphor for but the basis of a thoroughgoing critique of power–knowledge structures.[15] From this perspective, Moreau's subsequent filmic evolution from vivisector to geneticist appears to involve not simply translating evolutionary theory into the language of genetics but more pointedly questioning the social power of (human) genius and (animal) genes in regulating differences within and across species. This interrogation starts perhaps with the nested narrative structure of Wells's novel, which compounds as it frames the problems of narrating life forms emerging through scientific practice even as it tells the story of the isolated Dr. Moreau's surgical transformation of animals into humans. Filtered through the perspective of a multiply traumatized narrator named Prendick,[16] the story follows the scientific production, liberation, and final degeneration of the Beast People (as the novel terms Moreau's chimeras), whose shifting status as creatures "tainted with other creatures" (Wells, 1988, p. 195) mirrors the profound uncertainty of their narrative context. This troubled framing contributes as well to the overall generic instability of the tale,[17] and together the structural and generic flux amplifies the thematic uncertainty about what distinguishes the human from other animals, leading Fried to conclude that the novel stands out not just from Wells's other novels but from all other novels as "one of the most unpleasant books to read in world literature" (Fried, 1994, p. 101).

The displeasures of the text converge in its account of response to pain as the rational, scientific measure of a creature's attainment of humanity. Different definitions of the human—particularly the humanity of the experimental animal subject—compete at the core of this novel and these conflicts show how this quality specific to human species being cannot be detected, let alone experimentally reproduced. Especially through the filmic versions, this struggle over what counts as human gives way to a particular idea of cognition and science as a relationship of genius to genes when the creatures use language to negotiate their own roles in these aesthetics.

But in the novel, the conditions of possibility for this approach to language lie in another aspect of this experimental aesthetic. For Wells's Moreau, qualities like capacity for speech and physical proximity to a homunculus prove insufficient bulwarks against biological inscription. He tries to prove that he can "make a rational creature" (Wells, 1988, p. 120) by "find[ing] out the extreme limit of plasticity in a living shape," and his proof lies in the creature's response to being vivisected. Moreau's equation of rationality and humanity in opposition to atavistic guidance by "pleasure and pain," to him "the mark of the beast" (p. 115), indicates that by doing rather than being a body affirms or betrays its own humanity. Given the anachronistic absence of anesthetic in these experiments (Fried, 1994, p. 110), it is what a creature does in response to Moreau's surgical infliction of pain that elevates it (or not) to the realm of human genius. More than just acting like a human, the creature has to act on a human capacity for rational thought, more specifically with a rational repudiation of Moreau's own ideas of the creatures as beasts.

Pitting his own idealized rational mind against what he terms the "beast flesh" he succeeds in pushing the limits of the plasticity of the flesh but ultimately sets his own failure in motion. In light of Wells's predictions of what he termed in an 1895 treatise the "artistic treatment of living things" as the future of science (Wells, 1975, p. 39), according to Suvin (1973, p. 341) this use of surgical pain in defining the human signals the novel's figuration of a farther reaching "mutation of scientific into aesthetic cognition." Because Moreau's final experimental subject only problematically realizes this mutation, the novel creates the conditions for the proliferation of genetic aesthetics, even though it falls short of modeling nonhuman cognition.

First Mutation: Revenge of the Beast People

In the novel, the last Beast Person created by Moreau starts out as a puma and this brown, female, Native American cat is in many ways aligned with his other animal experiments, whose similar dark coloring is described by Prendick (who sometimes sees them as human) in racializing terms. The narrator first encounters her in a filthy and close confinement, aboard a ship full of similarly suffering animals in a scene that as it evokes the trade in exotic animals flourishing at that time[18] underscores the mixed-species social contexts that animals like her traverse en route to Moreau's lab and come to embody through their sufferings there.

Like all of Moreau's creatures, the puma is vivisected in an experiment to create cognitive behavior. Unlike the others, however, who endure their tortures only to be disposed of as failures, the puma escapes by ripping her chains out of the laboratory

Fig. 1 Moreau (Charles Laughton) and the Panther Woman (Kathleen Burke) in still from The Island of Lost Souls (1933) (credit: http://www.moviematerials.com/)

wall and, when pursued, fatally turns on Moreau. Significantly, she doesn't kill him like a puma might, that is, she doesn't claw or bite him to death; instead, when Moreau's body is later found alongside hers, the narrator ascertains that "his head had been battered in by the fetters of the puma" (1896, p. 165). The special attention that Moreau boasts he paid to working on "her head and brain" (p. 122) appears to backfire in an account that suggests that his death comes at the hand of a rational being rather than at the paws of a Cartesian machine.

But the status of this cognitive achievement is undermined by the fact that the narrator never processes it as such. This female, New World-native, and possibly rational animal bringing down the white European scientist may ground an image of the overthrow of social oppression but the critical power of this scene becomes mitigated by the question of whether she remains an animal. And, even as the novel reconstructs this scene through the narrator who finds the corpses, it further defers the chimera's social agency by casting this action in the passive voice—again, "his head had been battered in by the fetters of the puma." The creature's action may be based on rational thought but this creation or isolation of what is here the defining quality of human species being in an experimental animal remains a fatal anomaly, one that leads directly to Moreau's death and her own. The film versions draw out the novel's implied social critique of this experiment and its setting but in a way that builds on the sense of genetic and genomic connectedness that develops more broadly in the interim.

All of the film versions diverge sharply in the details of their depictions of Moreau's demise. While they are consistently more squeamish than even the novel as they shy away from showing the particularities of this image (the camera always cuts away on contact), all as well develop further the novel's suggestion that a Beast Person could act on a rational decision to murder Moreau. What is more, in each film the final experiment is bifurcated; it is always a group of "failed" chimeras who stage revenge kills while a single, humanlike (humane?), and sexy cat-woman pledges allegiance to the other humans, often sacrificing herself to save them. And it is through this division that the possibility of not just nonhuman cognition opens out to a vision of transgenic community in the most recent film.

Because the killer of the scientist in the film versions is never his final experiment, the cat-woman who loses this capacity gains another as the storylines instead link her character to a different sort of final solution. Albeit with different cats, all the films assume that Moreau's last experiment succeeds at least in creating an intelligent, beautiful creature who acts human, so much so that the nature of the experiment changes from a Frankenstein-like creation of the human to a eugenic breeding project involving this singular chimera and an unwitting human man. Instead of persuading the Prendick character to accept Moreau's new logic (his misunderstood scientific genius), the film versions posit Prendick as lured into the experiment by the seductive cat-woman (Moreau's misguided genetic aesthetics).

Long before the patenting of animal transgenics, these new Moreaus grapple with the social implications of genetically blurred human-animal distinctions: does the man's sexual desire demonstrate the cat-woman's humanity (as Moreau argues) or the man's animality (as he fears)? Though never clearly answered, these questions alone clarify how the context of genetic science alters the experimental objective.

Manipulations of individual bodies become the pretext for chromosomal interventions into heredity and the worst fears of the novel's narrator become realized as he becomes the human who is indistinguishable from the experimental animal. But what becomes of the puma?

In a broader film tradition of hybrid sex-kittens and feline femmes fatales that includes *Cat People* (1942 and 1982) and *Catwoman* (2004), it may be more surprising that the possibility of becoming human (again the quality that distinguishes the novel's puma experiment) remains, however, strictly limited. In every Moreau film, the catwomen's only rational actions are at once altruistic and self-destructive. Stereotyped as a dark, so-called native woman, the films never imagine her as capable of single-handedly ending the experiment and instead she remains beholden to Moreau, alternately cast as her creator, savior, tormentor, and even father. In this way, the cat-women pick up the "quasi-human" baggage of experimental and factory-farmed animal subjects but in doing so they lift this burden from the remaining Beast People, who together are then free to take on the task of finishing off Moreau and his experiments.

Instead of the one rogue creature, the films depict the chimera collectively negotiating this cognitive leap and acting on it. They do so by staging deliberate conversations among the Beast People, who work together not only to demonstrate but also to make rational judgments concerning Moreau's fallibility and mortality. Whereas in the novel, this knowledge comes after his death and leads to chaotic reversion to a bestial state of mutual predation, the films imagine the Beast People deducing it and then using it to avenge their torment.

The 1933 film *The Island of Lost Souls* spells this out most clearly in a monosyllabic exchange among the Beast People, who together think aloud their own liberation and in this way, according to Sherman (1995, p. 869), become more "horrific at the expense of Moreau." Here the eugenic Moreau's sadistic monstrousness is subjected to the vengeance of the bestial revolutionaries, who according to Showalter (1992, p. 79) introduce "the iconography of class revolution" to the story of his downfall by tearing from him the white suit, whip, and other talismans of his rule and then turning on him the surgical instruments in his laboratory or House of Pain before destroying it. The 1977 film makes the kill scene itself more impulsive, with the Beast People quickly dispatching Moreau with their teeth and claws, yet sustains their rationality further through the remorseful reflection that they have thereby proven themselves no better than him. By the 1996 film, this image of the collective negotiation of cognition mutates again to outline an alternative to alienation from species life in what gradually emerge as transgenic Beast People.

This profound shift in the staging of Moreau's bad end I understand in terms of the development of what Varela calls the "selfless or virtual self," which derives from Varela's notion, developed in the context of cognitive science, of action as not just demonstrating but constituting cognition. Situated within a society of agents, the selfless or virtual self demonstrates cognition when it "gives rise to what appears to an observer to be a purposeful and integrated whole without the need for central supervision" (Varela, 1999, p. 52). The model that Varela offers to illustrate this concept is the social insect colony, a "superorganism" in relation to which the perception of a personal "I" emerges as an interpretive narrative of some of the everyday activities that go into its production (p. 61). As his example suggests, animal subjectivity gains complexity through this approach, offering a means by which I hope to elaborate the development of Moreau's murder as a collective nonhuman cognitive action in film, one that gives rise to a social vision of transgenic animals beyond service to humanity.

Cognition in this approach proceeds without consciousness so that the discrete self and, more importantly, the representational model through which it gains significance appear to be only one possibility among many at the intersections of bodies and language.[19] As I suggested above, Wells's novel conceptualizes the Beast People's use of language as deeply ambiguous, a sign of their humanity to Prendick and an instrument of their repression by Moreau, who mocks and abuses the Law that they memorize and chant. As the films transform the Beast People into what acts more like a selfless or virtual self, the new Moreaus ground genetic aesthetics in more than alienated individual forms. Conceptualizing continuities across ideas of humans and animals as subjects, this vision accounts for how the status of knowledge shifts as it is multiply (and in multiform ways) embodied through transgenics. The multiplicity of the virtual self, figured at first as inimitable phenomena and later as transgenic animals, serves not simply as a point of contrast to the singularity of the human "I" but as its sustaining context.

Next Mutation: Transgenic Communities

At least, this is the significance I see in Frankenheimer's 1996 film adaptation of *The Island of Dr. Moreau*, which depicts the isolated scientist who literally anthropomorphizes animals in ways that amplify the problems of genetically updating any animal narrative. By translating this singular tale of the transformation of beasts to humans into the contemporary scientific language of genes, Frankenheimer's film does more than simply reflect changes to the world in and around both its literary source and its filmic progenitors. Bringing the interconnected questions of gendered and sexual agency, racial identity, and species demarcation characteristic of the earlier Moreaus into an unequivocally transgenic context, the film suggests that cognition beyond the human model is an inevitable consequence of contemporary recapitulations of social difference in terms of genetics.

Thus updating as well as situating Moreau in the late twentieth century with this latest incarnation, the film narrative's environment shifts from a vivisectionist nightmare involving painful cross-species grafting to a eugenic transgenic experiment involving primarily subcutaneous interventions: manipulations of protein strands to design creatures, followed by injections of endorphins and hormones to prevent their regression as well as the implantation of electrodes to stabilize the social interactions of these new forms. Moreau literally gets under the skin of his creations in multiform ways here, all designed to augment his control over organisms and failing in the end to do so. And the film makes clear that this shift in the experimental aesthetic reflects an acute and current anxiety about the porousness of species boundaries.

Moreau, as geneticist, abandons the representational model of the human to make way for a socially transitional role for these experimental life forms: he says that he works not on animals to make them human but on "animals that have been fused with human genes" in order to make something that is "better." This new Moreau pursues his version of genetic aesthetics not to locate or preserve but rather to extend and ideally to improve upon human species being. Mixing the terms of culture and science as much as of humans and animals in his description of his own experiments, he claims that he battles with a "tiresome collection of genes" as part of a "process of the eradication of destructive elements found in the human psyche" in order to produce a "divine creature that is pure [and] harmonious," by which he means, "absolutely incapable of malice."

The narcissism of Moreau's genius extends through the now conventional trope of genetic immortality via heredity, for apparently his own genes are the primary source of this human infusion into the animal. The scientist's biological inscription into his parthenogenic children pointedly conflates his sense of self, his aesthetic, and his practice as geneticist in both social process and product. To adapt Varela's phrasing, this Moreau emerges as a selfish and material "I" through these attempts to elevate himself genetically beyond the contexts that sustain this self-perception. Through the doctor's approbation of the beauty of the transgenic children who favor him against his brutal intimidation of the bestial and deformed failures, this film begins to separate the scientist's transgenic eugenics and the scientific creations' very different genetic aesthetics.

Moreau's articulation of his values as geneticist more broadly suggests how conceptions of the creative potential of transgenics formally rely on the isolation of one successful body, what transgenic artist Eduardo Kac terms the "beautiful chimera," (Kac, 1998) in order to keep at bay the social threat posed by the multiplicity of what in contrast appear to be "ugly" bodies involved in the production of the singularly successful (here pure and harmonious) body. Like the new Moreau, Kac, a contemporary artist who views transgenic creatures as media, abstracts the successful individual in his GFP Bunny projects involving the transgenic animal he calls Alba, an otherwise ordinary lab rabbit whose DNA has been modified to express throughout her body a version of a green fluorescent protein (or GFP) found in deep-sea jellyfish. Though apparently at odds with the scientists who produced her and therefore falling far short of the fictional Moreau's range of control, Kac's aesthetic similarly has built in strict limits to the transgenic chimera's social potential. Socialization, a central objective of Kac's original project, means for him that this rabbit, who luminously glows in blue light, will live in Kac's home along with his wife and child. Although his aesthetic vision of her life beyond the laboratory may never be realized, his "parental" plans (Lynch, 2003, p. 75) for this beautiful chimera follow the same logic by which Alba's less successful prototypes will never leave the lab in the name of art or science.

If genetic technologies succeed in blurring conceptual divisions of species, their popular, commercial, even scientific success hinges on their ability to control the transgenic animal's social circulation and, what is more, perceptions of the unsuccessful experiments leading up to it. Early cloning methods, for instance those used in the production of Dolly the sheep, were premised on the creation and summary execution of hundreds of severely deformed and disabled creatures for every successful one. Like their bodies, representations of these unsuccessful experimental animals or ugly chimeras remain largely confined to the lab, even in popular horror film images such as *Alien Resurrection* (1997), where a walk past specimens of the failed clones of herself terrifies the main character Ripley. This rare scene of the beautiful (humanoid) chimera haunted by the remains of her ugly (alien animal) prototypes makes plain how, in the context of overwhelming (and overwhelmingly murky) genomic overlap, genes take on the power to define individuals as social and species beings only through alienation from mixed-species contexts. In this way, the bodies of transgenic animals take on the slippery work of managing ideological and (in the case of race now more clearly as) genetic distinctions in and across species.

Again the multiplicity of transgenics—as of the animal more generally, according to Deleuze and Guattari[20]—is the menacing or threatening element. In this model, the end, even if it is only the one beautiful thing, justifies a means involving the production of a multiplicity of others, the "unstable phenomena," freaks, or mistakes against whom the transgenically enhanced beautiful one at once emerges as a success and struggles for preeminence on the model of the individual human subject. In contrast to both his literary progenitor, who abandons his mistakes to fend for themselves, as well as his actual genetic scientist counterparts, who customarily destroy their mistakes along the way, Moreau as geneticist construes himself as a father in relation to all, even to those who fall short of his ideal, and so maintains a mixed community of chimeras as an extended family group.

In this way, the genetic narrative of Moreau introduces the context that even self-styled socially sensitive transgenic artists like Kac disavow, namely the mixed social setting through which the successful individual is both isolated and sustained. Bringing together all of the products in his lab, if only in an isolated island community, the collective conceptualization of Moreau's transgenic chimeras creates a setting in which the ugly masses can confront not just their creator but also his beautiful one (the cat-woman). And, in the aftermath of this confrontation with their own physical repression as well as ideological interpellation by Moreau's aesthetic, still another possibility emerges in which transgenics develop life beyond Moreau in the form of a self-sustaining community, thereby opening up a narrative space in which Moreau's transgenic eugenics become a posthuman genetic aesthetic in contrast to their own nascent theriocentric aesthetics.

In other words, in this evolution of *The Island of Dr. Moreau*, genetic technologies herald the birth of (at least) twin possibilities for transgenics, on the one hand products controlled by humans and on the other self-directed social groups. As befits this scion of Faust, the premise in the 1996 film remains the same, namely that the doctor's immediate success entails his ultimate failure. However, this time the creatures live collectively beyond the experimental context, suggesting that the doctor's eugenics and not necessarily the creatures' transgenic status provide the basis of failure. The film reinvents its title character, no longer an ostracized vivisector, as a Nobel-prize-winning recluse, whose death becomes a necessary step toward the liberation of his transgenic "children" rather than its climactic signal of their chaotic reversion to animality. If the devolutionary trajectory of Wells's novel closed this circle with the death of Moreau at the hands of the creature in whom he creates rationality, then the genetic mutations of Frankenheimer's narrative recast the death of Moreau as a key part of the longer development of the creatures' social life, suggesting that the problems lie with ideas about (and not the fact of) genetic modification.

At every level, the 1996 film signals this significant alteration in scientific and (arguably as) social vision. Avoidance of animal rights activism remains the motive for retreating to the island lab facility here, but the taint of professional violation has vanished from Moreau's story. Supported rather than betrayed by his colleagues, Moreau appears instead to be a product of rather than a recluse from the world. And the island retreat itself suggests not so much an isolated fortress as a microcosm of the converging military-bio-industrial forces of globalization. Early on, its mélange of architecture and equipment is explained through its multinational history of settlement: first a Dutch coffee plantation, then an American military facility, next a failed Japanese resort, and finally the geneticist's exclusive laboratory, this vision of the island clarifies the long and involved human cultural history within which Dr. Moreau's experiments transform animals. Corresponding with this globalized context and Moreau's mutation here into former international hero, the narrator too shifts from a shipwrecked scientific colleague and fellow countryman to a U.N. inspector gone astray. Conflicting with him as world citizen, the Prendick character of this film also foregrounds the multinational, if not multicultural, effects of the new Moreau's genetic experiments. Even more explicitly the creatures this time are genetically altered hybrids, recasting the central tension from animals evolving into or devolving from human nature toward the developments of intersecting and separate transgenic cultures, unintelligible in strictly anthropocentric terms.

Onscreen, the Beast People again talk through their collective overthrow of Moreau's economy of pain but this time divestment also becomes possible. While earlier versions end with an unequivocally violent physical overthrow of Moreau, Frankenheimer's film stages an initially peaceful scene of the Beast People's confrontation with their creator. Significantly Moreau himself activates this potential in the chimeras by introducing the concept of multiple aesthetics, although in a vain attempt to justify the scientific hierarchy on which Moreau's power depends, by invoking his own aesthetic. Moreau, playing a piano to illustrate, likens himself to George Gershwin and the creatures to Arnold Schoenberg as he makes a final and fatal attempt to universalize his notions of beauty. The ugly chimeras, deferring to him up to this point, slowly recognize that they do not share this aesthetic and, as in previous films, together tear Moreau apart, gratuitously destroying his piano along with the rest of his palatial home. What this staging clarifies is that they are not simply reverting but acting on a collective, cognitive deliberation that involves violently rejecting his containment of their aesthetic within the formalist ghetto of experimentalism as part and parcel of his authority as father and ruler.

What makes this film so unique is that these creatures then, in turn, face collective opposition within the broader community of transgenic Beast People. Moreau's killers savagely turn on each other as well as the other humans and Beast People who are not involved in the murder, and the latter's effective resistance opens up another possibility for the transgenic chimera's life beyond their creator's control. The Beast People's consolidation as a group leads immediately to a violent overthrow of Moreau's regime but not to a long-term repetition of (or chaotic reversion to) Moreau's brutal order. Those among them who try to assume the dictatorial human role are, like him, killed. And so is the beautiful chimera, the cat-woman who dies protecting the Prendick character, leaving behind a group of transgenic creatures whose stability derives from their own collective awareness of interdependence and its fragility.

Frankenheimer's film is, in this respect, a far cry from the original Moreau, in which all of the Beast People, betrayed by their animalistic biological origins, revert quickly from their postoperative state to social fragmentation, isolation, and mutual predation. Volatilizing the narrative of biodetermined culture that more pervasively forecloses the social potential of GMOs, Moreau's creatures, here "tainted" by a species intermixture that involves humans, come to cognition with a biological claim on human status but with no need for individual consciousness or identity on a human model. Instead they reject human identification outright; they ask the one remaining human to leave and not to return with other humans. In this emergent genetic aesthetic, human proximities ultimately prove threatening to the consciousness of transgenics as selfless or virtual selves.

In the case of the genetic Moreau, one Beast Person's final observation, "we have to be what we are, not what the father tried to make us," announces neither a forgetting of nor a morbid fixation on their creator but an active choice to disidentify with his values. The film closes with the promise of a new social order of Beast People involving "no more scientists, no more laboratories, and no more experiments." The narrator, as ever the sole human survivor of the island of Dr. Moreau, for once is depicted at peace with these chimeras and agrees to leave them alone. The future of these genetic laboratory failures remains uncertain but, significantly, they have a future. In this way Frankenheimer's film figures transgenics across a genetic aesthetic spectrum edged by histories of human eugenics on the one side and the potential for transgenic cultures on the other.

Permutations: Multiple Genetic Aesthetics

With the rapid development and proliferation of biotechnologies through the century since the novel's publication, HG Wells's story of Dr. Moreau's invention of the Beast People itself undergoes mutations in film versions that respond to shifting notions of aesthetics along with scientific technologies. In this way, the new Moreaus of film ultimately illuminate the critical role of animal subjectivity in the developments of genetic aesthetics. Where vivisection worked for Wells and his contemporaries as a scientific mechanism for social dominance, ensuing versions of the Moreau story over the past 100 years have come to position eugenic breeding and, most recently, transgenic splicing as the trope for playing out the central cultural work of ordering species in distinguishing human species being. Animals remain the medium in which these struggles are (re)enacted, but their transgenic forms enable an investigation of how this process comes to define the human as genetically 0.1% removed from the animal. Extending the longstanding human traditions of species hybridization in agricultural plants in ways that also call into question genetic knowledge of humans and, more to the point, the social consequences of these emerging knowledges, transgenic animals have thus become sites for working out not just the logistical problems but also more broadly cultural anxieties about GM technologies. Overwhelmingly presented as evil in popular films, animal transgenics work as scapegoats for not just the potential ecological catastrophe threatened by plant-shaped GMOs and their industries but also for the medical reinvention of the human promised through gene therapy and pharamacogenomic products.

As the Moreau narratives suggest, however, animals in the process become a means of interrogating the relationships of genetic science to cultural representation, modeling other aesthetics of genetics that challenge the premise of human singularity. As the complex social lives characterizing single-and mixed species communities that have been documented by ethologists, cognitive scientists, and animal studies scholars threaten the foundations of humanist ideologies, the questions of nonhuman subjectivity profoundly disrupt the ideological alienation required of new commercial interests in transgenic life. This tension begins to suggest why the comparatively limited numbers of GM animals at present are at odds with their high public visibility. For the problem of perceiving transgenic bodies may not be so much a defining quality of the contemporary human individual as it

Fig. 2 Moreau losing control of the Beast People in still from The Island of the Lost Souls (1933) (credit: http://www.moviematerials.com/)

is a problem of seeing populations as more than a mere background or contrast to the human individual. Transgenics visualized as animals thus threaten to displace the transcendent human individual as a primary mechanism of social power at the same time that they locate competing interests in human and other animal bodies in the new global regimes of biopower.[21]

To be clear, my point here is not that genetic scientists or transgenic artists have become the new Moreaus (whatever that might mean) but rather that the mutating images of Wells's scientist and scientific chimera that emerge through film raise timely questions about the singularity of a genetic aesthetic that uses animals as containers for GM technologies already being implemented on a massive scale in plant forms. Genetic manipulation and genomic overlap become two sides of the same coin only within an aesthetic of controlled (and controllable) distance in terms of human genetic singularity. Situated amid theriocentric aesthetics, the anthropocentric values guiding genetic science appear more clearly (if less coherently) to respond to genetic inscriptions of past, present, and future mixed-species communities. Recalling how Jack London's fictional Buck turned away from the human to gain a new life in a multispecies social context in the Yukon woods by hearing "the call of the wild" in wolf howls, I ask, what does the genetically discrete human stand to gain (as well as lose) training its ears to hear only the chromosomal call of the other 0.1%? At stake in the genetic humanist ideology taking shape in such stories is no less than the social agency of bioengineered life, denied at human peril.

Notes

1. I acknowledge that this use of "transgenic organism" throughout to refer to genetically modified agricultural food products is itself controversial, as illustrated by Haraway's treatment of one of the earliest of such products to be approved by the US Food and Drug Administration: "Although [its patent owner] Calgene has claimed the Flavr Savr [a GM tomato variety] is not a transgenic organism, since the gene normally responsible for decay is genetically engineered to be reversed and so nonfunctional in the new product, I consider Flavr Savr strictly transgenic because it bears a gene for a bacterial enzyme inserted to act as a marker [in what has become a common practice] to verify successful insertion of the altered functional gene of intent." Haraway's assertion that this tomato is transgenic tactically revisits the strategic positioning of products that are at once appealing to knowledge of existing taxonomic categories (Flavr Savr as a kind of tomato) and "blasting" conventional notions of species distinctions (a tomato as a patented product of genetic lab science) (Haraway, 1997, p. 56). This distinction has become all the more contested with genetic modifications to plants to make them internally manufacture *Bacillus thuringiensis*, a bacterial toxin commonly referred to as Bt, which also is topically applied as an organic insecticide.
2. The Committee on Environmental Impacts acknowledges that Flavr Savr was a commercial failure but points to the "rapid and broad use of Bt-expressing cotton and corn" as subsequent evidence of "the commercial success" of transgenic crops more generally (2002, 221), in spite of the problems that stem from lack of knowledge about the natural ecology of this bacterium (p. 162).
3. Mitchell (1998, p. 221), relating especially the large, carnivorous dinosaur to modern historical patterns as an enduring image of "the spectacle of consumption itself", argues that the attraction of this narrative concerns its "redemption of big patriarchal capital" (p. 225). In contrast, Watkins (1995, p. 191) reads it as primarily concerned with "the thematics of

postmodern 'downsizing'" by positioning the dinosaur as "a kind of symptomatic monitor" of global economic patterns; Watkins relates how the fast, smallish velociraptors figure a shift in public perceptions of dinosaur life to the stripped-down corporate model of mobile capital rising in popularity in the 1990s, a reading that gains an eerie resonance in fallen corporate giant Enron's use of the name Raptor for an offshore (and allegedly money-laundering) subsidiary designed for this very purpose. Reading the animals as clones, Turner (2002, p. 903) argues compellingly that because of their "circulation as commodities within the same mass culture that maintains the ideology of DNA as a consumable object" the Jurassic narratives themselves fail "to carry out an unimplicated critique of the bioinformatics economy [...] *Jurassic Park* the novel is made into *Jurassic Park* the movie, with the inevitable book and movie sequels duly following."

4. For instance, identification with animal forms justifies the continuing imposition of an "evolutionary tree" model on systems that have evolved not simply through linear–hierarchic but also lateral–horizontal gene transfer (Venter, 2003), obscuring not only these emerging methods of communicating and influencing development across species lines but also the basic conditions of transgenic crop commercialization, which proceeds rapidly in spite of the fact that little is known about the long-term effects on consumers, ecosystems, or even the plants themselves. Venter's (2003, p. 52) conclusion is enigmatic: "The concern is that [lateral gene transfer] might be very extensive in the plant world, which has a lot of implications in terms of biotechnology in plants." The Committee on Environmental Impacts Associated with Commercialization of Transgenic Plants (2002) suggests that these concerns are further muddled by agricultural implementations of horizontal gene manipulations in crop plants, which may have previously obliterated "the line between conventional crop breeding and the creation of transgenic crops" (p. 48).
5. Controversies surrounding studies of wild-breeding transgenics complicate this situation further. Editors of the premiere science journal *Nature* published and in a rare move quickly retracted one such study, in which Quist and Chapela (2001) claimed to detect the markers of altered genes moving across generations within native corn and its uncultivated relatives in the Mexican state Oaxaca, a region that as Dalton (2001, p. 413) elaborates is part of the presumed ancestral origin of this plant and the "global center" of its genetic diversity. Quist and Chapela (2001, p. 542) conclude their article with a call for long-term studies that similarly trace "the gene flow from hybrids to traditional landraces in the centers of origin and diversity of crop plants" but the scandal surrounding *Nature*'s retraction makes this kind of work less viable. At issue is how to trace the environmental impact of altered genes and their possible mutations, a situation in which the illegal planting of transgenic seeds already contaminates possible scientific controls. Indicating that such "empirical inference" about this level of transgenic pollution is now noncontroversial, Metz and Fütterer (2002, p. 601) cite studies that document it through illegally grown transgenic soy and cotton, respectively, in Brazil and India, even as they refute Quist and Chapela's findings. This controversy illustrates how ideas of ownership, power, and science converge to bracket off the agency of transgenic organisms.
6. Reading Moreau's creatures as "animals artificially elevated to 'human' status," McCarthy (1986, p. 27) concedes nonetheless the complexity of their "case, one that almost completely blurs the line between man and beast." McCarthy's development of this novel's overlap with Conrad's *Heart of Darkness* (1899) gains significance with Marlon Brando's depiction of Moreau in the 1996 film version, where the actor who famously portrayed the Kurtz figure in *Apocalypse Now* (1979) again is swathed in white, surrounded by modern military gear, and presented as both insane in his excesses as well as symptomatic of empire.
7. Analyzing the political response of animal rights activists to patenting organisms—"the ultimate manipulation of being 'made-into'"—Sabloff (2001, p. 141) points out that strategies premised on a "flight into the human realm" underscore the overarching assumption that even these "quasi-human" animals exist outside of human history (p. 132), an assumption that I hope here to challenge by revisiting the notion of species being.

8. Schwab (quoted in Sabloff, 2001, p. 127) uses this vivid phrase to call attention to the hormone, antibiotic, and other drug use endemic to factory farming. But I think that it resonates with the uses of GM life forms to produce pharmaceuticals for humans as well as factory-farmed animals because this approach to microbial life has already become standard practice. Pharmaceutical giant Monsanto became "the biotech frontrunner" in the 1990s by aggressively developing agricultural GMOs, creatures like bacteria that produce bovine somatropin (BST), the first bovine growth hormone to be marketed, as well as acquiring companies with the rights to them, such as Calgene (Haraway, 1997, p. 291, n. 64).
9. For Noske (1997, pp. 15–16), the "genetic engineering" of animals as "specialized workers"—such as the featherless chickens recently developed by Israeli researchers to withstand the high heat of confinement facilities in the desert—demonstrates not only how genetic science extends a Taylorist division of "animal 'skills' and bodies" but also, in the ways in which such genetic interventions enable animals to be "deprived of their own society" and even human company, how such practices relocate the "species alienation" that Marx diagnosed in human workers into nonhuman bodies (p. 20).
10. See Haraway (1997, p. 79), who develops this point about the irreducibility of the facts and fictions surrounding OncoMouse, which is not only "the first patented animal in the world" but also a figure that enables people metaphorically and materially to "inhabit the multibillion dollar quest narrative of the search for the 'cure for cancer.'"
11. Ringel (1989, p. 64) points to this fixation on "the boundaries between the animal and the human" as she accounts for why in the 1980s there were more objections to "research in recombinant DNA than [... to] all the atomic research of the 1940s and 1950s." In the 1990s, species crossings continue to be the focus of critical attention. According to Schaal (2002, pp. 110–111), "The greatest area of concern at this time remains direct genetic modification—the insertion of a gene from one species into the genome of another—that is, genetic engineering or the production of GMOs." She points to anxieties about species boundaries to account for why public debate about biotechnology in agriculture is "extremely active and visible," "international in scope," and "often exceedingly bitter" (p. 109).
12. Recent discussions of Heidegger's focus on the Geschlect or species being of humanity call into question whether such distinctions inevitably reinforce a humanist teleology, endlessly deferring questions of animal subjectivity. Derrida (1989, p. 56) argues that such a conflation "has remained up till now [...] the price to be paid in the ethico-political denunciation of biologism, racism, naturalism, etc.", leaving open the possibility of distinguishing human being from anthropocentrism, a separation of the discourse of species from the institution of speciesism that Wolfe suggests Derrida pursued (if never quite accomplished) "all along" (Wolfe, 2003, p. 216, n. 24). The question remains whether it is the humanist insistence on being or the deconstructive focus on language—what is more, the inherent contradictions within language or human perceptions of them—that endlessly defers such a project.
13. Though not referring directly to Derrida, Agamben contends that the use of language "to isolate the nonhuman within the human" gets in the way of understanding how it works in the service of an "anthropological machine" to produce the human as a site of endless negotiation over such divisions (Agamben, 2004, pp. 37–38). Pointing to the influence of zoologist von Uexküll's "unreserved abandonment of every anthropocentric perspective in the life sciences" on Heidegger's as well as Deleuze's work (39), Agamben shows how von Uexküll's distinction of species members through their Umwelt or perceptual worlds defers the possibility of any absolute distinction between man and animal. Agamben refuses the absolute division within language and instead posits an ongoing negotiation at the site of the human so, for him, Heidegger's insistence on the distinctiveness of human species being in terms of the possibility of metaperception of animal perceptual worlds not only points the way to the conclusion of metaphysics but also anticipates a new purpose that has emerged for (and at the same time closes down) this sense of being: "Genome, global economy, and humanitarian ideology are the three united faces of this process in which posthistorical humanity seems to take on its own physiology as its last, impolitical mandate" (p. 77).

14. According to Showalter (1992, p. 80), "The all-male community [. . .] raises disturbing possibilities and fantasies of homosexuality and bestiality that add to its unpleasantness for readers" but I see the connection as far less "subtle."
15. See Foucault's (1978, p. 43) discussion of the discursive transformation of the homosexual into a "species" at the beginning of the twentieth century, a form of clinical sexual agency which involves a precarious transcoding between medical and animal science.
16. Especially in the gaze of the Leopard Man, the first of the Beast People to revert to killing and to be killed for it, Wells's Prendick intuits "the fact of its humanity" though he admits that he "cannot explain the fact" (Wells, 1988, p. 147), a perception that as (he calls himself) a scientific man he assumes will not convince the reader any more than his scientific human companions. Structurally, such narratorial insistence on transcendent humanity as a fundamental difference may bookend Moreau's insistence that the creatures are all failures but Prendick's reliability in making these claims in turn is surrounded by doubt. The many layers of reference to cannibalism in the framing of the story invite the dismissal of Prendick's narrative altogether, as Showalter (1992, p. 81) argues, "as a hysterical hallucination," but this interpretation precludes the significance of the novel's elaborate deferral of the vivisected chimera's human status. What I want to develop here instead is how, as Fried (1994, p. 204, n. 19) indicates, Moreau's approach to vivisection is deconstructive avant la lettre: "When Moreau concludes that he might just as well have made sheep into llamas or llamas into sheep, he is in effect expressing a commitment to difference (or say representation) within beasthood itself."
17. Vacillating across beast fable, scientific romance, castaway thriller, and the sort of "technonovel" popularized generations later by Crichton, Wells's text exhibits characteristics of multiple novelistic genres without finally establishing one as its dominant model (Krumm, 1999, p.51). Sherman (1995, p. 869) contends that this categorical difficulty, though one of the distinguishing features of *The Island of Dr. Moreau*, also in part accounts for why it is the "least widely read and taught" of Wells's popular scientific romances.
18. Although European stories of the people who captured animals were highly romanticized, Rothfels (2002, p. 68) shows how they became part of the larger historical narrative of empire, for "in most cases this business had a ruthless side in its dealings with animals, people, and environments which was almost a necessary component of the trade."
19. As Varela writing together with Maturana elaborates, this "ongoing descriptive recursion which we call the 'I'" is produced through the interface of human bodies and language: "the self, the 'I', arises as the social singularity defined by the operational intersection in the human body of the recursive linguistic distinctions in which it is distinguished" (Maturana and Varela 1992, p. 231). The ambiguous status of language that emerges here is central to Wolfe's deconstructive argument about animal subjectivity more generally; to him, Varela and Maturana's "processive, recursive, antirepresentational account of the relation between material technicities, linguistic domains, and the emergence of subjectivities" becomes a means of evading altogether "the blind alleys of 'intention' or 'consciousness' (or what amounts to the same thing on methodological terrain in the sciences, 'anthropomorphism')" that trouble prior accounts of animal in relation to human subjectivity (Wolfe, 2003, p. 87). The limitations of language, however, seem to weigh on this argument, for as Wolfe notes, Varela and Maturana refute the characterization of bee communication as a "language" and insist instead that "it is a largely fixed system of interactions whose stability depends on the genetic stability of the species and not on the cultural stability of the social system in which they take place" (2003, p. 218). In this context, it is all the more curious that Varela later returns to social insects as a model of the selfless or virtual self.
20. Moreau's transgenic children offer a stark delineation of how any animal can "be treated in all three ways" distinguished by Deleuze and Guattari: the beautiful chimera, kept in Moreau's house and most frequently referred to as his daughter is akin to the "individuated animals, family pets, sentimental, Oedipal animals, each with its own petty history"; this cat-woman later fights to the death with members of the largely unindividuated crew who kill him, who

are "more demonic animals, pack or affect animals that form a multiplicity"; and, in the ruins of Moreau's experiment, all who remain appear to be "animals with characteristics or attributes" in this case of genetic modification moving beyond the realm of human control, "archetypes or models" of genetic aesthetics in defiance of the Law of their shared father (Deleuze, 1987, pp. 240–41).

21. Foucault (1978, p. 139) develops this term to distinguish the "anatamo-politics" or production of the individual human body and identity through disciplinary structures from "biopolitics," the mechanisms by which populations are configured through "biopower." On the connections of biopower with globalization and empire, see Hardt and Negri (2000, xv).

References

Agamben, G. (2004). *The open: Man and animal,* Stanford: Stanford University Press.

Comar, J.C. (Director) & Kane, B. (Writer). (2004). *Catwoman.* United States: Warner Brothers.

Committee on Environmental Impacts Associated with Commercialization of Transgenic Plants. (2002). *Effects of transgenic plants: The scope and adequacy of regulation.* Washington, DC: National Academy Press.

Conrad, J. (1902) *Heart of darkness.* Edinburgh: Willliam Blackwood and Sons.

Coppola, F. (Director/Writer) & Milius, J. (Writer). (1979). *Apocalypse now.* United States: Paramount Pictures.

Crichton, M. (1990). Jurassic park. New York: Knopf.

Dalton, R. (2001). Transgenic corn found growing in Mexico. *Nature,* 413, p. 337.

Deleuze, G. & Guattari, F. (1987). *A thousand plateaus: Capitalism and schizophrenia.* Minneapolis, MN: University of Minnesota Press.

Derrida, J. (1989). *Of spirit: Heidegger and the question.* Chicago: Chicago University Press.

Foucault, M. (1978). *The history of sexuality, Vol. 1: An introduction.* New York: Random House.

Frankenheimer, J. (Director), Stanely, R. & Hutchinson, R. (Writers). (1996). *The island of Dr. Moreau.* United States: New Line Studios.

Fried, M. (1994). Impressionist monsters: HG Wells's *The Island of Dr. Moreau.* In S. Bann (Ed). Frankenstein, creation and monstrosity. London: Reaktion Books.

Haraway, D. (1997). *Modest_Witness@Second_Millennium.* FemaleMan©_Meets_OncoMouse™. New York: Routledge.

Hardt, M. & Negri, A. (2000). *Empire.* Cambridge: Harvard University Press.

Jeunet, J. (Director) & Whedon, J. (Writer). (1997). *Alien resurrection.* United States: Twentieth-Century Fox.

Kac, E. (1998). Transgenic art. *Leonardo Electronic Almanac,* 6, p. 11.

Kahn, J. (2004). How a drug becomes "ethnic": Law, commerce, and the production of racial categories in medicine. *Yale Journal of Health Policy, Law and Ethics,* 4(1), 1–46.

Kenton, E. (Director), Young, W. & Wylie, P. (Writers). (1933). *The island of lost souls.* United States: Paramount Pictures.

Krumm, P. (1999). *The Island of Doctor Moreau,* or the case of devolution. *International Review of Science Fiction,* 28(75), 51–62.

Lynch, L. (2003). Trans-genesis: An interview with Eduardo Kac. *New Format* 49, 75–90.

Maturana, H. & Varela, F. (1992). *The tree of knowledge: The biological roots of human understanding.* Boston: Shambhala Press.

McCarthy, P. (1986). *Heart of Darkness* and the early novels of HG Wells: Evolution, anarchy, entropy. *Journal of Modern Literature,* 13(1), 37–60.

Metz, M. & Fütterer, J. (2002). Suspect evidence of transgenic contamination. *Nature* 416, 600–601.

Mitchell, W. J. T. (1998). *The last dinosaur book: The life and times of a cultural icon.* Chicago: Chicago University Press.

Noske, B. (1997). *Beyond boundaries: Humans and animals*. New York: Black Rose Books.
Quist, D. & Chapela, I. (2001). Transgenic DNA introgressed into traditional maize landraces in Oaxaca, Mexico. *Nature*, 414, 541–543.
Ringel, F. (1989). Genetic experimentation: Mad scientists and the beast. *Journal of Fantastic Arts*, 2(1), 64–75.
Rothfels, N. (2002). *Savages and beasts: The birth of the modern zoo*. Baltimore: Johns Hopkins University Press.
Sabloff, A. (2001). *Reordering the natural world: Humans and animals in the city*. Toronto: University of Toronto Press.
Schaal, B. (2002). Genomics and biotechnology in agriculture. In M.Yudell & R. DeSalle (Eds.), *The genomic revolution: Unveiling the unity of life*. Washington, DC: Joseph Henry Press.
Schrader, P. (Director/Writer) & Ormsby, A. (Writer). (1982). *Cat people*. United States: Universal Pictures.
Sherman, K. (1995). Are we not men? *Queen's Quarterly*, 102(4), 869–875.
Showalter, E. (1992). The apocalyptic fables of HG Wells. In J. Stokes (Ed.), *Fin de siècle, fin du globe: Fears and fantasies of the late nineteenth century*. New York: St. Martin's.
Spielberg, S. (Producer/ Director) & Crichton, M. (Writer). (1993). *Jurassic park*. United States: Universal Pictures.
Suvin, D. (1973). *The Time Machine* versus *Utopia* as a structural model for science fiction. *Comparative Literature Studies*, 10(1), 334–352.
Taylor, D. (Director), Ramrus, A. & Shaner, J. (Writers). (1977). *The island of Dr. Moreau*. United States: MGM.
Tourneur, J. (Director) & Bodeen, D. (Writer). (1942). *Cat people*. United States: RKO Pictures.
Turner, S. (2002). Jurassic Park technology in the bioinformatics economy: How cloning narratives negotiate the telos of DNA. *American Literature*, 74(4), 887–909.
Varela, F. (1999) *Ethical know-how: Action, wisdom, and cognition*. Stanford: Stanford University Press.
Venter, J. C. (2003). Whole-genome shotgun sequencing. In M. Yudell & R. DeSalle (Eds.), *The genomic revolution: Unveiling the unity of life*. Washington, DC: Joseph Henry Press.
Watkins, E. (1995). The dinosaurics of size: Economic narrative and postmodern culture, *Centenniel Review*, 39(2), 189–211.
Wells, H. G. (1931). *Introduction. The Atlantic edition of the works of HG Wells, Vol 2*. London: Fisher Unwin.
Wells, H. G. (1975). The limits of individual plasticity. In R. M. Philmus & D.Y. Hughes (Eds.), *H. G. Wells: Early writings in science and science fiction*. Berkeley: University of California Press.
Wells, H. G. (1896/1988). *The Island of Dr. Moreau*. New York: Penguin.
Wilson, E. O. (1984). *Biophilia*. Cambridge: Harvard University Press.
Wolfe C. (2003). *Animal rites: American culture, the discourse of species, and posthumanist theory*. Chicago: Chicago University Press.

Landseer's Ethics: The Campaign to End "Cosmetic Surgery" on Dogs in Australasia

David Delafenêtre

Some years ago a fellow came to my office and insisted that he needed to dock the tail of his dog. I asked, "How will your dog be better when it doesn't have a tail?" He replied, "Well, that's the standard of gundog owners associations around the world." I said, "Who cares?" Why would an owner chop off a dog's tail simply so that the dog can compete at the Royal Easter Show or some other show in Brussels or Paris where judges can say, "Look, it's a one-inch tail. That's perfect." and give a tick, so that the dog can win a ribbon and the owner can prance around and claim all the credit? It is a joke—and not a very good one. (Alan Ashton, New South Wales, Hansard, April 06, 2004)

Abstract In Victorian Britain, Edwin Henry Landseer refused to paint cropped dogs for ethical reasons. As such, the artist became one of the first persons to challenge a centuries-old tradition inflicted upon the *canis familiaris*. The manipulation of dog breeds for aesthetic purposes on the basis of pseudo-scientific reasons has an ancient history. In international cynology, the term "cosmetic surgery" refers to a series of practices that lead to the permanent mutilation of companion animals. In the nineteenth century, the growing popularity of dog shows contributed to the institutionalization of cosmetic surgery. Today tail docking and ear cropping are the most common procedures performed on certain dog breeds. The late twentieth century and early twenty-first century have seen increasing efforts to ban both practices in Australasia. Led by the veterinary profession and animal protection groups, the campaign to end non-therapeutic surgical operations on the canine species has met with a great deal of resistance in the world of purebred dog breeding. This chapter looks at the way the activists' ethical principles have fared during attempts by Australasian parliaments to legislate a ban on cosmetic surgery. It will be argued that while the legal outcome of the political debate varied between the Australasian jurisdictions, all are likely to continue to work towards a harmonization of their legislations. This chapter also shows how the liberationist nature of the campaign has

D. Delafenêtre (✉)
Simon Fraser University, Burnaby, BC, Canada
e-mail: caodaserradeaires@caodaserradeaires.ca

C. Gigliotti (ed.), *Leonardo's Choice*, DOI 10.1007/978-90-481-2479-4_11,

contributed to the non-speciesist vision of the animal protection movement. As it happens the enactment of bans against ear cropping and tail docking coincides with the emergence of genetic technologies that have actually revived threats to the bodily integrity of dogs. This case study addresses the ethical implications of transgenics in the context of anthrozoology.

Keywords Cosmetic surgery · Mutilation · Dogs · Animal protection · Genetics

Introduction

The term "cosmetic surgery" refers to a series of practices that lead to the permanent mutilation of companion animals. Tail docking and ear cropping are the most common procedures performed on certain dog breeds.[1] The late twentieth and early twenty-first centuries have seen increasing efforts to ban both practices in Western countries. By deliberately interfering with the normal physiological, behavioural, and anatomical integrity of the animal, these purely cosmetic surgical procedures can be compared to genetic manipulations. Therefore this anthrozoological study seeks to inform the debate on bioethics and animals by examining how publicly elected institutions have come to legislate a ban against unethical procedures. The bulk of the scientific literature has convincingly demonstrated that contemporary ear cropping and tail docking are mostly done for aesthetic reasons (Broughton, 2003). Led by the veterinary profession and animal advocacy groups, the international campaign to end non-therapeutic surgical operations on the canine species has met with a lot of resistance in the world of purebred dog activities. The concept of cognitive dissonance has been successfully applied to those who continue to deny the overwhelming evidence against non-therapeutic surgical procedures on companion animals (Bennett and Perini, 2003b). The shared past of New Zealand and Australia offers an interesting setting to present a comparative perspective on a legal aspect of animal protection history.

The parallels between the Australian and New Zealand campaigns to end cosmetic surgery are so numerous that it can be said that they are completely interrelated. This chapter looks at the stage of the transnational campaign when it reached the formal political arena. So our aim is not to discuss the social dimensions of the activism involved in the pro- and anti- cosmetic surgery lobbying but rather to examine how the debate was translated at the legislative level. It is possible to identify three stages in that process. The chapter starts with a discussion on the historical background to the Australasian campaign. It shows that the British legacy over ear cropping influenced the way Australian and New Zealander policy makers approached a ban on cosmetic surgery. We then proceed with a discussion of the Australasian parliamentary debates that took place over the issue of tail docking in both Trans Tasman nations. The tail wagging bills that emerged between 2000 and 2004 reflected the complexity of legislating uniform regulations compatible with the ethical principles of the campaign. Finally, we reflect on the broader implications of the Australasian campaign for the anti-speciesist movement. It is argued that

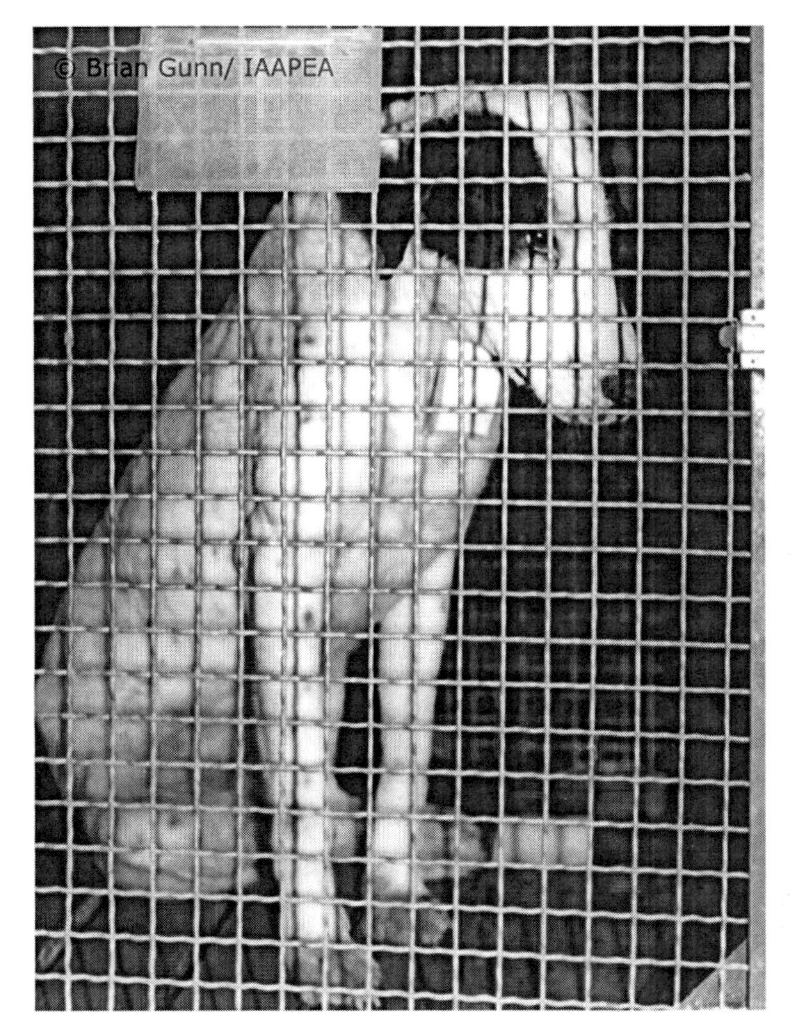

Fig. 1 Dog whose vocal chords have been severed to keep him quiet (credit: © Brian Gunn/IAAPEA)

the historical, political and sociological findings of the present study should prove useful to address the controversy over the use of genetic technologies on animals.

Lending an Ear

Archeological evidence suggests that the canine species was the first domesticated animal by humans. By the same token selective breeding has ancient origins and is clearly visible in the variety of characteristics (size, shape, colour, hair) of the *canis familiaris*. A huge diversity of breeds are included under the term as the result of artificial selection. Humans have a long history of altering the natural aspect of dogs by performing mutilations. Ear cropping and tail docking were the most common practices inflicted on the canine species. Those mutilations were thought to prevent disease or injury. Sometimes removing body parts was actually legislated such as the British law that used to require working dogs to be tail docked to avoid a tax. This law was repealed in 1796 but mutilations continued for certain breeds because they were so ingrained in anachronistic traditions. In Australasia, a formal campaign to end cosmetic surgery started two centuries later. The Australian Veterinary Association [AVA] officially launched a formal campaign in 1998 by issuing the following statement:

> Cosmetic tail docking and ear cropping are unnecessary, unjustified surgical alterations and are detrimental to the animal's welfare. The Australian Veterinary Association therefore recommends that: Kennel Control Councils throughout Australia phase tail docking requirements out of the relevant breed standards and Federal and State authorities and animal welfare organisations have tail docking and ear cropping declared to be illegal acts in all States and Territories except where performed for therapeutic reasons. (AVA, 1998)

This position statement sums up very well what the aims of the natural lobby group were at the end of the twentieth century. It also reflected the specificity of the Australasian situation. At the end of the nineteenth century, Australasian canine associations had removed the ear cropping requirements from their standards following a similar decision by the Kennel Club in Britain. The latter had declared cropped dogs ineligible for competition from 9th April 1898 in response to a 1895 letter from the Prince of Wales expressing his dislike of the practice (The Kennel Club). As Patron of the Kennel Club, the views of the future King Edward VII held some weight particularly since the Royal Family had been involved in the development of purebred dog activities. The British monarch was not the only celebrity of the time to take offense at ear cropping. The famous animal painter of the Victorian era, Sir Edwin Henry Landseer would not paint an animal with cropped ears: he considered the animal injured "in point of health and beauty" (as cited in Niven, 1967, p. 77). His contemporary, lay veterinarian William Youatt called tail docking a "barbarity" (Morton, 1992, p. 2). However outlawing those mutilations never seems to have been a priority of animal protection societies at that time. Concerns over ear cropping and tail docking developed indirectly since both were routine practices used in dog fighting circles. But it is the exploding dog fancy of the second half of the nineteenth century that would legitimize cosmetic surgery.

In fact, nineteenth century Britain became the first country where ear cropping and tail docking were formalized. Breeding dogs became a more popular activity by mid-nineteenth century when prosperous and leisured Victorians developed a passion for the exhibition of purebred dogs (Ritvo, 1987). The term purebred is misleading in that respect because all the registered breeds of canine organizations today are historically the result of mixed breeding. In the second half of the nineteenth century though, dog fanciers would establish the foundations for genetic "purity." As a result, certain procedures such as ear cropping and tail docking eventually became "trademarks" of several breeds. Through the adoption of written standards, mutilations were therefore institutionalized at conformation events held by canine organizations. The British passion for purebred dog activities was very quickly transplanted in the Australasian colonies. In 1862, the first conformation event took place in Hobart, Tasmania. The traditions and rules of the Kennel Club founded in 1873 were strictly followed in the seven Australasian colonies.

So it is not surprising to see Australasian canine associations adopt the British model. Yet the British attitude towards ear cropping was then the exception rather than the rule in international cynology. In the rest of Europe, canine organizations continued to standardize both ear cropping and tail docking. In Australasia only the second of the twin standards of cruelty, tail docking, became officialized. Neither the Kennel Club nor its Australasian followers can be credited for having banned ear cropping but over time the frowning upon the practice influenced community attitudes. By the time the campaign to end cosmetic surgery was launched in Australasia, ear cropping was considered an unethical and unacceptable surgery as illustrated by the debate that took place within the New Zealand veterinary profession in the 1980s (Gumbrell, 1984). The 1998 position statement of the AVA also took into account the more recent developments that had challenged cosmetic surgery in other parts of the world. (AVA, 1998).

In 1987, the European Council passed the European Convention for the Protection of Pet Animals [ECPPA]. Article 10 specifically dealt with surgical procedures and was worded so as to place the emphasis on the prohibition of surgical interventions which were mainly carried out for aesthetic reasons (Council of Europe, 1987). However, in the lead to its ratification, the ECPPA was amended to allow signatories to make a reservation on tail docking.[2] This confirmed the fact that traditionally the case against tail docking has proven more difficult to make than the one against ear cropping. Internationally though, the ECPPA did a great deal to support the anti-cosmetic surgery campaign. By the time the ECPPA came into force in 1992, Australasia had already experienced efforts to alert politicians on the issue of cosmetic surgery. For example, the debates over Queensland's green paper on animal welfare mention a call to ban tail docking (Queensland, 1991a, p. 6257; Queensland 1991b, p. 8312). However, there was still the belief that puppies a few days old did not experience pain during the procedure. This resulted in legislation limiting tail docking but not banning it. For example, the Animal Welfare Act passed by the Australian Capital Territory [ACT] in 1992 allowed tail docking to be performed on puppies under 10 days of age (ACT, 1992a). In June 1997 New South Wales parliament discussed a ban on cosmetic docking (New South Wales, 1997, p. 11177) but tail docking continued to be allowed for puppies under 5 days of age. The involvement of the New Zealand Veterinary Association [NZVA] started around the same time with preparations for the Animal Welfare Bill. 1997 also marked the creation of the New Zealand Council of Docked Breeds [NZCDB].

In the first phase of campaigning, it seems that the animal advocacy groups had an earlier involvement or at least were more vocal than the veterinary profession (Clarke, 2001). Faced with the possibility of a ban on tail docking in 1998 the Australian National Kennel Council [ANKC] responded by reminding the major players of its opposition to ear cropping:

> The Australian National Kennel Council and its Member Bodies abhor the practice of cropping of ears of any breed of dog. The act of ear cropping is an offense under the Rules and Regulations of the Australian National Kennel Council and its Member Bodies (ANKC, 1998).

Needless to say, the New Zealand Kennel Club was prompted to do the same when in December 1997, the Animal Welfare Advisory Committee had expressed its opposition to the practice of tail docking in the Code of Recommendations and Minimum Standards for the Welfare of Dogs (New Zealand, 1998). Published in 1998, the code had been influenced by the position of the NZVA. During the consideration stages of the Animal Welfare Bill, the Primary Production Committee considered at length a proposal to prohibit tail docking but it was unable to agree on the inclusion of a clause that would have prohibited the docking of dogs' tails from 1 January 2003. In 2004, the Minister of Agriculture, Jim Sutton, recalled the circumstances,

> When that Act was in the course of passage, the issue of docking dogs' tails was debated before the Primary Production Committee, and in the House at some length. I must say that while the weight of the evidence from the veterinary profession and animal welfare groups was strongly in favour of the prohibition of dog tail-docking for any reason other than therapeutic reasons—that is, for an injured, malformed, or diseased tail, or in the interests of the animals—the scientific evidence in respect of the effect of docking on the animals themselves was equivocal at that time. So the committee, of which I was a member, came to quite a finely balanced decision, which the House supported, to allow the status quo to prevail. The status quo was passionately supported by a number of breed organisations that claimed that tails could get damaged when hunting. (New Zealand, 2004a)

The lobbying of the artificial breeding community had won the first round on the tail docking issue in New Zealand even though the Code of Welfare of Dogs had explicitly issued recommendations against tail docking. In this respect, New Zealand was the first Australasian jurisdiction to consider a complete ban on cosmetic tail docking. Nevertheless the Animal Welfare Act adopted just before the defeat of the National government had the merit of providing a definition of ear cropping: "Crops, in relation to the ears of a dog, means the performance on the pinnae of the ears of the dog of a surgical procedure that is designed to make the ears of the dog stand upright" (New Zealand, 1999). The need to legally define cosmetic surgical procedures would become clear in the aftermath of the tail wagging bills discussed below. At any rate, New Zealand was seen then as the leader of the Australasian campaign to end cosmetic surgery on dogs.

Across the Tasman the tail docking campaign also became associated with efforts designed to have the ear cropping of dogs made illegal in all Australian States and Territories. Overall, the campaign to ban ear cropping succeeded because like in New Zealand the practice had been severely frowned upon for a century. In 2001, a member of the Legislative Assembly of Queensland expressed surprise that

> animals in Australia could ever be cropped. We owned Dobermans years ago and they were never allowed to be cropped, but the Americans think that cropping looks good for some reason. Therefore, I was surprised to see that it is still listed in the bill, but if it were not listed some dog owners may wish to have it done because they like the look of the dogs in America when they stand to attention on guard. (Queensland, 2001a, p. 2938)

While the campaign to outlaw ear cropping was a relatively easy venture due to the support of the purebred canine associations, the first phase of the campaign to end cosmetic surgery revealed certain elements of concern.

In general the consensual opinion over ear cropping made legislators overlook certain aspects of the legislation designed to ban the practice. For example, the Queensland Care and Protection Act of 2001 states that "A veterinary surgeon must not crop a dog's ear unless the surgeon reasonably consider the cropping is in the interests of the dog's welfare" (Queensland, 2001b). This wording could be interpreted as allowing prophylactic ear cropping even though it was considered an offense of cruelty in the previous legislation, the Animals Protection Act (Queensland, 1995). The same problem of wording applies to the Northern Territory's legislation passed in 2000 that restricts the procedure to a veterinarian "if he or she is of the opinion that it is reasonable and necessary to do so" (Northern Territory, 2000). According to the Animal Welfare Acts of Western Australia and Tasmania the act of mutilating an animal is considered an offence but contrary to other Australasian jurisdictions no provision is specifically made to deal with ear cropping. For its part, the New South Wales prohibition of ear cropping in place in 2000 was the closest to the one adopted in New Zealand. The same year the South Australian government had restricted ear cropping to therapeutic purposes. Finally in 2001 the ACT legislation imitated the provisions of the Victorian Prevention of Cruelty to Animals Act that restricted ear cropping to prophylactic and therapeutic reasons. In both jurisdictions, the prophylactic purpose was only removed in 2007.

In the end, the legal confusion existing over ear cropping can appear insignificant due to the generally accepted view held in Australasia that the practice is considered unacceptable except for therapeutic purposes. Limited public pressure had achieved legislative change without much political debate. However, the realization that legislation banning ear cropping differed across Australasia could have helped the anti-cosmetic surgery lobby during the second phase of the campaign that was completely dedicated to achieving a ban on tail docking.

The Tail Wagging Bills

The campaign to ban non-therapeutic tail docking in Australasia from 2000 to the present differed substantially from the one that ended ear cropping in Australasia. For one, this time, tradition was on the side of the artificial lobby. Typical of the climate of the moment was the claim the NZKC made in 2004 that Queen Elizabeth II, Patron of the RSPCA and New Zealand Head of State who owns and breeds Pembroke Corgis (a traditionally docked breed) had stated, "As dog breeders we have been given a charter to maintain the appearance of the breeds as handed down by our forebears through the various breed standards." The NZVA was quick to respond that "Whatever HRH's personal opinions are, the law will reflect New Zealand society's values, not any one individual's" (NZVA, 2004). The second and more significant distinction was that between 2000 and 2004 all the Australasian parliaments would discuss bills dedicated specifically to the issue of tail docking. In Australia, this came as the result of a federal governmental initiative that had been absent during the campaign to ban ear cropping.

In 2000, the AVA issued a new policy statement similar to that of their New Zealand counterparts and focusing exclusively on tail docking. To the New Zealand veterinary profession, it was a question of unfinished business that needed to be revisited. Both the AVA and the NZVA built on the momentum of the first phase of the campaign this time assuming its leadership in close cooperation with the RSPCA and other animal protection groups in both countries. Both the ANKC and NZKC proceeded with challenging the AVA and NZVA by arguing there was nothing in their rules or regulations to prohibit a dog of a customarily docked breed from being exhibited with an entire tail. Most standards for traditionally docked breeds were amended to allow the non-docked option. So, on the face of it, the ANKC and the NZKC did try to send the message that they were open to two versions of the same breed. Therefore, they could not be accused of favouring docking. This was in line with the central argument used by the pro-docking lobby, that is, the so-called "freedom of choice" of breeders. Yet, it is easy to see how those measures eventually backfire for the artificial breeding community. By accepting to have the natural appearance of traditionally docked breeds standardized, those in favour of docking indirectly admitted that it was after all perfectly fine not to dock the very breeds that had been customarily docked. This in itself undermined the basic prophylactic argument that leaving a dog undocked was a threat to his/her welfare. To the opponents of docking, the move to allow both options confirmed what the scientific community had stated for some time, that is, docking was a purely cosmetic surgery without any benefit to the dog. So rather than appease the call for a ban, the "freedom of choice" campaign provided more ammunition for those opposed to cosmetic tail docking.

The AVA adopted the approach used by the NZVA in 1997 by working closely with the national animal welfare body. The National Consultative Committee on Animal Welfare [NCCAW] was the Australian equivalent to New Zealand AWAC that had been replaced in 1999 by the National Animal Welfare Advisory Committee [NAWAC]. The NCCAW had been successfully persuaded to recommend against cosmetic tail docking by the AVA (Clarke, 2001). In October 2000, its revised position stated that routine tail docking of puppies served no practical purpose and that the practice should be banned (Australia, 2000). The NCCAW envisioned a phase-out of tail amputation in all Australian jurisdictions. By then, the Australian Capital Territory successfully passed the Animal Welfare Amendment Bill. The ACT initiative to prohibit docking of dogs' tails for cosmetic reasons received bi-partisan support. Yet the legislation did not ban prophylactic docking (ACT, 1992b). Meanwhile, in 2001 the Convention for the Protection of Companion Animals [CPCA] adopted by the World Small Animal Veterinary Association of which both the AVA and NZVA are members called for legislation to be enacted to prohibit surgical operations for the purpose of modifying the appearance of a companion animal for non-therapeutic purposes (WSAVA, 2001). As for the European Council's legislation, the CPCA allowed signatories to make a reservation on tail docking.

Nevertheless, the Australian government took the initiative at the federal level to push for a ban on cosmetic docking.

In May 2002 tail docking of dogs was on the agenda of the first meeting of the Primary Industries Ministerial Council [PIMC] in Hobart (Australia, 2004). Interestingly, this federal body established in 2001 also includes New Zealand even though their representatives do not seem to have taken part in the discussions over a proposed national ban on routine tail docking of dogs introduced by the Victorian Minister for Agriculture. A background report on tail docking prepared by the NCCAW was circulated to all the ministers and in October 2002 the PIMC considered the recommendation to ban tail docking with a phase in period during which the procedure should be conducted by a veterinary surgeon. It was noted then that the ANKC had resolved to amend all breed standards for breeds with docked tails to have the words "preferably docked" removed. This move was undoubtedly designed to appease the overwhelming opposition to cosmetic tail docking acknowledged by the PIMC. The task of developing a nationally consistent approach to routine tail docking of dogs for cosmetic purposes was rendered more difficult since under the current Australian constitution, legislative responsibility for animal welfare within Australia rests primarily with state and territory Governments. Therefore all states and territories had to include the tail docking ban in their respective animal welfare legislation and were expected to use a legal mechanism similar to the one applied in the Australian Capital Territory in 2000. More importantly, the question was whether the rest of the Australian jurisdictions would adopt a tail docking amendment close to the model adopted by the ACT.

Despite the strong scientific case against cosmetic docking, some jurisdictions expressed a desire to consider their positions on the issue. In April 2003, the AVA accused Agriculture Ministers from New South Wales and the Northern Territory of stalling on the proposed tail docking ban (AVA, 2003a). Midway through 2003, the case put forward by the AVA and animal welfare organisations received a boost with the publication of a substantial study on tail docking. Bennett and Perini's (2003a) article published in the April 2003 issue of the *Australian Veterinary Journal* represented an updated and overwhelming case against tail docking. After a round of consultation, the PIMC decided in October 2003 that non-therapeutic tail docking of dogs would be prohibited and therapeutic tail docking of dogs would only be able to be carried out by a veterinarian (Australia, 2004). At first hand, this seemed consistent with the position advocated by the AVA but the Council's failure to define what it meant by non-therapeutic docking soon became problematic. Certain jurisdictions had actually proceeded to legislate a ban before the PIMC finalized its position. Provisions for prophylactic tail docking were also made without the benefits of the conclusions made by Bennett and Perini (2003a).

In November 2002, on the heels of adopting its new Animal Welfare Act, Western Australia had passed its own piece of legislation (Western Australia, 2002a). Inspired by the ANKC Code of Practice for the Tail Docking of Dogs it allowed a veterinarian to dock the tail of a dog if it was deemed necessary for therapeutic or prophylactic purposes only (Western Australia, 2002b, p. 2851). By June 2003, the Northern Territory confirmed it would move to ban cosmetic tail docking following intense pressure from the AVA at its annual conference in Cairns (AVA, 2003b). However, the debates that took place in the Legislative Assembly indicate that the Northern Territory changed its initial commitment and eventually proposed a less restrictive legislation that would have included a prophylactic or preventative clause (Northern Territory, 2004). Subsequently in February 2004, the Northern Territory passed regulations that could be interpreted as making prophylactic tail docking legal. This conformed with the regulations in place in the ACT and confirmed that preventative or prophylactic tail docking was not interpreted as equal to cosmetic surgery in some jurisdictions.

The confusion over whether prophylactic docking legally constituted cosmetic docking was best illustrated in the case of Queensland. Interestingly enough, Brisbane had been leading the push for a national ban and was quick to pass regulations effective 27 October 2003. However, the Queensland legislation was a setback due to the wording on the restrictions on tail docking: "A veterinary surgeon must not dock a dog's tail unless—(a) the surgeon reasonably considers the docking is in the interests of the dog's welfare..." (Queensland, 2001b). The phrasing was not clear and it came as a disappointment to the AVA who had fully endorsed Queensland's call for a national ban earlier in the process. According to the then National President of the AVA, Dr. Jo Sillince: "Unfortunately in some States there is a loophole in legislation that allows tail docking for preventative reasons and it must be closed otherwise it makes the gesture meaningless" (AVA, 2003c). Consequently, the AVA worked hard on lobbying the other Australian jurisdictions that had yet to legislate a ban.

While comprehensive, the consensus reached by all Australian jurisdictions in 2003 did not deliver a coherent ban. The PIMC's failure to properly coordinate across all Australian states and territories ruined the objective of having a consistent ban. The ACT legislation had set a precedent that influenced the early parliamentary debates. Warning against such loopholes became a priority of the AVA which was more successful with the pressure it exerted on the South Australian government to allow tail docking for therapeutic purposes only. As in other jurisdictions, the debate in South Australia highlighted the fact that the pro-docking lobby represented a minority voice:

> The very few people who make a lot of noise about continuing this barbaric practice are mainly animal breeders. I saw a sticker on the back of a car the other day that said, 'I'm pro tail docking.' I do not think the owner of the car was speaking for the dog in the back. It is barbaric to chop off a dog's tail; this procedure has no place in the year 2003. It has been the tradition in the past to dock dogs' tails for a number of reasons, all of which can be discounted. There is no health reason; there is no safety reason; there is no valid reason whatsoever for chopping off a dog's tail. (South Australia, 2003b, p. 750)

South Australia made it illegal for any person to dock the tail of a dog for non-therapeutic purposes. However, a mechanism allowing a review of the general prohibition with respect to individual breeds was adopted. If valid statistics can demonstrate satisfactorily to a properly constituted panel that a particular breed should be allowed to be docked because of hygiene or injury concerns, that breed may be docked as a preventive measure. The AVA turned its attention to Victoria where despite strong lobbying from the pro-docking breeders, the government eventually adopted a piece of legislation similar to that of South Australia but without the possibility for breed clubs to seek an exemption. The Victorian Canine Association submitted its recommendations for acceptance of the proposed Western Australian legislative model whereas the Dog Body, another pro-docking group, made a less restrictive proposal (Victoria, 2003, p. 1324). But where the tail docking bill enjoyed bi-partisan support in Victoria, it became a battleground in the state of origin.

Early on New South Wales had made it clear that they would initiate a period of consultation before agreeing to pass the necessary legislative mechanisms to enact a ban. New South Wales was therefore the last jurisdiction to commit to the national ban. This was reflected in the sophistication of the debate. The government's position reflected the dilemma legislators had faced in other jurisdictions. Speaking for the bill, Mr. Ashton insisted that

> ...therapeutic docking has been defined as clinical treatment in response to a pre-existing and otherwise untreatable injury or ailment and that the term "non-therapeutic docking" includes tail docking for any other purpose, including prophylactic, cosmetic, or breed standard purposes. (New South Wales, 2004a, p. 8075)

This understanding corresponded to the one that prevailed in both South Australia and Victoria and assumed that the term therapeutic excluded prophylactic docking. Like the Queensland's piece of legislation though, the wording of the tail docking amendment in New South Wales was not specific enough. In particular, the terminology "in the interests of the dog's welfare" was not defined. Nevertheless, the government had managed to pass the bill despite the strong opposition of the Liberal–National coalition.

Tasmania ended up being the last Australian state to implement the ban. In June of 2003, the Tasmanian government had decided to follow whatever would come as the result of the PIMC process. Therefore it proceeded with a ban on therapeutic tail docking only after New South Wales' legislation came into force. The Tasmanian regulations that came into force in July 2004 were similar to the ones adopted earlier by Victoria and confirmed that if Australia had now a national ban on tail docking, each state and territory had a different version. So, after two years of debates, the eight Australian jurisdictions had complied with the federal move. While the result fell short of a completely satisfactory result, it did at least help the anti-docking lobby in New Zealand on two levels. First, the Australian scene provided additional evidence that tail docking was not justified. Since restrictions on tail docking had to be passed into law, the parties in power had to argue their case and a close examination of the Australian parliamentary debates show that they were very articulate in doing so. Second, the limitations of the Australian bills suggested that although it had been possible to put the specific issue of tail docking in politics, the wording of the legislation needed to be carefully studied. Even before the tail docking ban was enacted in several parts of Australia, the NZVA had managed to convince a Labour MP to launch a private members' bill on tail docking. New Zealand became then the ninth Australasian jurisdiction to consider banning non-therapeutic tail docking.

The extent of the divide between the natural and artificial lobbies would again be reflected in the political arena but this time, Labour was in power.

The "tail wagging" bill as it became known was launched in Auckland on 1 February 2004 and, compared to the Australian experience, it represented the best worded version of the intent of legislation: Section 17 (2) of the Animal Welfare Act was to be amended by adding after the words "interests of the animal" the words

> which, in the case of tail docking, is based on an informed veterinary opinion that the procedure is necessary for the welfare of the animal where the tail has been damaged by injury or disease, and is not being performed for cosmetic or prophylactic purposes. (New Zealand, 2004b)

MP Dianne Yates introduced the restriction on Docking of Dog's Tails bill in Wellington on 5th August 2004. She insisted on the need to be in line with the Australian legislations. On the occasion, the spokesperson for NZVA Animal welfare, Dr. Virginia Williams, referred to the AVJ paper in her support for the proposed legislation to ban cosmetic and prophylactic tail docking in dogs. In fact, the proposed bill supported the veterinary Council of New Zealand Code of Professional Conduct which states that veterinarians should not perform surgical procedures for purely cosmetic reasons. The first reading took place on 21 October 2004 and the matter was referred to a select committee for review. As in New South Wales, the Official Opposition made it clear it would not support the bill. The National Party along with a number of other parties was against the proposed amendment to the Animal Welfare Act arguing that the issue had been put to rest in 1999.

Eventually the bill was sent to the Government Administration Committee instead of the Primary Production Committee as in 1999. However, a quick look at the composition of the committee shows that the bill did not stand much chance of returning to the House (New Zealand, 2007d). While the six-member committee was equally divided between Labour and National, it was chaired by Shane Arden, one of the National MPs who had spoken against the bill during its first reading. Shane Arden had also held several positions with Federated Farmers. What could be characterized as the second phase of the campaign to end cosmetic surgery in Australasia thus ended ironically. The Bill that reflected more accurately the objectives of the anti-docking lobby was to languish in a committee for three years during which the Labour Party became embroiled in another controversy over a legislation dealing with the micro-chipping of dogs. By the time the New Zealand Select Committee returned its decision not to proceed with the tail wagging Bill in August 2007, the campaign had already moved into another phase in Australia. In the ensuing 4 years after its enactment, the national ban was put to the test. The eventual defeat of the New Zealand bill seems to have motivated the pro-docking lobby in certain Australian jurisdictions. Several Australian states and territories had to revisit their tail docking regulations to fix the legal loopholes that came to threaten the very spirit of the tail wagging Acts.

A Tale Worth Hearing

Tasmania was the first Australian jurisdiction to have its legislation challenged even though it was the last to pass it. In November 2004, the Legislative Council of Tasmania heard an attempt to have the tail docking regulations dismissed. At least, the move had the benefit of initiating a debate in a state where the tail wagging bill had received little attention. Once again, the natural lobby found a politician, this time in the person of Ms Ritchie, to point out the illogical nature of the pro-docking argument:

> One cannot argue that the creation of a distinctive looking animal by surgical means can then be called natural or traditional even if the process did start over 2000 years ago... I do not believe that we have the right to physically manipulate an animal. (Tasmania, 2004, p. 76)

In Tasmania, a review of the Animal Welfare Act, 1993 published in July 2006 actually further tightened regulations on docking dogs' tails. It recommended the offense be one of docking a dog's tail or permitting, allowing, or causing a dog's tail to be docked (Tasmania, 2006). It is noteworthy that most jurisdictions would respond to challenges to their tail docking regulations by improving and clarifying their legal extent. Since 2004, there has been no reversal of existing legislation on tail docking even if the less restrictive regulations have also been targeted by the artificial lobby.

In April 2004, the Minister responsible for the Animal Welfare regulations of 2003 in Western Australia reiterated his government's position that the decision to dock was left to the veterinary profession and that the State had been at the leading edge of Australia on tail docking (Western Australia, 2004, p. 1645). At least, Perth resisted the calls of the Canine Association of Western Australia to review the tail docking legislation. CAWA reminded dog owners that there had been no ban on tail docking in Western Australia, just restrictions, and added that

> We therefore wish to see the regulations changed so that these qualified people may do this crucial animal husbandry procedure without the threat of action taken against them and in the case of vets, misconduct charges hanging over their heads. (CAWA, 2004)

For its part, in 2007, the ACT modified the 2001 amendment to its Animal Welfare Act to reflect not only the ban initiated by the PIMC but the developments that had taken place since 2004. In addition to removing the prophylactic option, a clause stating that "a veterinary surgeon must not carry out a medical or surgical procedure on an animal for a cosmetic purpose only" (ACT, 1992c) was added. While the Northern Territory did not change its legislation, it is worth noting that it was one of the few that had included a definition of docking. The lack of such provision meant that some tail wagging Acts would be tested in the courts.

It appears that New South Wales expected legal challenges to its legislation even before it came into force. Therefore Sydney undertook an information campaign coupled with a policy of enforcement. The New South Wales government proceeded to clarify its legislation: "Veterinarians, breeders and owners should be aware that the intention of the legislation is to ban tail docking which is done for 'routine', 'prophylactic' or 'cosmetic purposes, as is traditional with many dog breeds" (New

South Wales, 2004a). The intensity of the New South Wales debates over tail docking discussed earlier prompted the government to publish a second information sheet detailing the ban and its rationale. In particular, New South Wales pointed to the combination of two factors: the weight of scientific opinion about the negative effects of tail docking on dog welfare and changing community expectations about unnecessary surgical procedures on animals (New South Wales, 2004b).

Nevertheless, the pro-docking lobby was quick to seize on the weaknesses of the legislation. Several court cases in New South Wales gave an opportunity to those opposed to the turning point of 2004 for challenging it. In November 2006, one case in particular drew the attention of Dogs New South Wales. A judge argued that "Parliament did not intend to include banding in the ban of docking within legislation which has the prime purpose of preventing cruelty" (Dogs NSW, 2007). In 2006 Dogs New South Wales enlisted the support of the official opposition by trying to make a distinction between tail docking and banding. The latter refers to the use of a tight rubber band that cuts off blood circulation and causes the tail to die. As such, it is plainly another form of tail docking since it leads to the amputation of the tail. Yet, the fact that the procedure was promoted by the NZKC in its submission to the New Zealand parliamentary committee in 2005 gave another reason for New South Wales' Deputy Opposition leader Duncan Gay to express his support for tail docking (Dogs NSW, 2006). In a position statement issued on January 31, 2007, the president of Dogs New South Wales proposed a change to the legislation adopted in 2004 by allowing tail banding. Along the same line, the ANKC issued a statement on tail shortening on 15 December 2006 stating that it regarded "regulated, professional tail shortening of the newborn pups of certain dog breeds as not being cruel" (ANKC, 2006). This new terminology was being used to avoid the negativity attached to docking. The lack of definition of the latter in several pieces of legislation has made it easier for the pro-docking lobby to challenge the ban passed in 2004. Yet, these attacks on the national ban incited some jurisdictions to go back to the drawing board and improve their legislation. For example, New South Wales, in June 2007, amended its tail docking legislation by defining tail docking as the removal of all or part of the tail, by surgical or other means. This rendered illegal the act of docking a dog's tail using a rubber band. Tail docking was also added to the list of procedures that veterinarians need to record in a registry (New South Wales, 2007).

Obviously, due to the similarity in the wordings of their legislation, it would be possible for Queensland to adapt its legislation to that of its southern neighbour. Despite winning an Animal Welfare League Award for its stand on tail docking, the Queensland government was forced in 2005 to issue a guide to its ban on tail docking dogs that stated that: "Veterinarians are being advised that it is generally considered inappropriate to dock a healthy tail on the basis of a possible future event. Only therapeutic tail docking is justifiable" (Queensland, 2005). Yet the new regulations are still left to legal interpretation: "If a case comes to court, the decision on whether or not an offence has been committed will be decided by a magistrate" (Queensland, 2005). However, there is no indication that a court case has yet challenged the need to define the term "docking" in Queensland. In the meantime, it seems that the Queensland government is hoping that the prospect of enforcement will be sufficient enough to deter any attempt to contravene the law: "although each case will be considered on its merits, animal welfare inspectors would normally investigate any docking of a healthy tail with a view to potential prosecution" (Queensland, 2005). While the legislation as it stands leaves too much room for possible challenges, one can assume that the experience of other jurisdictions will eventually incite Queensland to review its tail docking regulations. The lack of a definition of docking does not seem to have been problematic either in South Australia.

As we have previously seen, the South Australian legislation was accompanied by the creation of a tail docking committee whose functions were to advise the Minister on any matter relating to tail docking and to consider submissions from persons who consider that for prophylactic purposes a specific breed should be docked. In four years only one breed club, that of the Old English Sheepdog sought a prophylactic exemption and it was rejected (D. Kelly, personal communication, May 28, 2008). Because the tail docking policy is not a regulation, it is possible to envision the prospect of easily removing it altogether. Considering the instruction that any recommendation to exempt a breed from the prohibition of tail docking would have to be based on statistically sound evidence, not anecdotes, policies or personal beliefs, it seems unlikely that the five-member tail docking advisory panel could issue an exemption. Nevertheless Adelaide could possibly improve its legislation based on the experience of its eastern neighbour.

Since 2004 the Victorian government has had to pass several amendments to its tail docking restrictions to make the regulations more enforceable. In 2007, Melbourne adopted the Animals Legislation Amendment (Animal Care) Act 2007 introducing the offence to show or exhibit an illegally tail-docked dog (Victoria, 2007). This came as a result of routine inspections performed at conformation events that indicated that tail docking was still being carried out in violation of the 2004 legislation. Dogs Victoria responded in such manner as to remain within the parameters of the state law by issuing a Docked Tail Dog Identification Card to its members (Dogs Victoria, 2007). This card is to provide documentation that their dog's tail was docked for medical/therapeutic reasons in accordance with the local jurisdiction of the state where it was whelped. However, the Victorian amendments which came into force in March 2008 do not address the issue of docking tourism. For dogs who have been docked outside of Victoria, allowance has been made for the dog to be shown provided the procedure was done in accordance with the legislation of the jurisdiction (country/state/territory) in which the procedure was carried out. A more aggressive attitude would have imposed a complete ban on the showing of tail-docked dogs born after the regulations were put into place regardless of the place of birth.

Four years after the national ban was supposed to take effect it is fair to say that the campaign to end cosmetic surgery in Australia is not over. The loophole for prophylactic docking in Western Australia and the lack of clarity of the Northern Territory's legislation has enabled pro-docking breeders to circumvent the ban in other Australian jurisdictions. In many respects, the Queensland and New South

Wales legislations are not as restrictive as the South Australian and Victorian pieces. For a national ban on tail docking to be effective it would require all states and territories to harmonize their legislation with a wording similar in intent to the New Zealand bill of 2004. The nationally coordinated state and territory ban failed to deliver a coherent legislation across Australia because it did not pay enough attention to the definition of the procedure itself. Yet, recent developments point to the fact that community attitudes are changing due to the simple reality that more and more undocked dogs of previously docked breeds are being seen in the streets and parks of Australia.

On the whole, the anti-docking lobby should remain vigilant and monitor the situation as well as put pressure on the relevant administrations to make sure that docking tourism is targeted. Hopefully the Australian Animal Welfare Strategy (AAWS) whose task is the development of nationally consistent welfare policies in all Australian states and territories will contribute to harmonize state and territorial legislation on tail docking. At its meeting in Canberra on June 8, 2007, the new National Consultative Committee on Animal Welfare [NCCAW] discussed the issue of prophylactic tail docking of dogs, specifically in Western Australia (Australia, 2007, p. 8). It is time to eliminate the discrepancy between NCCAW guidelines and state and territory legislations. Since 2004, the Australian experience has continued to influence the debate across the Tasman. In a press release dated 05 November 2006, the NZCDB hailed the New South Wales court ruling as a "landmark decision" that augured well for New Zealand pro-docking dog owners (NZCDB, 2006). Yet the fact that New Zealand currently remains the only Australasian jurisdiction without a ban on tail docking does not mean its proponents have been silenced. An attempt is currently under way to make New Zealand legislation compatible with Australian regulations on tail docking.

After the failure of the proposed bill presented by Dianne Yates in 2004 the issue has resurfaced with the draft Code of Welfare for Dogs released for comments on September 17, 2007. NAWAC initiated a period of consultation that closed on 1 November 2007. At the time of this writing, public submissions were being summarized for the new code to be presented in the second half of 2008. Significantly, the draft code was written by a group convened by the New Zealand Companion Animal Council, an anti-docking group, in consultation with NAWAC (New Zealand, 2007a). In announcing the release of the draft code for comment, Dr. Peter O'Hara, Chairman of NAWAC indicated that

> Tail docking is likely to attract interest. NAWAC felt that, in light of the Private Member's Bill recently being dropped, there was a need to introduce some regulation around this practice. Essentially, the potential for pain might not be outweighed by non-medical reasons for removing the tail. (New Zealand, 2007b)

Needless to say, the NZKC voiced its opposition to the proposed draft and accused NAWAC of hijacking the consultation process that led to the draft. However, while calling disingenuous NAWAC's announcement that the NZKC had been involved in the preparation of the draft, the NZKC added that NAWAC and proponents of a ban on docking had failed to promote voluntary changes by persuading anybody to not dock and had not "considered the feasibility and practicality of effecting a transition from current practices" (NZKC, 2007). This denotes an awareness of the possible implications of having Australian states and territories getting their acts together on tail docking. This would definitely put pressure on New Zealand to review its legislation. The minimum standard proposed in New Zealand is actually more specific than the different Australian legislations: "Tail docking must not be performed on puppies or dogs, except where required to manage existing damage or disease to the tail or malformation which harms animal welfare" (New Zealand, 2007c). In addition, NAWAC comments are clearly geared towards the opponents of a ban: "The alternatives to tail docking are amending breed and judging standards to accept entire tails, removal of the tail only following any injury or disease, and attention to hygiene and regular care" (New Zealand, 2007c). That brings us back to the original aim of the campaign that all canine organisations in Australia phased out any recommendations for tail amputation from their breed standards. None of the Australian tail wagging bills address the issue of canine associations still having tail docking in their standards which *de facto* contravenes the very spirit of the tail docking ban. Meanwhile, it is too early to determine the contents of the final draft of the New Zealand Code of Welfare for Dogs and, more importantly, whether it would be accepted by the Regulations Committee. The prospect of a National government at the next election could ruin the efforts of NAWAC due to that party's pro-docking stand.

In order to understand the opposition the tail wagging bills received in Australia and New Zealand, one needs to go beyond the issue at hand. Part of the objection raised against banning tail docking in dogs had to do with the implications the legislation could have on farming practices. In New Zealand, the Nationals accused the Labour government of having ulterior motives:

> Can they (Labour MPs) stand in this House and say to the farming industries of New Zealand that this is not a Trojan Horse to bring in some back-door legislation that will put pressure on sheep farmers for docking sheep, dairy farmers for docking cows, and horse breeders for docking horses? (New Zealand, 2004a)

What seems to have worked against the New Zealand bill is the ability of the pro-docking interests to lobby the official opposition. The New Zealand failure to pass its tail wagging bill can be attributed to the coalition that the NZKC was able to build with the hunting and particularly the farming interest groups. Parliamentary opposition to the bill went further than repeating the arguments of the artificial lobby. If the issue of tail docking became so divisive in Australasia it is not only because the conservative pro-docking lobby did not want to have the activity of breeding regulated or even questioned. The reactionary attitude displayed by conservative interest groups such as the farming and hunting lobby in the parliamentary debates was reflective of the fear from those groups that banning tail docking on dogs was the beginning of an attack on their "rights" to exploit non-human animals. That is why the Australasian tail wagging campaign is relevant for the animal advocacy movement.

Indeed, if governments in charge of the tail wagging bills took pains at rejecting the charges of trying to expand the legislation to other animals, it would be a mistake to overlook the ideological implications of the legislations. In Australasia, animal rights groups such as Animals Australia have been prominent in their support of the campaign against cosmetic surgery. As a result, the lobbying has been more powerful and has brought significant results. The key here is that while anti-speciesist groups are well aware of the welfarist nature of animal welfare governmental institutions or even of the veterinary profession, they have come to understand the broader picture. By getting a ban on cosmetic surgery for dogs, they can and have challenged similar practices inflicted on other animals. Similarly, to preserve the bodily integrity of dogs is not limited to phasing out mutilations inflicted upon them but it includes their right to life since some countries still abuse them for food or fur. Also, the achievements of the anti-cosmetic surgery lobby in Australasia ultimately have some positive consequences for dogs subject to similar mutilations in other parts of the world. By preventing humans from altering the physical aspects of animals one indirectly goes to the heart of the problem of exploiting animals for food, fur, and a myriad of other uses. . . As one New South Wales MPs put it: "The bottom line is that we should leave their bottoms alone" (New South Wales, 2004c).

The case of cosmetic surgery is particular in the sense that dogs are not considered food animals in Western societies. What has happened over the past two decades or so is a recognition that the physical appearance of dogs should not be altered. It is pretty obvious that the manual mutilations of the past are connected to the genetic manipulations of today. The reality is that canine organizations are no longer seen as the legitimate representatives of the dog lobby and have been successfully challenged by other organisations such as the NZCAC.

At the end of the day, what could have initially appeared as a mostly welfarist form of activism falls within a liberationist framework similar to the one conceptualized by Munro in his comparative study of the animal advocacy movement in Australia, the United Kingdom and the United States. In this regard, the Australian sociologist reminds us of the anti-speciesist framework of animal protectionism: "While the movement ideologically is divided between welfarist, liberationist and rightist traditions, these different strands are held together by the movement's central campaign against speciesism" (Munro, 2005, p. 189).

Conclusion

This chapter has shown how the ethical principles of the campaign to end cosmetic surgery on dogs have been translated into law. It remains to be seen whether the activists will completely achieve their goal in Australia but the lessons learned there could help New Zealand achieve a more effective ban on cosmetic tail docking. Recent studies should also help the arguments of the natural lobby (Leaver and Reimchen, 2008). At any rate, this case study should serve as a reminder that genetic manipulation has an old history in cynology. The abuse resulting from recessive gene technology used to produce bobtail specimens did not escape the attention of politicians when they discussed the "dog fancy":

> There are genetic deformities and usually there is a really good reason for a dog having a very shortened tail or, in some cases, appearing to have no tail. The Australian Shepherd and the Pembroke Corgi are two breeds that appear to have no tail, but they have very deformed and very short tails. If you X-rayed them, you would see deformed coccygeal vertebrae, the remnant of where the tail should be; most of it has gone because of the selective breeding. (South Australia, 2003a p. 751)

Interestingly the campaign to ban cosmetic surgery coincides with the emergence of genetic technologies that have actually revived threats to the bodily integrity of dogs. The cloning of dogs (Best, 2009) is now a reality and it is possible to imagine how genetic manipulation could produce tailless dogs or breeds with artificial erect ears to bypass any legislation designed to protect their rights to keep their natural tails and ears. More generally, the ethical implications of domestication and breeding for "purity" in the present context need to be addressed (Spencer et al., 2006). Concerns over the bodily integrity of companion animals can and should be extended to other domesticated animals. In this regard, the ability to mobilize public opinion is critical in the sense that the establishment of worldwide standards for the protection of companion animals is the desirable outcome. Yet, one needs to look beyond the short-term implications of a ban on ear cropping and tail docking and realize that ultimately the campaign to end cosmetic surgery on dogs is part of a non-speciesist vision. As Munro (2005) put it,

> Ironically, the contentious issue of genetic engineering may be the trigger that will eventually turn this vision into a reality. The spectre of genetically engineered animals in laboratories, on farms and in the wild (not to mention the dinner table) may so alarm ordinary people that a new breed of animal activist will emerge to challenge what the conservationist John Muir denounced as "Lord Man's creativity". (p. 190)

Clearly the advances made in Australasia covered in this chapter are linked to international cooperation. This should not come as a surprise because ultimately the fight to protect the bodily integrity of dogs needs to be global since it is imperative to have universal standards of care, protection and rights. As Sue Kedgley (Green Party of New Zealand) insists,

> There are wider issues here, as well. There has been an attitude for many, many years that we humans are entitled to do whatever we like to animals. We can cut off their tails, manipulate them, put them in cages, clone them, genetically engineer them, patent them—do whatever we like, no matter how cruel, providing it makes them more efficient producers of meat, eggs, and milk for human consumption. This bill challenges that view, just to a small extent. It challenges the view that we are entitled to do whatever we wish to animals, no matter how cruel. It is a small step towards regarding animals as living, sentient beings, having intrinsic value in their own right and not simply existing as the property of humans, for us to do whatever we wish. It is a small step also to accepting that animals have a right to natural forms of behaviour. ... (New Zealand Parliamentary Debates, Hansard, 20 October 2004)

Notes

1. The removal of dew claws is another widespread mutilation inflicted on dogs. However, it has not been the object of a campaign due to the fact that most veterinary associations accept the prophylactic arguments put forward by canine associations.
2. Some European countries such as the Netherlands are signatories but have not ratified the convention. However ear cropping and tail docking are already banned in the Netherlands and the Raad van Beheer (Dutch Kennel Club) has been pro-active in promoting the natural appearance of breeds.

References

Australia. (2000). Department of Agriculture. Fisheries and Forestry. Retrieved May 18, 2007 from: http://www.daff.gov.au/animal-plant-health/welfare/nccaw/guidelines/pets/taildocking

Australia. (2004). Primary Industries Ministerial Council. Retrieved April 16, 2008 from: http://www.mincos.gov.au/__data/assets/pdf_file/0008/316088/pimc_res_04.pdf

Australia. (2007) National Consultative Committee on Animal Welfare. June 08, 2007 Meeting. Retrieved April 16, 2008 from: http://www.daff.gov.au/__data/assets/pdf_file/0005/372929/nccaw-session1-minutes.pdf

Australian Capital Territory. Department of Territory and Municipal Services, Code of practice for the welfare of dogs. Retrieved April 14, 2008 from http://www.tams.act.gov.au/_data/assets/pdf_file/0016/13642/dogwelfare-codeofpractice.pdf

Australian Capital Territory [ACT]. (1992a). Animal Welfare Act. Retrieved April 18, 2008 from: http://www.legislation.act.gov.au/a/1992-45/19921102-4648/pdf/1992-45.pdf

Australian Capital Territory [ACT]. (1992b). Animal Welfare Act. Retrieved April 18, 2008 from: http://www.legislation.act.gov.au/a/1992-45/20010912-607/pdf/1992-45.pdf

Australian Capital Territory [ACT]. (1992c). Animal Welfare Act. Retrieved April 18, 2008 from: http://www.legislation.act.gov.au/a/1992-45/current/pdf/1992-45.pdf

Australian National kennel Council [ANKC]. (1998). Retrieved March 10, 2004 from: http://www.ankc.aust.com

Australian National Kennel Council [ANKC]. (2006). Retrieved April 10, 2008 from: http://www.ankc. org.au

Australian Veterinary Association [AVA]. (1998). Retrieved March 08, 2004 from: http://www.ava.com.au

Australian Veterinary Association [AVA]. (2003a). Retrieved May 18, 2007 from: http://www.ava.com. au/news.php?c=0&action=show&news_id=60

Australian Veterinary Association [AVA]. (2003b). Retrieved May 18, 2007 from: http://www.ava.com.au/news.php?c=0&action=show&news_id=55

Australian Veterinary Association [AVA]. (2003c). Retrieved May 18, 2007 from: http://www.ava.com.au/news.php?c=0&action=show&news_id=32

Bennett, P. C., & Perini, E. (2003a). Tail docking in dogs: A review of the issues. Australia Veterinary Journal, 81 (4), 208–218.

Bennett, P. C., & Perini, E. (2003b). Tail docking in dogs: Can attitude change be achieved? Australian Veterinary Journal, 81 (5), 277–282.

Best, S. (2009). Genetic science, animal exploitation, and the challenge for democracy. In C. Gigliotti (Ed.), *Leonardo's choice: Genetic technologies and animals*. Dorchedt: Springer.

Broughton, A. L. (2003). Cropping and Docking: A Discussion of the Controversy and the Role of Law in Preventing Unnecessary Cosmetic Surgery on Dogs. Michigan State University. Retrieved May 14, 2007 from: http://www.animallaw.info/articles/dduscropping.docking.htm

Canine Association of Western Australia [CAWA]. (2004). Retrieved April 14, 2008 from: http://www.cawa.asn.au

Clarke. R. (2001). Cosmetic Surgery: Legislation in the Australia/Pacific Region. Proceedings of the 26th World Small Animal Veterinary Association Congress. Retrieved May 17, 2007 from: http://www.vin.com/VINDBPub/SearchPB/Proceedings/PR05000/PR00015.htm
Council of Europe. (1987). European Convention for the Protection of Pet Animals. Retrieved March 13, 2007 from: http://conventions.coe.int/treaty/en/Treaties/Html/125.html
Dogs New South Wales. (2006). State Opposition Supports Dogs NSW. Retrieved March 23, 2008 from: http://www.dogsnsw.org.au/util/doc.jsp?n=Duncan%20Gay-20%20Dec%2006-Tail%20docking_1166679916888l9.pdf
Dogs New South Wales. (2007). Shortening Dogs' Tails by Banding. Retrieved March 23, 2008 from: http://www.dogsnsw.org.au/util/doc.jsp?n=Tail%20Banding%20Position%20 Statmment_117219946637691.pdf
Dogs Victoria. (2007). Application for Docked Tail Dog Identification Card For Dogs Whelped in Australia after 12.12.2007. Retrieved April 14, 2008 from: http://www.vca.org.au/ assets/docked%20card%20after%20dec07.pdf
Gumbrell, R. C. (1984). Canine Ear Cropping. *New Zealand Veterinary Journal*, 32 (7), 119
Kennel Club. Retrieved May 20, 2007 from http://www.thekennelclub.org.uk/item/343
Leaver, S., & Reimchen, T. (2008). Behavioural responses of Canis familiaris to different tail lengths of a remotely-controlled life-size dog replica. *Behaviour*, 145 (3), 377–390.
Morton, D. (1992). Docking of dogs: Practical and ethical aspects. *The Veterinary Record*, 131, 301–306.
Munro, L. (2005). Confronting Cruelty: Moral orthodoxy and the Challenge of the Animal Rights Movement. Leiden: Brill.
New South Wales. (1997). Parliamentary Debates. Legislative Council. Hansard, June 26, 1997.
New South Wales. (2004a). Information for dog breeders on tail docking. Retrieved June 16, 2007 from: http://www.agric.nsw.gov.au/reader/aw-companion/tail-docking-info-to-breeders.htm
New South Wales. (2004b). A Guide to the NSW ban on tail docking of dogs. Retrieved April 14, 2008 from: http://www.dpi.nsw.gov.au/agriculture/livestock/animal-welfare/ general/other/dogs-horses/dogs/tail-docking-ban
New South Wales. (2004c). Parliamentary Debates. Hansard, April 6, 2004.
New South Wales. (2007). Prevention of Cruelty to Animals Act 1979 No200. Retrieved April 14, 2008 from: http://www.austlii.edu.au/au/legis/nsw/consol_act/poctaa1979360/
New Zealand. (1998). Code of Recommendations and Minimum Standards for the Welfare of Dogs. Retrieved May 18, 2007 from: http://www.biosecurity.govt.nz/animal-welfare/codes/ dogs/index.htm
New Zealand. (1999). Animal Welfare Act. Retrieved April 14, 2008 from: http:// www.legislation.govt.nz/act/public/1999/0142/latest/DLM49664.html?search= ts_act_Animal+Welfare+Act+1999&sr=1
New Zealand. (2004a). Parliamentary Debates. Hansard, October 20, 2004.
New Zealand. (2004b). Animal Welfare (Restriction on Docking of Dogs Tails) Bill. Retrieved May 17, 2007 from: http://www.vets.org.nz/News/Public/tailDock/DogsTailsBill.pdf
New Zealand. (2007a). Letter Accompanying Draft Code of Welfare for Dogs. Retrieved April 14, 2008 from: http://www.biosecurity.govt.nz/files/biosec/consult/draft-code-of-welfare-dogs-letter.pdf
New Zealand. (2007b). Draft code of welfare for dogs released for comment. Retrieved April 14, 2008 from: http://www.biosecurity.govt.nz/media/17-09-07/dogs-code
New Zealand. (2007c). Animal Welfare (Dogs) Code of Welfare 2007. Retrieved April 14, 2008 from: http://www.biosecurity.govt.nz/files/biosec/consult/draft-code-of-welfare-dogs.pdf
New Zealand. (2007d). Select Committee Report. Government Administration Committee. Animal Welfare (Restriction on Docking of Dogs' Tails) Bill (169-1). Retrieved April 14, 2008 from: http://www.parliament.nz/en-NZ/SC/Reports/e/8/a/48DBSCH_SCR3856_1-Animal-Welfare-Restriction-on-Docking-of-Dogs-Tails.htm
New Zealand Council of Docked Breeds [NZCDB]. (2006). Media Release 5th November 2006. Retrieved April 14, 2008 from: http://www.nzcdb.co.nz/toc.htm

New Zealand Kennel Club [NZKC]. (2007). New Zealand Kennel Club comment on Animal Welfare (Dogs) Code of Welfare 2007 Public Draft. Retrieved April 15, 2008 from: http://www.nzkc.org.nz/article_pdf/epwc1j6ad5w1hs13.pdf

New Zealand Veterinary Association. (2004). The Proposed Bill to Ban Tail Docking: The Facts. Retrieved March 18, 2008 from: http://www.vets.org.nz/News/Public/TailDock/TailDockingNZVAResponse2.pdf

Niven, C. D. (1967). *History of the Humane Movement*. London: Johnson.

Northern Territory. (2000). Animal Welfare Act. Retrieved April 10, 2008 from: http://www.austlii.edu.au/au/legis/nt/consol_act/awa128/

Northern Territory. (2004). Parliamentary Debates. Hansard, March 30, 2004

Queensland. (1991a). Parliamentary Debates. Hansard, February 20, 1991

Queensland. (1991b). Parliamentary Debates. Hansard, May 31, 1991

Queensland. (1995). Animals Protection Act 1925. Retrieved April 12, 2008 from: http://www.legislation.qld.gov.au/LEGISLTN/REPEALED/A/AnimalsProtA25_01.pdf

Queensland. (2001a). Parliamentary Debates. Hansard, October 17, 2001

Queensland. (2001b). Animal Care and Protection Act. Retrieved April 12, 2008 from: http://www.legislation.qld.gov.au/LEGISLTN/CURRENT/A/AnimalCaPrA01.pdf

Queensland. (2005). Department of Primary Industries and Fisheries, Animal welfare and Ethics, A Guide to Queensland's ban on tail docking dogs, Retrieved June 14, 2007 from: http://www2.dpi.qld.gov.au/animalwelfare/13771.html .

Ritvo, H. (1987). *The Animal Estate: The English and Other Creatures in the Victorian Age*. Cambridge, Mass.: Harvard University Press.

South Australia. (2003a). Parliamentary Debates. Legislative Council. Hansard, November 12, 2003.

South Australia. (2003b). Parliamentary Debates. House of Assembly. Hansard, November 21, 2003.

South Australia. Department for Environment and Heritage. Tail Docking. Retrieved April 14, 2008 from: http://www.environment.sa.gov.au/animalwelfare/companion/index.html

Spencer, S., Decuypere, E., Aerts, S., & Tavernier, J. (2006). History and Ethics of Keeping Pets: Comparison with Farm Animals. Journal of Agricultural and Environmental Ethics, 19 (1), 17–25.

Tasmania. (2004). Parliamentary Debates. Hansard. Legislative Council, November 23, 2004.

Tasmania Department of Primary Industries and Water. (2006). Animal Welfare Act Review Report. Retrieved April 14, 2008 from: http://www.dpiw.tas.gov.au/inter.nsf/Attachments/SROS-6UN4RG/$FILE/Animal%20Welfare%20Act%20Review%20Report.pdf

Victoria. (2003). Parliamentary Debates. Hansard. Assembly, October 29, 2003

Victoria. (2007). Department of Primary Industries. Tail Docking of Dogs FAQ'S. Retrieved April 14, 2008 from: http://www.dpi.vic.gov.au/DPI/nrenfa.nsf/LinkView/128D4E82FC7215F2CA2573F3000EEB042B72296A5108C4FFCA25734F0009F96F/$file/Taildocking%20Dogs%20FAQ%27s%20Dec%2007.pdf

Western Australia. (2002a). Department of Local Government and Regional. Development.Retrieved April 14, 2008 from: http://www.dlgrd.wa.gov.au/Legislation/AnimalWelfare/Act2002.asp

Western Australia. (2002b). Parliamentary Debates. Hansard, November 12, 2002.

Western Australia. (2004). Parliamentary Debates. Hansard, April 2, 2004.

World Small Animal Veterinary Association [WSAVA]. (2001). Convention for the Protection of Companion Animals. Retrieved April 13, 2008 from: http://www.wsava.org/Conventi.htm

Adoration of the Mystic Lamb

Kelty Miyoshi McKinnon

Abstract Using the metaphor of the sheep, and the hefted sheep in particular, this chapter examines the impacts of genetic technologies on animals and the resulting deterritorialization of both animal and human from the three ecologies outlined by anthropologist Gregory Bateson: environment, social relations and body/mind. Due to their flocking instinct, dominance hierarchy and similar physiology to humans, sheep were one of the earliest animals to be domesticated by humans, and have been manipulated to provide everything from wool, milk and meat to tissue, biopharmaceuticals and "human" stem cells. Humans have deliberately altered biological constitutions to optimize animal-use value through crossbreeding and selective culling for millennia, but this chapter argues that there is a fundamental difference between the commingling of bodies that traditional animal husbandry involves and the distancing abstraction of contemporary genetic manipulation. The sheep as an animal of sacrifice is discussed in relationship to its promise of eternal youth (via the development of pharming) and redemption (from the current mass extinction via the promise of cloning).

Keywords Genetic · Ecology · Hefted sheep · Cloning · Dolly

From an environmental perspective, the complex entangling of sheep with environmental ecologies is implicated in Scotland's Dissolution of the Monasteries, England's Enclosures Acts, and the British tradition of the Romantic and Picturesque landscape. The multifaceted social relations of sheep amongst sheep and humans amongst sheep are examined from varying points throughout history, and the ecology of body/mind analysed as a dissolution of integral bodies into a fluid dispersal of genes comprising unlimited hybridomas and chimeras. The deterritorialization from these three ecologies speaks to humanity's faith in total biological control, and the intensifying struggle against the loss of this control.

K.M. McKinnon (✉)
School of Architecture and Landscape Architecture, University of British Columbia, Vancouver, BC, Canada
e-mail: keltymc@gmail.com

C. Gigliotti (ed.), *Leonardo's Choice*, DOI 10.1007/978-90-481-2479-4_12,

Fig. 1 Detail from Hubert and Jan van Eyck's adoration of the mystic lamb (credit: Image courtesy of Scala/Art Resource, NY)

At the National Museum of Scotland the glass encased body of Dolly the Sheep rotates atop a grass green plastic plinth. Her stance evokes a strange doppelganger to Hubert and Jan van Eyck's "Adoration of the Mystic Lamb," the iconographic 1432 polytych that centres upon a holy sheep atop an elaborately carved altar. A strong vertical axis locates the animal as the pivotal point of redemption between the heavenly powers above and the healing waters of the verdant landscape below. From the chest of the Mystic Lamb flows a cascade of sacrificial blood into a golden chalice. Below the chalice is the Fountain of Life, promising redemption with the inscription, "This is the fountain of the water of life proceeding out of the throne of God and the Lamb."

Atop her rotating altar, Dolly is both the life blood and sacrificial lamb taken quite literally from her "mother"s' breast. In 1996 at the Roslin Institute at the University of Edinburgh, a maternal mammary cell nucleus was injected into a denucleated egg

Fig. 2 Dolly on display at the National Museum of Scotland (credit: Photo by Dick Warren. Courtesy of the National Museum of Scotland)

cell to create the first mammal to be cloned from an adult somatic (non-sex) cell. Dolly survived for 6 years, less than half the expected life expectancy of a healthy sheep. In 2003 she was euthanized due to progressive lung disease, a common disease in sheep. Now immortalized within her rotating glass case, instead of angels, Dolly is surrounded by a circle of computer screens with such interactive activities as "Spot the Clone", "Design a GMO Crop" and "Who's Like You".

A year before Dolly's birth, scientists at the Roslin Institute cloned sheep Megan and Morag from embryonic cells. Until Dolly was born, it was believed that embryonic cells were the only cells that could regenerate a new organism. As animals grow and cells differentiate, the DNA is altered in each cell type as genes are chemically activated or silenced. Although each adult cell still carries a full complement of DNA, normal growth and development over time was thought to progressively constrain genetic potential. The successful birthing of Dolly proved it was possible that the genomes of differentiated adult cells were still able, under certain conditions, to generate an entirely new animal. For many, the creation of Dolly was akin to the discovery of the Fountain of Life, replete with the promise of eternal youth and a solution to the current extinction crisis.

> After this I beheld, and, lo, a great multitude, which no man could number, of all nations, and kindreds, and people, and tongues, stood before the throne, and before the Lamb, clothed with white robes, and palms in their hands; And cried with a loud voice, saying, Salvation to our God which sitteth upon the throne, and unto the Lamb. And all the angels stood round about the throne, and about the elders and the four beasts, and fell before the

> throne on their faces, and worshipped God, Saying, Amen: Blessing, and glory, and wisdom, and thanksgiving, and honour, and power, and might, be unto our God for ever and ever. Amen. (Revelation 7:9)

Two of Dolly's three mothers, for she was born of breast, egg and womb, were Scottish Blackface Sheep, or Blackies. The breast cell nucleus which contained the nuclear DNA was taken from a Finn Dorset sheep, transplanted into the denucleated Blackie oocyte, and then planted into the womb of a final surrogate Blackie. This particular breed is known as "hefted", a word that refers to their strong territorial instincts and extreme hardiness. The hefted sheep is bonded to specific territories that they have come to know for countless generations and will never willingly leave. Because of this unwillingness to stray, they are highly adapted to their local environments, and do not need walled enclosures. The hefted sheep, taken out of its flock and territory, is both behaviourally and ecologically utterly lost.

A year after Dolly's death, British artist Damien Hirst created *Away from the Flock*, a steel and glass box encasing a Scottish Blackie lamb preserved in formaldehyde. The significance of the lamb, at once separated from its mother, its flock and its territory, speaks of the deterritorialization of something-once-hefted, and once again brings to mind the image of van Eyck's sacrificial mystical lamb. One of the early warning symptoms of an ill sheep is isolation. Sheep are strong flockers, social by nature, and require the presence of at least four or five sheep when grazing to feel relaxed enough to feed. While grazing, sheep maintain a constant visual link with one another, and the slightest sudden movement will cause the group to react quickly and synchronously. While hefted sheep are highly adapted to survive in harsh, cold, nutritionally poor landscapes, they are not adapted to single life away from their home turf. Van Eyck's lone lamb, once promising salvation and eternal youth, is altered by Hirst into a memento more after Dolly's untimely death and the Foot and Mouth crisis of 1991 that resulted in the culling of over ten million sheep and cattle in the UK.

Over a decade later, Hirst's lamb is an even weightier memento of the ecological pickle we are in. The latest report of the United Nations International Panel on Climate Change, created by 2500 of the world's scientists and their governments, states that without significantly altering our current way of life and industry, global warming will lead to the extinction of over one quarter of the earth's species (Fischlin et al., 2007, pp. 211–272, Podger, 2002). Climate change is now a greater threat to species survival than habitat destruction and modification, with one overtly reinforcing the destructiveness of the other. That humans are just another species that will fight for survival is evidenced in migrations already forced by climatic and environmental catastrophe, particularly in the developing world.

The American Museum of Natural History (1997) conducted a nationwide survey of biologists entitled *Biodiversity in the Next Millennium.* 70% of the scientists surveyed agreed that we are in the midst of a mass extinction of living things, and that this loss of species will pose a major threat to human existence in the next century. Unlike pre-historic mass extinctions which were blamed on natural phenomena, this current one is the result of human activity. Since the explosion of the human population following the invention of agriculture around 10,000 years ago, our demand for and impact upon biological resources has dramatically intensified. These same scientists also claim that during the next 30 years as many as one-fifth of all species alive today will become extinct, while 30% of those polled dramatically increased this number to half of all species. This current mass extinction is in fact the fastest in Earth's 4.5-billion-year history, faster than the last great extinction 65 million years ago that wiped out most of the world's dinosaurs (Ayres, 1998). We are quickly realizing the extreme connotations of the fundamental conjoinment of the human with environmental and animal ecologies, and the shared vulnerability that links us all.

Dolly and a subsequent brood of genetic experiments continue to offer the hope of technological salvation from our collective afflictions. Cloning is now considered a viable tool in the fight to preserve endangered species. "Frozen Zoos" cryogenically store animal DNA, sperm, eggs and embryos. Both the San Diego Zoo and the Audubon Center for Research of Endangered Species in New Orleans are systematically preserving genetic materials with the expressed goal of eventually reviving future extinct species through cloning, artificial insemination, embryo transfer and in vitro fertilization. Dr. Betsy Dresser is Senior Vice President of the Audubon Center for Research of Endangered Species. Her research for the Center's "Species Survival Plan" involves taking tissue samples from animals, growing thousands of cells from them, and preserving them in liquid nitrogen to store their DNA. Dr. Dresser calls it a "living library for the future", and in a recent interview on the BBC's *Planet Earth* series she states,

> We have the DNA- we have the science behind us to be able to at least bring the numbers up of this species so they don't go extinct... Man played God a long time ago...I believe God gave us stewardship over these animals, and what we're doing is using the capabilities that we have as humans to not destroy the animals any longer, but to protect them and preserve them and bring them back....Someday we may have to populate another planet. What better way than to take animals with us in frozen form...to the moon, to mars. ...If we can't release animals back in the wild today because of our shrinking habitats, maybe there will be another option one day. What we do in the laboratory is a safety net. (as cited in Beeley, 2006)

Most animal conservationists quickly point out that cloning, necessarily done on an individual basis, monetarily exceeds conventional populating techniques such as embryo transfer or captive breeding, and is rarely worth the cost. Dolly was the anomaly in 277 attempts to create viable life from adult somatic cells. The remaining 276 Dollies were lost at various stages of experimentation. Dolly herself suffered from inexplicable obesity and arthritis before developing progressive lung disease at the early age of six (the average life span for a healthy sheep is between 12 and 18 years). Subsequent high profile cloning successes such as Snuppy, the dog (BBC, 2005) in 2005, ANDi, the monkey (Trivedi, 2001a, b) in 2001 and CC, the cat (BBC, 2002) in 2001 have occurred after hundreds if not thousands of attempts. Snuppy, a clone of an Afghan hound carried to term by a Golden Retriever, was the sole success of over 1,000 embryo transplants. The trail of failed experiments include everything from inviable embryos and stillbirths to deformed babies, aberrant life forms and immense animal suffering.

It is not surprising that Dr. Dresser's "Species Survival Plan" is sponsored by the American Zoo and Aquarium Association. The desire to "save" or "create" individual species through genetic technologies comes out of an attitude of taxonomic reductionism that separates complex animal relationships into hierarchical taxa and distills the Animal into an informational code. As Jeremy Rifkin (1998) writes,

> Living beings are no longer thought of as individual entities, as birds or bees, foxes or chickens, but rather as bundles of genetic information. All living beings are emptied of their substance and transformed into abstract messages. Life becomes a code that is waiting to be deciphered. It is no longer thought of as sacred or specific. (p. 282)

Fetishization decontextualizes. The isolated lamb, whether Van Eyck's, Hirst's or Dolly, speaks volumes of this need to "know", which translates into a need to control, and ultimately to distill and reduce. In each case the sheep is framed, elevated and separated from the earth, a tripled isolation: the removal from flock, territory, and body itself. Anthropologist Gregory Bateson writes about the inseverability of these three realms, environment, society and mind, in his influential series of essays, *Steps to an Ecology of Mind.* Ecology for Bateson is a complex, immersive condition of environment, social relations and subjectivity stressing pattern and differentiation over individual materialities. Culture, nature and body are intertwined into a basic unit of survival, "The unit of survival [or adaptation] is organism plus environment. We are learning by bitter experience that the organism which destroys its environment destroys itself" (Bateson, 1972, p. 483). Bateson's *Ecology of Mind* encourages a relational understanding of environment, where species are not isolated individuals within a larger system, but are unfolding from and within that system—indeed they are immanently part of that territory (Guattari, 2000). The hefted sheep demonstrates this state of mutual embeddedness, simultaneously unfolding from, interacting with and intimately transforming its much localized environment. Gilles Deleuze (2001) relates this tripartite "Ecosophy" to the idea of immanence,

> Absolute immanence is in itself: it is not *in* something, *to* something; it does not depend on an object or belong to a subject. . . .When the subject or the object falling outside the plane of immanence is taken as a universal subject or as any object to which immanence is attributed, . . . immanence is distorted, for it then finds itself enclosed in the transcendent. (pp. 26–27).

Environment

The etymology of the word "sheep" is a direct statement on the inseverability of animal and environment. A historical tracing of the word "landscape" leads one to sheepish etymological roots: sceap, scep and scap in Old English, schaf in Germanic and schaap in Dutch. This linguistic connection is significant, as sheep have been intimately tied to massive historic transformations of the physical environment into a landscape aesthetic that is deeply engrained in our consciousness. By the late 1700s, the word "scape" was defined as "scenery or view", concurrent with the English rise in landscape aesthetic theory as defined by Romantic and later Picturesque writers, painters and philosophers See (Price, 1796) and (Repton, 1794).

Roman settlements in Britain during the Iron Age resulted in both rising sheep numbers and massive ecological change (Franklin, 2007, p. 93). Trees were cleared from large tracts of land to make room for livestock, and were used for fuel, industry, and the building of ships and settlements. Mass deforestation and livestock occupation resulted in the erosion of high quality soils into low lying areas, and the development of extensive marshlands and heather-covered moorlands. Sheep were ideal inhabitants for the resulting harsh, open, nutritionally poor landscapes due to their hardiness and capacity to improve poor soils. From Roman times on, the value of sheep to the British economy, and the effects their proliferation have had on the British landscape have been inestimable.

Initially, the wool boom was dominated by the Cistercian monasteries (Franklin, 2007, p. 96). The Cistercian monks lived according to the rules of St. Benedict, emphasizing *ora et labora*, prayer and manual labour. In British monasteries this involved farm work and the keeping of large manorial flocks for the trading of wool. Over 2 million acres were owned by over 800 monasteries, with tens of thousands of tenant farmers working the lands. This land was communally accessible to residents for fuel, water, and grazing livestock. In the 1500s, King Henry the VIII ordered the Dissolution of the Monasteries, where the majority of monasteries, nunneries and friaries were confiscated by the monarchy for redistribution to aristocrats and private merchants with vested interests in the rising price of wool. Vast numbers of shepherds and peasants were forced from their homes to make a living in urban centres or the British colonies. The Enclosure Acts of the eighteenth century continued this transfer of previously open, communally accessible fields over to privately enclosed properties. In the process, entire communities were dislodged, villages moved or burned, and massive amounts of arable lands were converted to enclosed pastures for sheep farming. Dispossessed peasants fled to the cities, creating a labour surplus to fuel the advent of the industrial revolution. Thomas More's (1516) *Utopia* describes the massive influx of sheep onto once common lands, and the displacement of entire communities in the interest of profit:

> ....the increase of pasture...by which your sheep, which are naturally mild, and easily kept in order, may be said now to devour men and unpeople, not only villages, but towns; for wherever it is found that the sheep of any soil yield a softer and richer wool than ordinary, there the nobility and gentry, and even those holy men, the abbots not contented with the old rents which their farms yielded, nor thinking it enough that they, living at their ease, do no good to the public, resolve to do it hurt instead of good. They stop the course of agriculture, destroying houses and towns, reserving only the churches, and enclose grounds that they may lodge their sheep in them. (p. 9)

Both the Dissolution of the Monasteries and the Enclosures Acts asserted a class based deterritorialization of the shepherd from his flock, and the flock from the larger landscape.

The deforested, industrialized landscape, with its rolling green hedgerowed and stonewalled fields and valley bottoms and open high fells, has become the image of

Fig. 3 Sheep in the Lake District (credit: Photo by Tina Mahoney, Courtesy of www.english-lakes.com)

the pastoral ideal, the quintessential British landscape. Romantic and Picturesque conceptions of nature were developed by William Wordsworth and William Gilpin, both born in the heart of sheep country, the Lake District of Northern England. Due to their steep terrain, shallow soils and harsh weather, much of the Lake District and the Highlands of Scotland were not enclosed, and have been maintained as common lands. Romanticism gained strength during the Industrial Revolution as a reaction to the industrialization and rationalization of nature, focusing instead on emotion as a source of aesthetic experience.

Both Wordsworth and Gilpin believed that the hierarchies and separations imposed by the Industrial Revolution were the primary cause of the alienation of people from their essential "purity" due to the effects of being separated from nature. Only through a "return" to the land could this essential "goodness" be retrieved. While Wordsworth lamented the urbanization and industrialization of rural agrarian life, he nostalgically posited the agrarian peasant as a symbol of an harmonious and simpler lifestyle that was innately tied to the land. The transformation and demarcation of the landscape during this period was critiqued as an extension of the "deformities" created through the fascination with the science of animal husbandry. Richard Payne Knight (1806) in his treatise on aesthetics foreshadows this "indulgence" with scientific rationalism:

> ... (He) indulge(s) those partial and extravagant caprices of his taste, which he has so abundantly displayed in the productions of his own art and labour. As far, however, as he has been able, he has done it most profusely. At one time, he crops the tail and ears of his dogs and horses, and at another, forces them to grow in forms and directions which nature

> never intended: his trees and shrubs are planted in fantastic lines, or shorn into the shapes of animals or implements; and all for the sake of beauty. Happily for the poor animals, it has never appeared possible to shear or twist them into the shapes of plants, or it would, without doubt, have been attempted; and we should have been as much delighted at seeing a stag terminating in a yew tree, as ever we were at seeing a yew tree terminating in a stag. (pp. 7–8)

Despite lamenting the loss of the intimate knowledge of land and animal ecologies, Wordsworth and Gilpin's theorizing ironically resulted in the extreme aestheticization and objectification of the natural. William Gilpin (1792), one of the primary originators of Picturesque landscape theory, defined the Picturesque as a series of principles based primarily on landscape painting. His approach to landscape was overtly scenographic in that he sought to compose particular views from key vantage points within the landscape. For him, landscape was primarily for the visual enjoyment of a "unified whole" that incorporated the animal as a slowly moving graphic counterpoint to the carefully composed framework of foreground, middle ground and background:

> ....In pastoral subjects, sheep are often ornamental, when dotted about the sides of distant hills. Here little more is necessary, than to guard against regular shapes- lines, circles, and crosses, which large flocks of sheep sometimes form. In combining them, however, or, rather scattering them, the painter may keep in view the principle, we have already so often inculcated. They may be huddled together, in one, or more large bodies, from which little groups of different sizes, in proportion to the larger, should be detached. (p. 255)

Asymmetric, broken, rough, intricate and sunken lines were selected to counterbalance the visual effects of enclosure. These lines were extended from what existed on the site- rutted paths, old trees and rugged slopes.

Interestingly, the development of Picturesque theory led to a thriving landscaping industry as land owners sought to enhance the picturability of their properties for status. Landscape became a means of expressing wealth through the impression of size and importance. Strategically cut vistas for "borrowed views," ha has (sunken fences), curved approach drives, and the careful placement of sheep flocks in fore, middle and background gave the visual impression of limitless territory. The animal body not only became a mobile counter to the static ground of landscape, it also became a visual indicator of size and extent,

> ...the eye forms a very inaccurate judgement of extent, unless there be some standard by which it can be measured; bushes and trees are of such various sizes, that it is impossible to use them as a measure of distance; but the size of a horse, a sheep or a cow, varies so little, that we immediately judge their distance from their apparent diminution, according to the distance at which they are placed; and as they occasionally change their situation, they break that surface over which the eye passes, without observing it, to the first object it meets to rest upon. (Repton, 1840, pp. 348–349)

The transformation of "schap", "schaap", "sceap", "scep" and "scap" into "scape" was complete. From relational animal subject *of* the land, the sheep was deemed an independent object to be placed *on* the land.

But despite the human rhetoric and objectification, sheep themselves continued to proliferate, metamorphisizing in response to landscape and human intervention while simultaneously transforming landscape. More so than in other countries, British sheep breeds are as diverse as the microclimates they inhabit. The name of each breed echoes the territory it constitutes, and from which it is constituted. The Dorset, Leicester, Suffolk, Lincolnshire, and Scottish Blackface are all uniquely adapted to their particular ecologies (soil types, plants, water drainage, solar and wind orientation, pathogens and other species), as well as the diversity of ways that they are put to human use, such as breeding, improving the land, and the production of wool, meat and milk. The sale of a farm in Britain will often include its sheep as part of the transaction, so tied to the land are the flocks. The Cheviot for example developed a strong constitution, well developed mothering instinct and durable rot resistant wool in the harsh winters of the Cheviot Hills bordering England and Scotland, while the Norfolk developed an extremely muscular body, adapted for traveling great distances for food through the rugged southeastern region of England where forage was sparse (Oklahoma State University, 1995). This diversity of British breeds is seen as a national boon, a genetic repository that can be accessed to respond to fluctuations in the market. A shift in consumer preference toward warmer wool, sweeter milk, or more fertile waste products can be accommodated through animal husbandry to maintain the overall British competitive advantage. The Lincoln breed, produced by crossing a Leicester and a native sheep of Lincolnshire, resulted in a higher quality wool and sturdier bodies. Each sheep bred is a complicated mix of both natural and inbred adaptations to climate, territory, and market, and each landscape develops a corresponding aesthetic and ecology, linked by sheep and shepherding paths that have today been formalized into highways, roads, and hiking trails. Many British place names from the Middle Ages ended in "scap" or "sceap", expressing the seamless connection between sheep and land.

The hefted sheep is an exaggeration of this territorialization in that it has been bred and encouraged over hundreds of years to become hefted, or heafed, to the landscape. Since at least the twelfth century the Scottish Blackface, or Blackie, was raised in British monasteries for their wool, and in the sixteenth century King James IV established an "improved" Blackie flock of about 5,000 sheep in the Ettrick Forest of Scotland. These flocks were bred to thrive in adverse weather conditions and the spartan landscapes of higher climes. Other breeds, such as the Hedgewick, Rough Fell and Dalesbred, have been hefted to rugged territories in the Highlands of Scotland, the Lake District, Howgill, the Pennines, Dartmoor and Northern Ireland. These hefted breeds are best suited to hill and mountain habitats due to their ability to thrive and reproduce in harsh weather, and their ability to survive on nutritionally poor forage. Their long coarse wool is quick to dry, an adaptation uniquely suited to protect from harsh winter winds, snow and rain in the most extreme locations. Paradoxically, these animals were domesticated and bred for their wildest traits: their self sufficiency, hardiness, independent lambing abilities, protective instincts and their strong territorial bonds. Hefted ewes are able to yield lambs, milk and wool on the most marginal of pastures, and will defend their young against predators if necessary. They pass intimate territorial knowledge on to their lambs, teaching them where shelter, water and shade can be sought and what plants are of nutritional and medicinal value (New Scientist Magazine, 2006; Lawrence, 2007). Hefted lambs are avid mimickers, and quickly inherit these maternal maps, which are then passed down through subsequent generations. Hefted sheep will not cross into other territories other than their own, and can be kept in unfenced areas without constant shepherding. The result for sheep farming has been an animal that needs minimal interaction, feeds itself, is able to birth and raise lambs independently, and is able to self-medicate for minor ailments.

The British Foot and Mouth crisis of 1991 revealed the devastating effects on both landscape and economies when sheep disappear. The culling of ten million British sheep and cows wreaked havoc on a nationwide level. Cumbria, the heart of Blackie country, was the worst affected area, with the mandatory culling of entire flocks of hefted sheep leaving no ewes to pass on territorial knowledge to future generations. The crisis is estimated to have cost the UK approximately $16 billion in the agriculture and tourism sectors. The significant conflation of sheep and landscape was made painfully clear as public rights of way across lands were closed and the tourist industries that thrived on sentimental yearnings for the picturesque were threatened with landscapes quickly transforming from easily accessed lush, close cropped fell sides to inaccessible bramble covered slopes. The open commons of the highlands, long made possible by bramble and scrub eating hefted flocks, was no longer feasible homeland for new communities of lambs that had not been hefted onto grounds through the intimate knowledge of their mothers. It is estimated that to re-heft a flock could take years. A full time shepherd is essential to keep sheep from straying, and is economically infeasible for the majority of farmers. As an alternative to this expensive option, some farms have strung up electric fences to teach the sheep the limits of their territory. The estimated time for sheep to reheaf based on fear of electric shock is five to ten years, and this only guarantees the heafing will regard limits to movement. The Foot and Mouth crisis serves to remind us of the conjoined states of environment, animal behaviour and human economy. Attempts to return the land and economy to a pre-Foot and Mouth state raises the issue of this complex interweaving and the absurdity of attempting to create a particular animal, with particular behaviors, relationships and subjectivites from genomic encoding alone.

Social Relations

The isolated animal is a voiceless one. Desocialized, its existence is no longer structured by daily interaction— it is truly "away from the flock". Like many prey animals, sheep have a strong flocking instinct, preferring to stay with the group rather than traveling individually. Bonding within a flock occurs through close body contact and social interaction and sheep not only tolerate, but thrive on the calming effect of close quarters. Because of their dependence on the flock for safety and comfort, they become highly agitated if separated from the group, and will actively struggle to rejoin their peers.

In 2001, behavioral scientist Keith Kendrick from the University of Cambridge completed a highly publicized study that showed strong evidence that sheep use faces to recognize one another in a very similar way to humans. By measuring activity in the parts of the sheep brain that are associated with visual recognition, they found that sheep have similar neural systems to humans in the temporal and frontal lobes for remembering particular faces, and that only after more than two years absence do the memories of these faces begin to fade. Kendrick (2007) explains,

> Neural networks within these two lobes are dedicated to processing faces, as opposed to other visual objects, just as they are in monkeys and humans. The neural networks involved are organized hierarchically, with components classifying identity of specific categories of faces and individual faces taking longer to encode face stimuli than those simply involved in the process of distinguishing face from non-face stimuli. (para 8)

The study found that sheep are able to remember up to 50 sheep faces from multiple directions, and that they are able to form a mental picture of particular flock mates in their absence (Kendrick et al., 2001).

Sheep form individual bonds, or friendships as Kendrick calls them, with one another that last a few weeks at a time. Other researchers, such as Thelma Rowell, have studied sheep sociality and how it is expressed through such behaviors as head rubbing or “affiliative reassurance behaviour”. Rowell has also observed that sheep will intervene in antagonistic interactions between pairs in order to maintain the larger social order of the flock (as cited in Franklin, 2007, p. 197). Keith Kendrick’s team found that a particular subgroup of cells that encode memory showed increased activity when pictures of the faces of sheep that were either currently or previously part of their immediate social group were shown. The sheep vocalized in the same way when looking at pictures of former members of their social group as they did when looking at current ones (Kelleher, 2002). Kendrick and his team extended their research in a 2004 study that concluded that pictures of familiar sheep faces, compared to similar images, reduce symptoms of stress and fear in sheep (Da Costa et al., 2004). When agitated lone sheep were exposed to images of either inverted triangles, goat’s faces, or sheep faces of the same breed, it was found that the sheep shown sheep faces returned to normal heart rates and blood hormone levels within 15 minutes. Those shown inverted triangles remained stressed with high heart rates and elevated adrenaline and cortisol levels, while the sheep shown goat faces were slightly less stressed. Previous research had shown Kendrick that sheep were more attracted to the faces of animals from the same breed than others. Further study showed that sheep also use visual cues to judge the emotional state of other sheep from their facial expression (New Scientist Magazine, 2004).

The initial establishment of a strong ewe–lamb bond is also essential to a lamb’s ability to thrive in a group. It is estimated that failed bonding is implicated in 80% of pre-weaning deaths (Dwyer, 2007). After birthing, a ewe is initially attracted to the smell of her own fluids spilled at the birth site and covering her lambs. She bonds with her young by vigorously licking these fluids. The higher the level of licking, the stronger the bond will be, and the better she is able to recognize her lamb by sight, smell and the sound of its bleats. There is a limited window of 30 to 60 minutes where the ewe will be receptive to forming a bond with a particular lamb or lambs, and afterwards she will only be bonded to them. If bonding fails to happen during this time frame, there is a very high chance of infant mortality (Scottish Agricultural College, 2006). Particularly in hefted sheep, the mother– child bond is the primary vehicle through which the lamb learns social skills and how to read the landscape, but they also form relationships and test their skills with other sheep and lambs in the flock.

In somatic cell nuclear transfer, the cloning technique used to create Dolly, the cloned animal is manually pieced together from a selection of biological parts taken from a number of individual animals. The DNA from one individual's nucleus is transferred into the denucleated cell of another individual, which is then inserted into a third individual—the surrogate mother. Born not from the coupling of a male and female to one "natural" mother, the creation of life instead becomes completely dependent on human intervention. In this case males were completely absent: all three sheep involved in Dolly's creation were ewes. Ian Wilmut, lead scientist in the Dolly project, describes Dolly as "the stuff of which myths are made. Her birth was other-worldly—literally a virgin birth; or at least one that did not result directly from an act of sex" (Wilmut et al., 2000, p. 233). Not only does the animal lose its agency in reproduction, it is reduced to an object or source of biological material. Gender is made irrelevant.

Dolly is appropriated from the idea of "mother" in multiple ways. She is given life from her biological mother's udder—not from her milk, but from a cell taken from her cryogenically preserved breast tissue. Not only is Dolly's mother dead before conception, but her breast, a symbolic source of life and nourishment, is no longer symbolic of "mother", nourishment or life. It is reduced to a generic cellular expression of "sheepness". Sheep milk has long been re-directed for human consumption, now it is redesigned transgenically to produce valuable human proteins that can be used in the production of pharmaceuticals. The breast is re-created as a "lactation system" for the production of drugs to treat human disorders such as diabetes, haemophilia and cystic fibrosis.

Human meddling and tampering with sheep life has been incredibly complex, emotional and tangled. Keith Kendrick's research found that sheep also remember and respond to familiar human faces, even after a separation of up to 12 months. Sheep have been closely affiliated with humans for thousands of years—archaeological evidence indicates that people were keeping sheep 8,500 years ago in the Neolithic Period. Mummified sheep have been found in Egypt, and there are several biblical references to the entwined conditions of humans and sheep.

> When he has brought his own sheep outside, he walks on before them, and the sheep follow him because they know his voice. They will never follow a stranger, but will run from him because they do not know the voice of strangers *or* recognize their call... I am the Good Shepherd. The Good Shepherd risks *and* lays down His life for the sheep. But the hired servant (he who merely serves for wages) who is neither the shepherd nor the owner of the sheep, when he sees the wolf coming, deserts the flock and runs away. And the wolf chases

> *and* snatches them and scatters [the flock]. Now the hireling flees because he merely serves for wages and is not himself concerned about the sheep. I am the Good Shepherd; and I know *and* recognize My own, and My own know *and* recognize Me. (John 10:4–18)

The heafing of the hefted sheep is the behavioural legacy of hundreds of years of good shepherding when shepherds lived with the sheep in the fells. Over generations, the behaviours taught by the shepherd to their flock were then passed from ewe to lamb. In non-hefted flocks, sheep are tended to intensively by humans for birthing, tail docking, worming, vaccinations, hoof trimming, feeding and watering, sheltering, weaning, dipping and shearing (for parasite control) and tupping. The Farm Animal Welfare Advisory Council and the Society for the Prevention of Cruelty to Animals (SPCA) recommend "five freedoms" that good animal husbandry should provide: freedom from thirst, hunger and malnutrition; freedom from discomfort; freedom from pain, injury and disease; freedom to express normal patterns of behaviour; and freedom from fear and distress. The intensive required interaction is described by many sheep farmers as a familial intimacy. The 1991 UK Foot and Mouth epidemic revealed, devastatingly, the extent of human "emotional attachment to sheep as embodiments of human labour, industry and accomplishment" (Franklin, 2007, p. 192). The alarm of Cumbrian farmers at the realization of the scope of their loss was expressed through the media: "We would rather die than let them kill our flock", "The compensation doesn't matter. They are our lives. . .our sheep are part of our family" (Franklin, 2007, p. 180). Prince Charles recognized the mutual "heftedness" of rural people to sheep and to a particular way of life: "family farms, with the people that run them, are in many ways hefted . . .and you cannot just start them up because you have lost centuries of tradition, knowledge on the local area and local conditions—these are absolutely crucial things" (BBC, 2001). Describing the required culling as a "national agony," he reflected the grieving of a nation mourning the loss of an entire way of life, one that was the very backbone of British identity. The Foot and Mouth crisis magnified the catastrophic emotional, cultural, social, environmental and economic impacts of the rapid disappearance of an animal with such an intimate relationship to humans.

Body and Subjectivity

Before Dolly was Dolly, she came into the world as most modern lambs do—as a number, Lamb 6LL3. Despite her status as the original transgenic, Dolly follows the legacy of the industrialization of livestock, and the tradition of the experimental redesign of animal DNA for capital expansion (Franklin, 2007, p. 6). The will to industrialize dissolves the individual animal subject into serial object, or specimen, that is then reduced again to its isolatable and crossable traits. This information

> is not concerned with the singular; it eliminates any trace of being. Its anthropology is in the process of becoming that of a meticulous physics of elements, whereby the genetic barriers between. . .individuals. . .are breaking down, since value and weight are only accorded to genetic crystallizations, which are themselves potentially provisional or modifiable. (Le Breton, 2004, p. 3)

The clone is reduced further in that it loses the guaranty of mixture and individuality that comes from sexual reproduction: the creation of a genetically unique individual from the mixing of a male and female's DNA. The offspring is an exact replicant of the donor, and diversity is lost. Genetic biodiversity is essential for long-term population viability, particularly in terms of maintaining high rates of fertility and resistance to disease. Maintaining a singular lineage, the clone moves outside of genealogy into a terrain where anything seems possible. As Sarah Franklin (2007) remarks,

> ...it is especially important to explore the implications of Dolly's cultured-up biology for the elementary formation of kinds and types, for example, by asking what happens to sex, breed, species, and reproduction when genealogy is retemporalized and respatialized. (pp. 31–32)

Dolly's bricolage biology heralds an "emergent system of categorical difference brought into being by her existence outside the genealogical lines and linearities that formerly mapped the interior of the biologically possible" (p. 32).

The disintegration of specificity also blurs the accepted divisions between breed, species, genus, and other taxonomic categories. Genetic technologies directly challenge the administrative categorical systems of Linnaean taxonomy. Plant, animal and human are no longer distinct modes of being. Recent developments in the genetic production of hybridomas and chimeras have tested the limits of the human imagination in terms of what the combination of plant, animal and human genes will produce. Hybridomas are the result of the fusion of two species' cells into a single cell, while a chimera results from transplanting the embryonic cells from one animal into the embryo of another. In 1984, a new kind of animal was invented by combining goat and sheep embryos to create a geep (Time, 1984). Subsequent experimentation has yielded rat/mouse chimeras, and a plethora of creatures crossed with the green fluorescent protein of the jellyfish including a zebrafish, rabbit, and monkey (Stewart Jr., 2006).

Plant and animal have also been merged in the pursuit of new species with traits beneficial to humans. The genetic coding for Arctic flounder has been merged with the tomato to attempt to create a plant hardier at colder temperatures (PBS, 2003). Corn, rice, soy, and tobacco, have been engineered to make products like growth hormones, blood clotting agents, vaccines, human antibodies, industrial enzymes and contraceptives (Ruiz-Marrero, 2005). In the transgenic experiment,

> ...all living beings appear to be constructed from the same components, which are distributed in different ways. The living world is a sort of combination of elements, which are finite in number and that resemble the product of a gigantic Meccano set, and this is Transgenic species- the result of the continual *bricolage* of evolution. (Jacob, 1998, p. 12)

The distinctions with which we define our humanity are increasingly undifferentiated.

The spectre of the chimeric human lingers amongst these new transgenic creatures. In 1997 ewes Polly and Molly were cloned from adult somatic cells, but with the insertion of human transgenes designed to express human clotting proteins in sheep milk. They were the first of hundreds of transgenic farm animals to subsequently be cloned from adult somatic cells. The production of pharmaceutical proteins for human consumption in livestock milk (and plants) has come to be known as pharming, and in August 2006, the European Medicines Agency approved the use of the first drug manufactured from the milk of a transgenic goat, ATryn. Now several farms across the United States and Europe have broods of transgenic animals expressing treatments for such ailments as tumours, haemophilia, and diabetes (Buchanan, 2006).

The biology of the domestic sheep makes it an "ideal host" for the production of pharmaceuticals and xenotransplantation. As Sarah Franklin (2007) states,

> more than any other domestic animal, the sheep so closely resembles humans that it has long been considered an ideal substitute for humans in medical scientific experimentation, particularly for questions related to reproduction and respiration. (p. 200)

Now we see the difference between human and sheep disintegrating in the production of human sheep chimeras. In 2007, researchers at the University of Nevada created a sheep that has 15% human cells and 85% animal cells (Joseph, 2007). Sheep are now reified as living factories for the production of tissue and organs for human implantation.

Transgenic species have become a way around the American moratorium on the use of discarded human embryos for the gathering of human stem cells. In 2003, scientists in Shanghai fused human skin cells with dead rabbit eggs to create the first human chimeric embryos. The embryos were harvested after a few days for their embryonic stem cells. In 2004 researchers at the Mayo Clinic created chimeric pigs with human blood, and in 2008, British researchers inserted human DNA into a denucleated cow's egg to create an embryo that was 99.9% human, and 0.1% bovine. While the embryos survived for only 3 days, the goal is to increase survivability to 6 days in order to harvest stem cells (Mott, 2005).

The creation of chimeric species raises a host of ethical issues; however to date, only Canada has passed legislation to ban the creation of human–non-human life forms. The Canadian Assisted Human Reproduction Act of 2004 however, does not ban the creation of animal to animal chimeras. There is strong opposition to the banning of chimeric research. Irv Weissman, director of the Institute of Cancer/Stem Cell Biology and Medicine at Stanford University, says,

> Anybody who puts their own moral guidance in the way of this biomedical science, where they want to impose their will—not just be part of an argument—if that leads to a ban or moratorium. . . . they are stopping research that would save human lives. (as cited in Mott, 2005)

Weissman has created human–mouse chimeras where human neurons are injected into the brains of embryonic mice. In 2005, Weismann created mice with 1% human brains. His ultimate goal is to create mice with 100% human brains.

This increasing overlap between animal, human and plant raises ethical questions about the definition of and rights allocated to the human and the animal. As the definition of humanness has included an argument about "self-awareness," if an animal is then deemed self-aware, are they then "elevated" to the status of human? Are they guaranteed the rights and freedoms accorded to humans in the Constitution of Canada and the Universal Declaration of Human Rights as the right to bodily integrity and security of person? The absolute inability for animals to control their biologies and indeed their integrity as sheep, chimpanzee or jellyfish corrupts all boundaries. In 2008 the Food and Drug Administration officially declared that it is safe for farmers to use cloned pigs, cow and goats in the production of meat and milk for human consumption (though, forebodingly, they have called for more testing on sheep) (FDA, 2008). Since there is no requirement at this time for these foods to be labeled accordingly, our own ability as humans to control what we put in our bodies is simultaneously diminished. The rapid global transfer of diseases between species, such as Foot and Mouth, BSE and SARS, is yet another manifestation and warning of these new proximities.

Deterritorialization, Thrice Hefted

Like the Blackface sheep, we are all thrice hefted, intimately immersed in environment, social relations and our own physicality and subjectivity. The turn of the century Estonian zoologist Jakob von Uexküll saw these "perceptual worlds" as infinitely varied and relatively exclusive, but "linked together as if in a gigantic musical score". Giorgio Agamben (2004) describes Uexküll's theory of "Umwelt" as the "environment-world that is constituted by a more or less broad series of elements that he calls "carriers of significance"... or... "marks," which are the only things that interest the animal" (p. 40). Deleuze and Guattari give an example of this blind unity in their writings about the Thynnine wasp (*Neozeleboria cryptoides*) and the wasp orchid (*Chiloglottis trilabra*). The petals of the wasp orchid resemble a female Thynnine wasp so closely that the male wasp will attempt to mate with it. As he does, he is covered with pollen which he then transfers to the next orchid (Genosko, p. 112). As Agamben (2004) describes,

> Everything happens as if the external carrier of significance and the receiver in the animal's body constituted two elements in a single musical score... though it is impossible to say how two such heterogeneous elements could ever have been so intimately connected. (p. 42)

The perceptual worlds of the wasp and the orchid are closely attuned to one another, though the wasp has no perception of the umwelt of the orchid and vice versa. Thus these two creatures are unwittingly hefted to each other's world.

To think of animals in the way of Uexküll and Agamben is to realize the absurdity of isolating traits and behaviours from the worlds that they unfurl from, but also the impact of deterritorialization on the larger collective environment. Throughout this essay the sheep has been a metaphor for the ways in which both animals and humans are hefted and deterritorialized. The sheep is thought to be the first domesticated animal, and has been living with humans for millennia. Humans have deliberately altered biological constitutions to optimize the animal's use value through cross-breeding and selective culling. But there is a fundamental difference between the commingling of bodies that traditional animal husbandry involves and the distancing abstraction of genetic manipulation. Traditional animal husbandry has involved a relational mutual understanding between human and animal, where the shepherd often lived with the sheep, travelled, ate, guided and cared for them. This involved humans who were familiar with animals and who maintained the wisdom which accompanies that familiarity. Contemporary genetic modification of animals involves a severance of ties, of both the animal and the human, from the environment and from direct human animal relations and communication. The separation and autonomy that is assumed in this model renders the environment, social networks and the mind into irrelevant anachronisms. Geneticist Francois Gros (2000) summarizes the goal of the Human Genome Project as "...(the) reduc(tion of) human behaviour and vital mechanisms to a gigantic algorithm for which the chromosome would be the programme, which we could more easily control if we could interpret it as computer data" (p. 220). The fate of the animal is the fate of humanity.

In 1894, British sheep historian William Youatt defined the sheep as "the measure of national prosperity or calamity". There is a keen echo of Youatt's wisdom in the absolute stillness of Dolly and the Hirst and Van Eyck sheep. A close inspection of Hirst's sheep reveals an array of inanimate air bubbles trapped in its preserved wool. Even air is isolated from movement " ...in a manner that neatly recapitulates the strict separations required to preserve a faith in "biological control" against contamination from its opposite, in the form of loss of control" (Franklin, 2001, p. 6). Heterogeneity, difference and multiplicity are sacrificed in the creation of placeless, voiceless, formless animals, captured in vials and stored in liquid nitrogen for future resurrection. The distillation of life into pure, abstract information and the simultaneous disappearance of thousands of species in this current epoch of extinction, speak to the immense sacrifice of life in the struggle for total biological control. This is the creation of the ultimate "still life", where the sacrificial lamb is divorced completely from context in the name of purity and righteousness. A seventeenth century poem in protest of the Enclosures Act is a germane comment on contemporary conditions:

> They hang the man, and flog the woman,~~~That steals the goose from off the common;~~~But let the greater villain loose,~~~That steals the common from the goose. (Anonymous)

Arguably, what is needed is a kind of re-hefting of humanity to the conditions that generate diversity, a recognition of the multiple ways in which we are entangled in the commons, and a re-acquaintance with the animal as something mutually and relationally embedded.

References

Agamben, G. (2004). *The open: Man and animal.* Stanford: Stanford University Press.

American Museum of Natural History & and Louis Harris and Associates, Inc. (1997). *National survey reveals biodiversity crisis- scientific experts believe we are in the midst of fastest extinction in earth's history.* Retrieved April 12, 2008 from http://www.amnh.org/museum/press/feature/biofact.html

Ayres, E. (1998, September 1). The fastest mass extinction in Earth's history [Electronic version]. *World Watch*. Retrieved April 12, 2008 from http://www.encyclopedia.com/doc/1G1-21111086.html

Bateson, G. (1972). *Steps to an ecology of mind*. San Francisco: Chandler Publishing Company.

BBC News Online. (2001, October 17). Prince Charles makes pledge to farmers [Electronic version]. Retrieved May 1, 2008 from http://www.bbc.co.uk/devon/news/ 102001/17/farming_prince.shtml

BBC News Online. (2002, February 15). First pet clone is a cat [Electronic version]. Retrieved May 1, 2008 from http://news.bbc.co.uk/1/hi/sci/tech/1820749.stm

BBC News Online. (2005, August 3. S Korea unveils first dog clone [Electronic version]. Retrieved May 1, 2008 from http://news.bbc.co.uk/2/hi/science/nature/4742453.stm

Beeley, F. (Producer) and Poland, S. (Narrator). (2006). *Planet Earth: The future-saving species* [Video]. London: BBC.

Boyd, C. (2007, October 31). Foot and mouth disease- the aftermath [Electronic version]. *BBC*. Retrieved May 10, 2008 from http://www.bbc.co.uk/insideout/ content/articles/2007/10/31/london_fandm_carl_s12_w7__feature.shtml

Buchanan, R. (2006, February 22). "Pharmed" goats seek drug licence [Electronic version]. *BBC News*. Retrieved May 12, 2008 from http://news.bbc.co.uk/2/hi/science/nature/4740230.stm

Da Costa, A.P., Leigh, A.E., Man, M.S. and Kendrick, K.M. (2004). Face pictures reduce behavioural, autonomic, endocrine and neural indices of stress and fear in sheep. *Proceedings of the Royal Society of Biology B*. 271, 2077–2084.

Deleuze, G. (2001). Pure immanence: Essays on a life. New York: Zone Books.

Dwyer, C. M. (2007). Genetic and physiological effects on maternal behavior and lamb survival. *SAC*, Edinburgh, UK. Retrieved April 29, 2008 from http://adsa.asas.org/ meetings/2007/abstracts/0458.PDF

Fischlin, A., Midgley, G.F., Price, J.T., Leemans, R., Gopal, B., Turley, C., Rounsevell, M.D.A., Dube, O.P., Tarazona, J. and Velichko, A.A. (2007). Ecosystems, their properties, goods, and services. In M.L. Parry, O.F. Canziani, J.P. Palutikof, P.J. van der Linden & C.E. Hanson (Eds.). *Climate change 2007: Impacts, adaptation and vulnerability. Contribution of working group II to the Fourth Assessment Report of the Intergovernmental Panel on Climate Change*. Cambridge: Cambridge University Press. 211–272.

Food and Drug Administration (FDA). (2008, January 15). FDA issues documents on the safety of food from animal clones. Retrieved on April 29, 2008 from http://www.fda.gov/bbs/topics/NEWS/2008/NEW01776.html

Franklin, S. (2001, June). Sheepwatching. Anthropology Today. 17 (3), 3–9.

Franklin, S. (2007). *Dolly mixtures: The remaking of genealogy*. London: Duke University Press.

Genosko, G. (2002). *Felix Guattari: An aberrant introduction*. New York: Continuum.

Gilpin, W. (1792). Observations, relative chiefly to picturesque beauty, made in the year 1772. London: Cumberland & Westmoreland.

Gros, F. (2000). L'ingenierie du vivant. Paris: Odile Jacob.

Guattari, F. (2000). *The three ecologies*. New Brunswick, N.J.: Athlone Press.

Jacob, F. (1998). *Of flies, mice and men*. Cambridge: Harvard University Press.

Joseph, C. (2007, March 27). Now scientists create a sheep that's 15% human [Electronic version]. Mail Online. Retrieved May 12, 2008 from http://www.mailonsunday.co.uk/news/article-444436/Now-scientists-create-sheep-thats-15-human.html

Kelleher, T. J. (2002). Remembering ewe – in sum – sheep intelligence. American Museum of Natural History. Retrieved May 15, 2008 from http://findarticles.com/p/ articles/mi_m1134/is_1_111/ai_82803320

Kendrick, K. (2007). Babraham Institute- cognitive and behavioural neuroscience. Retrieved May 27, 2008 from http://www.babraham.ac.uk/pjl_pages/kendrick/kendrick.html

Kendrick, K.M., da Costa, A.P., Leigh, A.E., Hinton, M.R. and Peirce, J.W. (2001). Sheep don't forget a face. *Nature* 414, 165–166.

Knight, R. P. (1806). An analytical inquiry into the principles of taste. London: Luke Hansard.

Lawrence, L. J. (2007). Being `hefted': Reflections on place, stories and contextual bible study. *The Expository Times*. 118 (11), 530–535.

Le Breton, D. (2004) Genetic fundamentalism or the cult of the gene. *Body & Society*, 10 (4), 1–20.

More, T. (1516). *Utopia*. New Haven: Yale University Press.

Mott, M. (2005, January 25). Animal-human hybrids spark controversy [Electronic version]. National Geographic News Online. Retrieved May 30, 2008 from http://news. nationalgeographic.com/news/2005/01/0125_050125_chimeras.html

New Scientist Magazine. (2004, September 4). A familiar face is all it takes to keep sheep chilled out [Electronic version]. 2463 (13). Retrieved May 30, 2008 from http://www.newscientist.com/article/mg18324632.200-a-familiar-face-is-all-it-takes-to-keep- sheep-chilled-out.html

New Scientist Magazine. (2006, April 26). Self-medicating sheep shake off the 'stupid' label [Electronic version]. 2549 (19). Retrieved May 30, 2008 from http://www. newscientist.com/channel/life/mg19025495.200-selfmedicating-sheep-shake-off-the-stupid-label.html

Oklahoma State University, Department of Animal Science (1995). Breeds of livestock, sheep. Retrieved May 24, 2008 from http://www.ansi.okstate.edu/breeds/sheep/

Podger, C. (2002, May 21). Quarter of mammals 'face extinction' [Electronic version]. BBC. Retrieved May 10, 2008 from http://news.bbc.co.uk/1/hi/sci/tech/2000325.stm

Price, U. (1796). *Essay on the picturesque, as compared with the sublime and the beautiful*. London: J. Robson.

Public Broadcasting Service (PBS) Online (2003). Gallery of genetic modifications: Genetically modified tomatoes [Electronic version]. Retrieved May 1, 2008 from http://www.pbs.org/wnet/dna/pop_genetic_gallery/

Repton, H. (1794). *Sketches and hints on landscape gardening: Collected from designs and observations now in the possession of the different noblemen and gentlemen, for whose use they were originally made: the whole tending to establish fixed principles in the art of laying out ground*. London: W. Bulmer and Co.

Repton, H. (1840). *The landscape gardening and landscape architecture of the late Humphery Repton, esq: being his entire works on these subjects*. London: printed by Longman and Co.

Rifkin, J. (1998). *The biotech century: Harnessing the gene and remaking the world*. New York: Jeremy P. Tarcher/Putnam.

Ruiz-Marrero, C. (2005, March 28). Biopharmaceutical crops are a disaster waiting to happen. Organic Consumers Association. Retrieved May 28, 2008 from http://www. organicconsumers.org/ge/biopharm32905.cfm

Scottish Agricultural College (SAC). (2006, August 8). Animal health and welfare. Retrieved May 28, 2008 from http://www.sac.ac.uk/research/themes/animalhealth/animalhealthwelfare/sheep/lambing/mortality/mismothering/ewebehaviour/

Stewart Jr., C. N. (2006, April). Go with the glow: Fluorescent proteins to light transgenic organisms. Trends in Biotechnology. 24 (4), 155–162.

Time Online (1984, February 27). It's a geep. [Electronic version]. Retrieved May 1, 2008 from http://www.time.com/time/magazine/article/0,9171,921546,00.html

Trivedi, B. P. (2001a, January 16). Introducing ANDi: The first genetically modified monkey [Electronic version]. Genome News Network. Retrieved May 20, 2008, from http://www. genomenewsnetwork.org/articles/01_01/ANDi.shtml

Trivedi, B. P. (2001b, November 7). Sheep are highly adept at recognizing faces, study shows [Electronic version]. National Geographic. Retrieved May 21, 2008, from http://news.nationalgeographic.com/news/2001/11/1107_TVsheep.html

Wilmut, I., Campbell, K., and Tudge, C. (2000). *The second creation: The age of the biological control by the scientists who cloned Dolly*. London: Headlines

Ending Extinction: The Quagga, the Thylacine, and the "Smart Human"

Carol Freeman

We played God when we exterminated [the thylacine]. I would like to think by bringing it back we are playing the role of smart humans.

Professor Mike Archer, former director of the Australian Museum, Sydney

Abstract In a museum in Cape Town, South Africa, a mounted specimen of a week-old quagga foal balances on slender legs. In the Australian Museum in Sydney, Australia, the four-month-old pouch-young of a thylacine (Tasmanian "tiger") floats in a fluid-filled jar. These extinct species are the subject of attempts at re-creation using genetic technologies: the quagga by selective breeding after tests on tissue samples from the taxidermy specimen and the thylacine by cloning from DNA extracted from the furless body in the glass container. While developments in reproductive technology involving human tissue and stem cell research have generated impassioned public debates, there has been relatively little discussion of cases where animals are the primary subjects. This chapter examines the way these two projects are represented on official websites, in newspapers and magazines, and in the Discovery Channel documentary *End of Extinction*, arguing that they draw heavily on and interact with fictional representations of cloning extinct species in the book and film *Jurassic Park*. In all these representations little concern is expressed for the animals involved in the research, or the potential well-being of those that may be produced if the programmes are successful.

The question then is are these projects in the interests of the creatures involved, or are attempts to retrieve extinct animals simply an expression of speciesism? I argue that revival efforts using genetic technologies should be critically examined on the websites of public institutions, questions should be asked about methods and claims, and alternative solutions to the problem of extinctions discussed.

Keywords Extinction · Cloning · Bioethics · Thylacine · Quagga

C. Freeman (✉)
School of Geography and Environmental Studies, University of Tasmania, Hobart, TAS, Australia
e-mail: carol.freeman@utas.edu.au

C. Gigliotti (ed.), *Leonardo's Choice*, DOI 10.1007/978-90-481-2479-4_13,

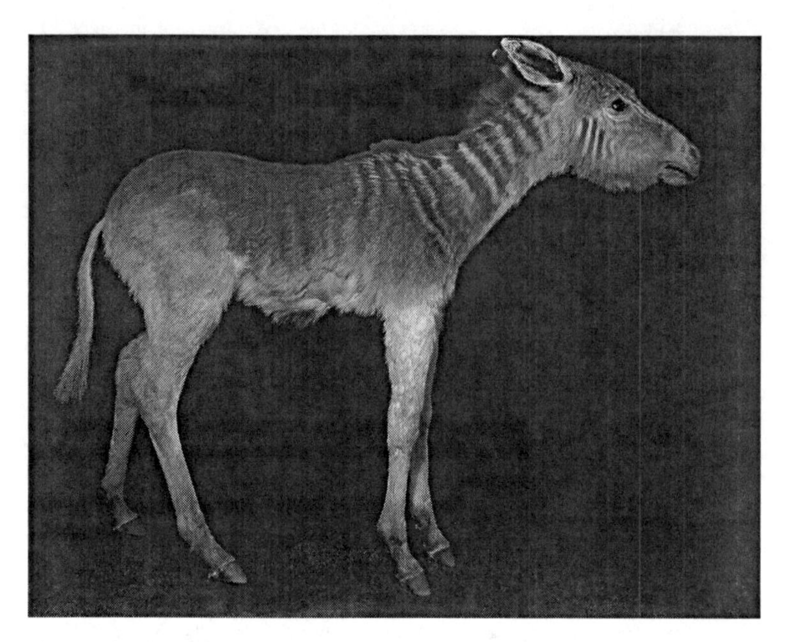

Fig. 1 Quagga foal in the South African Museum, Cape Town. DNA from tissue adhering to the skin of this taxidermy specimen was compared with DNA of the Plains Zebra to determine their relationship (credit: Photo: H. Mair, Iziko Museums of Cape Town)

Locked in a glass case in the South African Museum in Cape Town, a week-old female quagga balances on slender legs. The foal has a rough coat, associated with newborn animals, and faint stripes on her neck and upper body. In the Australian Museum in Sydney, the tiny, hairless remains of a pouch-young thylacine (Tasmanian "tiger") are curled in a glass-stoppered jar of alcohol. The wrinkled skin of the baby is pale, almost transparent, and she rests in a foetal position on the side of the jar. These "specimens" represent dozens of non-human animal species that became extinct in the nineteenth and early twentieth centuries, often as a result of colonial extermination policies, recreational practices and/or habitat disturbance. Now, nearly two centuries later, like the mosquito trapped in amber in Steven Spielberg's 1993 film *Jurassic Park*, they hold the key to attempts at rejuvenation of extinct animals and "sit at the junction of molecular biology, conservation ethics, and endangered-species politics" (Weidensaul, 2002). Indeed, information and publicity about "real life" projects to restore the quagga and clone the thylacine employ the kind of rhetoric found in the fictional representations in Michael Crichton's book and Spielberg's blockbuster film.

References to *Jurassic Park* by those involved in the quagga and thylacine projects are somewhat surprising, as there have been many actual attempts to revive extinct species in the last century. In the 1930s, German zoological park directors, Heinz and Lutz Heck, claimed to have "bred back"—that is, re-assembled the genes of an extinct animal thought to be present in the larger gene pool of interbreedable

Fig. 2 Pouch-young thylacine at the Australian Museum in Sydney. Tissue from this animal's body was the basis for a cloning attempt between 1999 and 2003 (credit: Photo by Stuart Humphreys, Australian Museum)

species—the ox-like auroch and horse-like tarpan, both of whom disappeared from Europe in the seventeenth century. The results of this selective breeding are called "Heck cattle" and their descendants survive in small numbers in Poland, Germany and the Netherlands (Ackerman, 2007; Vuure, 2005). Recently, attempts at cloning extinct animals have almost become commonplace, although few have been successful largely because of problems associated with degraded DNA. For instance, in 1997 DNA was successfully extracted from a tissue sample of a frozen woolly mammoth, extinct since 2000BC, found near Siberia. But the genetic material was so badly damaged—a jumble of small pieces that was nowhere near the full genetic code—that cloning was impossible.[1] There has been more progress with two species of moa, the giant flightless birds from New Zealand, who were the first extinct animals to have their complete mitochondrial genome sequenced by scientists at Oxford University in 2001 (Cooper et al., 2001). However, an entire nuclear genome needs to be sequenced to clone a species. Only the gaur, a rare species of wild ox from Asia, was successfully cloned in 2001 by implanting DNA from the frozen tissue of a dead male into the eggs of a cow. But within 48 hours of the birth the calf, called "Noah", died (Advanced Cell Technology [ACT], 2000; Lee, 2001).

This essay examines two recent projects, both using genetic technologies in attempts to resurrect extinct species, through an analysis of representations on official websites, in newspaper and magazine articles, and in a documentary film called *End of Extinction*. I note in particular how the aims, methods and results of each project are presented and take a close look at *Jurassic Park* in both novel and filmic forms and discuss how the projects interact with this fictional depiction of species' resurrection. I identify the motives behind the projects, assess their chances of success, and consider what Steven Best (2009), in this volume, terms "the potentially beneficial and perhaps destructive aspects of biotechnology". I focus chiefly on the well-being of the animals involved in the research, the prognosis for the lives of those produced, the costs of research, and the chances of their successful reintroduction into natural habitats. While there has been substantial discussion of genetic technologies in relation to human and domestic animal reproduction, especially their commodification by McHugh (2002), Haraway (2003) and Franklin (2007), the retrieval of extinct species has received less attention.[2] Do they present a different case? Are justifiable motives in operation? Are the animals produced as a result of genetic techniques viable or sustainable as species in a natural environment? Or are attempts to retrieve extinct animals simply an expression of speciesism?

Projecting the Quagga and the Thylacine

Quaggas (*Equus quagga quagga*) were similar in appearance to zebras but with stripes only on their head, neck and shoulders. They were indigenous to southern Africa, grazing in large herds in the Karoo and southern Free State of South Africa at the time of white settlement, but became extinct in the 1880s through recreational hunting and to eliminate competition with European livestock for sparse pasture. Quaggas were believed to be a genetically separate species, even regarded by some researchers as more closely related to horses than zebras. According to the South African Museum's web pages on the Quagga Project (Iziko South African Museum, 2007), questions about the species' genetic lineage were answered when Reinhold

Rau, a taxidermist at the Museum, decided to remount a crudely prepared quagga foal in 1969. Rau discovered flesh, veins and other tissue adhering to the skin and, in a far-sighted action, removed and preserved them. The web pages maintain that he remembered as a boy seeing an "auroch" bred by Lutz Heck at a circus near Frankfurt and dreamed of a breeding project that would reverse the extinction of the quagga. In the 1980s, the material from the South African Museum's quagga foal, together with tissue from specimens in Mainz, Germany, was analysed by Russell Higuchi and other geneticists at the University of California, Berkeley (Higuchi et al. 1984, 1987). In a well-publicized effort, the team managed to extract mitochondrial DNA from the quagga tissue and compare sequences of the code with those of the Plains Zebra. Finding they had a "close affinity," they concluded that the quagga was a subspecies of the prolific Plains Zebra (Iziko South African Museum, 2007, "Selective Breeding"; South Africa.info, 2006).

So, in March 1987 the Quagga Breeding Project was launched using nine Plains Zebras selected and captured at the Etosha National Park in South Africa and transferred to a "specially constructed breeding camp complex" at the Nature Conservation farm, Vrolijkheid, near Robertson in Cape Province (Iziko South African Museum, 2007, "Selective Breeding"). The first foal was born in December 1988 and further breeding stock was taken from Etosha and Zululand. The first foal of the second offspring generation was born in February 1997. In the meantime, due to costs of artificial feeding, the project at Vrolijkheid had to be abandoned and eventually all the zebras were moved to land near Cape Town and two other sites where there was sufficient natural grazing. In 2004, 83 animals in the Quagga Project were living at 11 different localities around Cape Town. Breeding is now in its twentieth year and coat patterns that resemble those of the extinct species—that is, with stripes gradually disappearing from the back portion of the body—are progressively emerging, although individual variations are still apparent (Iziko South African Museum, 2007, "Selective Breeding"). The photo gallery on the Project's web pages shows quite graphically how breeding is progressing in the second and third generation of offspring. It is believed that Reinhold Rau's work with the quagga provided the inspiration for the book *Jurassic Park* (Telegraph.co.uk, 2006; New York Times, 2006).

In contrast to the quagga, the thylacine was an animal with no close relatives. The species was unique to Australia, but confined to the southern offshore island of Tasmania since the waning of the Ice Age. These relatively shy, carnivorous marsupials were similar in appearance to large dogs, but with dark, transverse stripes across their backs and a backwards-facing pouch. They hunted smaller marsupials and birds and, although recent morphological analysis suggests they were highly unlikely to have attacked sheep or cattle, they became the focus of a colonial extermination programme, with private and government bounties in force at various times between 1830 and 1909.[3] The last time the species was seen was in 1936. The thylacine has been the object of scientific curiosity since Europeans settled in Tasmania in 1803; hence there is a large amount of tissue material and body parts in 103 museum collections in 21 countries (Sleightholme, 2006). Mike Archer, a paleontologist and head of the project to clone the thylacine at the Australian Museum in

Sydney, claimed that the idea of resurrecting the species was initially inspired by the pouch-young preserved in alcohol that he first saw in the Museum's collection in the 1960s.[4] When he became director of the Museum in 1999 and launched the project, he replied to initial objections: "We played God when we exterminated [the thylacine]. I would like to think by bringing it back we are playing the role of smart humans" (Benson, 1999, September 8). Later he envisaged that a cloned thylacine could become a pet—"just like the family dog" (Benson, 2001, March 28; Dayton, 2002, May 29). Meanwhile, a survey on the Australian Museum's website registered an "overwhelming majority of people" (2492 in favour and 276 against) supported cloning the species (Barbeliuk, 2001, October 30).

The progress of Archer's project was well publicized. In 2000 what was described as "good quality" DNA was successfully extracted from the preserved internal organs of the Museum's pouch-young and the bone, tooth and dried muscle of two others. It was planned to develop an embryo using either the Tasmania devil or numbat as surrogates (Brook, 2000, May 5). In 2002, geneticists Don Colgan and Karen Firestone at the newly formed Evolutionary Biology Unit at the Museum successfully replicated individual thylacine genes using polymerase chain reaction (PCR). This was described as "the most significant breakthrough" in the project because it showed that short fragments of the DNA were undamaged and that "there was no reason why these should not work in a living cell" (Australian Museum Online, 2002, May 28). The next step was to make large quantity copies of all the genes of the species so these could be used to construct synthetic chromosomes. But in 2003 the story of a huge specimen heist at the Australian Museum broke. Hank van Leeuwen, a pest exterminator and a skilled amateur taxidermist, had removed 2000 museum "objects"—skulls, mounts and other material—since he came to work at the Museum in 1996. Mike Archer, as the Museum's director, lost his job when it came up for renewal soon after, and the controversial cloning project was subsequently re-evaluated and scrapped. One of the major reasons cited by the new Museum director was that much of the DNA extracted from the baby thylacine was too degraded to be useful and there had been failure to sequence anything but short fragments of DNA (Australian Museum Online, 2002; Skatssoon, 2005, February 15).

A web of connections with fictional representations of cloning endeavours, particularly *Jurassic Park*, became evident when brothers Randolph and Owen Griffiths approached the Australian Museum with an offer of financial support not long after the project was announced. They joined with the Museum to establish The Rheuben Griffiths Trust, which would fund the expensive research. The Griffith brothers have filial links to Tasmania, but they also have an interest in reptiles and endangered species. Owen Griffiths owns and operates a crocodile park, La Vanille (named after the vanilla plantation it replaced), established in 1985 on the Indian Ocean island of Mauritius. In this beautifully maintained, tropical habitat Owen breeds Nile crocodiles and the giant tortoises, Aldabra and Radiata, and contributes to a captive breeding programme that helps prevent the latter from becoming extinct.[5] In an uncanny moment reminiscent of the scene in the movie *Jurassic Park* where Grant, Sattler and the children lunch in the Park's restaurant after having observed the feeding frenzy of velociraptors,[6] I was offered a meal in the leafy restaurant where crocodile was on the menu, after a tour of the grounds. But more disturbingly, a similar combination of ideas about island utopias, genetic technologies, consumerism and conservation was also apparent in a series of pages about the thylacine project on the Australian Museum's website in 1999 and 2002, which are now preserved on the National Library of Australia's Pandora Web Archive. Rather than having a consistent, conservative scientific style, scattered amongst the "official" information there are the kind of heroic pronouncements and dramatic expectations that appear in techno-thrillers or science fiction representations. For instance, when the Griffith brother's trust fund was announced, Randolph is reported as saying: "Hyde Park [in central Sydney] may never become Jurassic Park, but one day, the genetic imprint from this 130-year-old Tasmanian Tiger pup could help bring her species back to life" (Australian Museum Online, 1999, 2002, "Rheuben Griffiths Trust"). The rhetoric that is used or reported in media coverage for the quagga project also tends to segue into fictional accounts of extinction retrieval. These two attempts to resurrect extinct species, then, act as bookends to *Jurassic Park*: the first inspiring Michael Crichton's book; the second enacting Spielberg's movie.

Focusing on Jurassic Park

It is ironic how readily and optimistically the Australian Museum adopted the rhetoric that appears in *Jurassic Park* as Crichton's book, published in 1991, includes an introduction that warns against combining genetic technologies with commercial interests, and the central message of both book and film is that cloning enterprises could well end in disaster. Crichton's introduction explicitly comments on the trivial nature of much biotechnological research and notes that [in 1991 when the novel was published] there is "no coherent government policy, in America or anywhere else in the world" and "it is remarkable that nearly every scientist in genetics research is also engaged in the commerce of biotechnology ... everyone has a stake". Crichton calls the commercialization of molecular biology "the most stunning ethical event in the history of science" (p. viii). The narrative that follows is a critique of science and an elaborate and dramatic demonstration of the consequences of genetic manipulation for animals and for humans, where mathematician Ian Malcolm is a didactic voice stating, "scientists are actually preoccupied with accomplishment. So they are focused on whether they can do something. They never stop to ask if they *should* do something" (p. 284). Laura Briggs and Jodi Kelber-Kaye (2000) detect an anti-feminist message embodied in the exclusively female dinosaurs—animals that are violent, out of control and breeding wildly. As W. J. T. Mitchell (2003) puts it, "creatures who have taken over their own means of reproduction" (p. 498). In the book, the "unnatural" coupling of nature and technology is associated with the female. Perhaps by chance, the young quagga and the pouch-young thylacine from whom DNA is originally extracted are both female. But an animalist reading of *Jurassic Park* would also link the dinosaur's behaviour to "the demonisation of the animal as the monster or mysterious "outsider" ...[that, after Cavell] allegorizes the escape from human nature" (Wolfe, 2003, p. 5), a fear famously exploited in Edgar Allan Poe's story The Murders in Rue Morgue, in the film *King Kong*, and in countless other stories that feature animal "monsters".

Briggs and Kelber-Kaye suggest that one of the things science fiction narratives such as *Jurassic Park* do is show "how science should (or shouldn't) be done" (p. 93). Paleontologist Alan Grant represents courageous, responsible science, while the instigator of the Jurassic Park project, John Hammond, stands for research that is unregulated, commercialized and trivial. The end result of the venture is a theme park primarily for entertainment with little benefit beyond recreating the childhood fantasies of the men in the narrative. Briggs and Kelber-Kaye note that, in the book at least, the masculine scientist is hero, just as Rau and Archer take leading roles in the resurrection of the quagga and thylacine. In another vein, Robert Mitchell (2007) reads the film in relation to notions of sacrifice made explicit in the 1980 Bayh-Dole Act and a subsequent ruling in the biotechnology case of Moore vs The Regents of the University of California (1990). The Bayh-Dole Act permits a university, rather than the US government, to patent any invention produced using federal funds for research. It is intended to facilitate the commercialization of inventions. The result has been a surge in patents, an effect on trends in research and, particularly, problems with biomedical patents. In the Moore Case, a leukemia patient at the University of California Medical Center had his cancer developed into a cell line that was commercialized, but the court ruled that John Moore had no rights to profit from his own tissue. Rather, it was the work of research physicians that produced the cell line that constituted property, even though many of the procedures involved were concealed from Moore. Mitchell suggests that the case implied that repeated solicitations to 'nature' were necessary before rewards are forthcoming and finds the fictional creation of dinosaurs in *Jurassic Park* is an "already accomplished event" rather than the result of a period of intense and difficult labour. That is, there is no long sacrifice of time and travail in the face of "capricious nature," but a rapid innovation and achievement (pp. 130–131).

Archer's cloning project displayed these characteristics: hasty and dramatic announcements, few explicit details about the project, and an impression that results were imminent—elements that generated criticism, particularly from the scientific community. But he also implied that success would be achieved without the sacrifice of animal lives and well-being. Archer's comments and actions, particularly the way in which they were selected and reported, fit with Mitchell's notion of "innovation . . . understood in terms of immanence and linkage" (p. 134) that permeates the story of Jurassic Park and is inherent in the commercial *promise* of science fiction books and films. Archer's project was implicitly directed at the promise of a successful cloning and the recognition that it would gain for himself as director of the Museum, his Evolutionary Biology Unit, and Australia generally. Little was ever said about the practical problems of reintroducing the thylacine back into its habitat or the time, costs and difficulties of complex, untried cloning procedures. The commercialization of the project and its success were anticipated in the documentary film *End of Extinction*. The release of the DVD made it appear, as Dr. Malcolm in the *Jurassic Park* movie says: that "the scientific power that you're using here: it didn't require any discipline to attain it . . . you patented it, and packaged it, and slapped it on a plastic lunchbox, and now you're selling it" (quoted in Mitchell, 2007, p. 139).

On the other hand, the Quagga Project never expected quick success. Fifteen years after it began, Rau was saying that it might take 30 years to achieve a "perfect" quagga-like animal. While there has been some vagueness in the aims of the Project, the information on some of official web pages suggest that the goals are realistic. Rau conforms to the profile of the hard working, persistent scientist. After years of work on a range of projects for the Iziko-South African Museum, he is described as "more than your typical one-eyed conservationist, whose focus falls solely on a single species or on a single habitat" (Iziko South African Museum, 2007, "The Founder"). All the articles and websites about the Project that I accessed stressed his dedication and perseverance with the quagga project, which may account for the relative lack of criticism. Ironically, Rau, who died in 2006, bore a striking physical resemblance to Hammond, the entrepreneur described as "a potentially dangerous dreamer" (Crichton, 1991, p. 51), as played by Richard Attenborough in the movie *Jurassic Park*. Like Hammond, who says "if you want to do something important in computers or genetics you don't go to a *university*. Dear me, no" (Crichton, 1991, p. 124), Rau was not a scientist and his ambitions were initially regarded with suspicion at the South African Museum. However, Crichton transforms Rau's example of long, hard work into the model of unregulated and irresponsible science discussed by Robert Mitchell.

The central site of *Jurassic Park* the book, the movie, and the fictional Park itself is an island that is geographically outside mainstream institutions and politics. Islands, or exotic isolated and self-contained worlds, recur like a *leit motif* in the story of thylacine cloning: Archer suggests that his cloned animals might be placed on "a small uninhabited island" (he seems to have ignored any non-human animal population), while Owen Griffiths, one of the financiers of the project, already has a wildlife park populated by reptiles on a small and remote island (previously uninhabited by humans). The original habitat of the thylacine, an island that is promoted as a tourist destination based on the appeal of wilderness and ecologically unique biosphere, mimics locations in Spielberg's movie. In Tasmania, the local beer from Cascade Brewery introduced a dog-like image of the thylacine in print and television advertising and on the product's bottles. The landscapes in these images are crowded with ferns, exotic island tropes, and a trusting animal figure that gestures towards Eden and prelapsarian innocence—a nostalgic motif that embodies loss and longing. Images such as this that suggest resurrecting lost worlds and regaining innocence are often used in *Jurassic Park* and in reports about the two projects I am discussing.

Science Fact and Science Fiction

I will now take a close look at how the aims, methods and results of the Quagga Project and plan to clone the thylacine are expressed on official websites of institutions associated with the research, in press releases, and in newspapers reports and magazine articles. The dominant aim of both projects was to reverse extinction, to redress the human act of extermination, and restore each species to their habitat. The South African Museum's pages about the Quagga Project (Iziko South African Museum, 2007) include the banner "Saving the Quagga from Extinction". The Project's home page defines the venture's aims as "an attempt by a group of dedicated people in South Africa to bring back an animal from extinction and reintroduce it into reserves in its former habitat". As European settlers hunted the quagga to extinction, the website says, this will rectify "a tragic mistake made ... through greed and short sightedness." A *New York Times* article on the Quagga Project published in January 2006, a month before Reinhold Rau died, reported that he felt he had "a responsibility—even a destiny—to 'reverse this disaster'" (Max, 2006, January 1). In an obituary to Rau, the South African information site with government and commercial affiliations emphasizes his philanthropic intentions: "'There are many people all over the world who are concerned about the destruction mankind is doing to the world and are trying to stop it. I am one of them', Rau says" (South Africa.info, 2006, October 3).

At the peak of the thylacine cloning project, the Australian Museum's web pages titled "Australia's Thylacine" claimed similar aims for the more physically invasive techniques that were being used to resurrect the thylacine. In answer to the question "Should the thylacine be cloned?" Mike Archer and the staff in the Evolutionary Biology Unit were said to be attempting to "redress our immoral actions when we wilfully and wrongly exterminated this animal" (Australian Museum Online, 1999, 2002, "A conversation with Professor Mike Archer") Archer believed that "we have a moral obligation to try to bring the species back because it was early European settlers who directly caused their extinction" (Woodford, 2000, May 5). However, conscious of the impact of a successful clone, he also told a reporter on a national Australian TV news program that the project "is the equivalent, the biological equivalent, of that first step on the moon" (Mohr, 2000, May 4). Randolph Griffiths was equally extravagant: "it's a wonderful madcap venture ... research is not just focused on the end result but on uncovering all sorts of medical breakthroughs along the way". He felt that the project was about "the process" as well as the ultimate goal (Barbeliuk, 1999, September 9).[7] According to the Australian Museum website when it was reworked after Mike Archer departed, these processes were to "extract DNA of the highest possible quality from thylacine specimens" and to "make and distribute 'libraries' in bacteria or yeast, with complete coverage of the thylacine genome so that its genetic material can be maintained indefinitely" (Australian Museum Online, 2005). The major motivation put forward for retrieving these two species from extinction, however, was the *redemption* of human actions that resulted in their disappearance.

The degree to which genetic technologies are or were used in the two projects is substantially different. The Quagga Project uses DNA sampling and analysis as a tool to determine the genetic component of the quagga's lineage and its relation to that of the Plains Zebra, opening the way for genetic engendering by selective breeding. However, "breeding back" is controversial. The methods used in the Heck brothers' programme to breed the auroch and the tarpan are considered by many biologists to be problematic. They say a genetic lineage cannot be bred back because not all genes survive the passage of time. Although information about

Rau's scheme mentions a book by Lutz Heck, *Grosswild im Etoshaland* (1955), which suggested selective breeding of Plains Zebra on the basis of their brownish colour and/or reduced striping could produce an animal identical to the quagga (Iziko South African Museum, 2007, "Selective Breeding"), it does not mention connections between Lutz and Heinz Heck and Nazi political agendas, particularly eugenics. In fact, the re-breeding of tarpans was supported by Herman Goering and made possible by the removal of and experimentation on small primitive horses called koniks from Bialowieza National Park in Poland after German occupation. The Third Reich saw Bialowieza as potentially "its most splendid hunting ground . . . a great living laboratory of purely Teutonic species" (Schama, 1995, p. 71). Diane Ackerman (2007) writes that for years before World War II German scientists pursued "a fantastic goal: the resurrection of extinct species" (p. 103). But as the complete genetic blueprint of the auroch, the tarpan or the quagga is not known, so the basis of any reconstruction will always be appearance, and the animals produced may not necessarily behave like the species they represent. So, on the Genome News Network, Winstead (2000, October 20) is careful to state, "Rau and a group of researchers and conservationists are selectively breeding plains zebras that have *quagga-like traits*, an approach that livestock and horse breeders have used for centuries" [my italics].

Addressing this criticism further, the Quagga Project's web pages stress that, as DNA analysis has shown that the quagga was not a separate species of zebra but a *subspecies* of the Plains Zebra (*Equus Quagga*), "by breeding with selected southern Plains Zebras an attempt is being made to retrieve at least the genes responsible for the Quaggas *coloration*" [my italics] (Iziko South African Museum, 2007, Home). Indeed, according to Winstead "the relationship between the quagga and the plains zebra is still in dispute". Rau also admits that "we will never know how genetically similar the quagga is to the Plains Zebra because the entire genome of the quagga is not known" (Winstead, 2000, October 20). But in apparent contradiction, the Project's website claims that the genetic basis of the Quagga Project relies on research by Higuchi et al. (1987) suggesting that the mitochondrial DNA of the quagga is the same as the Plains Zebra, so the quagga should be considered merely a different population (deme) of the Plains Zebra. Another page stresses that extinction is final and the loss of a species is "irreversible", but goes on to state that, as the quagga was not a species of its own, the Project aims at "reversing" the extinction of the subspecies (Iziko South African Museum, 2007, "Extinction is Forever"). The aims of the project, therefore, are directly linked to the Museum's interpretation of the genetic associations the quagga has with Plains Zebra revealed by DNA analysis, and then with the selective breeding methods they use. The emphasis on the web pages is not on "re-creating" the quagga, but on "restoring" a subspecies.

In contrast to attempts to retrieve the quagga, the project to clone the thylacine aimed to reproduce a species through an entirely artificial process involving not just DNA analysis of dead tissue samples, but reproduction of cells in a laboratory to the point of creating a living organism or "clone" of an animal. Information about the now defunct project on the revised Australian Museum website consisted of science-based language, photographic magnifications, graphs, and other visual data. It relied on technical terms and was candid about the failures, as well as the successes, of the techniques. It stated that early work focused on extracting liver DNA from the ethanol-preserved thylacine pouch-young, but that "unfortunately subsequent tests (PCR amplification) showed that whilst there was some thylacine DNA in the extraction, the majority of the sample was from contaminating micro-organisms" (Australian Museum Online, 2005). DNA was then extracted from a variety of tissue sources including bone and teeth from two other thylacine specimens held in the Museum's collection, a male and an unsexed individual. Polymerase chain reaction (which makes many millions of copies of a specified sequence) was applied to these short DNA samples to confirm the presence of thylacine DNA. Samples from the bone and teeth showed very high similarity to the available thylacine DNA sequences for all of these amplifications. PCR using primer pairs designed to amplify longer stretches of DNA was not successful on any of the Museum material. The page admits that a number of attempts have been made at construction of genomic libraries from the thylacine DNA, but that none of these attempts have been successful. It concludes, "the lack of success with attempts to clone longer fragments hugely magnifies the difficulty of making a genetic map of the thylacine for the genomic sequencing planned as the third immediate goal of the project" (Australian Museum Online, 2005). The "facts" of this project are in conflict with the optimistic press releases and the DVD *End of Extinction*, discussed later in this chapter, that appeared as Mike Archer made the second major announcement of a "breakthrough" in 2002.

The website of a rival institution, Museum Victoria in Melbourne, identifies another potential problem with the methods proposed to clone the thylacine. It claims that, even if the entire genetic code had been extracted from several different thylacine specimens, finding a suitable host or surrogate in which to incubate an egg or nurture a pouch-young would be difficult. The thylacine's closest living relatives are the *dasyurid* marsupials (e.g. Tasmanian devil and the numbat) that are not only considerably smaller than the thylacine, they are not even closely related—"at least 25 million years of evolution separates these groups" (Museum Victoria, 2008; Barbeliuk, 1999, September 10).

The results expected of both projects are often unclear or extravagant although, because the Quagga Project is well advanced, plans for the herds seem more achievable. However, no time frame is mentioned for re-establishing the newly bred animals in the Karoo and, as a *New York Times* article points out, Rau's project raises a lot of questions. For instance, "what ... gives an animal its identity—its genetic makeup? Its history? Its behavior? Its habitat?" (Max, 2006, January 1, p. 2). The Quagga Project web pages argue, cunningly, that since it is now impossible to demonstrate whether there is evidence that the quagga had specific characteristics or not, these questions are "spurious" (Iziko South African Museum, 2007, "Criticism of the Quagga project"). On the other hand, Oliver Ryder, a geneticist from the Zoological Society of San Diego and an advisor to the Quagga Project, believes that the quagga was a genetically diverse species. He is quoted as saying that "re-breeding this animal can't establish the full extent of the genetic diversity." The project, he says, does not work towards the goals of conservation: "The purpose of conservation genetics is to give the future the robustness of genetic diversity minimally perturbed" (Winstead, 2000, October 20). These differing views call into question the future of the herd that will be eventually transplanted to the Karoo. Will it merely be a tourist attraction that will benefit South Africa's economy, or the basis for a viable population that will expand and fill an ecological niche and benefit other species over 100 years after the quagga disappeared from its habitat? Even giving the proponents of the scheme the benefit of the doubt, there is a sense in which this is a costly, if well-intentioned demonstration of how "smart" humans can be.

In the case of the thylacine, Archer's statements about the ultimate results of his project again recall the fiction of *Jurassic Park*: "[the thylacine's] environment is waiting for their return and, if the worst comes to the worst, then a small, uninhabited island could be found for the tigers until the community is ready to accept them as a legitimate part of Australia's fauna" (Woodford, 2001, November 24–25, p. 5). But challenging Archer's aim to build a viable population of thylacines, an Australian Government website comments "even if the difficulties with the technology were overcome, unless new genes could be artificially introduced into DNA from a thylacine in a museum, the only individuals produced from this alcohol-preserved specimen would all have exactly the same genetic make-up and would be the same sex" (Australian Government). A group of individuals cloned from even a small sample, that is, the bones, teeth and tissue of several other specimens, could not make up a viable population.

The End of Extinction

At the same time as Mike Archer announced a major "breakthrough" in the thylacine cloning project in 2002, Discovery Channel/TLC released a DVD called *End of Extinction: Cloning the Tasmanian Tiger* (O'Neill, 2002), which was publicized on the Australian Museum's website and was shown in 155 countries (Safe, 2002, June 1). The Australian Museum/Rheuben Griffiths Thylacine Trust Fund—Research Institute is listed after the producer's names, but the role of the Fund in the making of the DVD is not specified. The film traces the institution's efforts to clone the thylacine and the initial scenes are dominated by shots of locked steel doors, laboratory equipment, and procedures at the Museum that could only be achieved with the overwhelming cooperation of its director and staff.[8] The story is described as "one man's quest to bring back the Tiger" and the voice of Mike Archer is frequently heard, while music and sound motifs, visuals of animal skeletons, and the use of the camera are highly evocative of Spielberg's film. The DVD begins with compelling images of the pouch-young in its jar. The transparent container is highlighted against a black background that accentuates the pale skin and fine hairs of the immature creature. Like the mosquito in amber and the wobbly quagga foal, this image generates a message that is at once about the fragility of species, the vulnerability of nature, and the potential of reproductive technologies. Then Archer announces that there exists a "threshold of opportunity to question whether extinction is forever", although he admits there are "many challenges" before this can be achieved. The bases for his expectations are the successes that have already been realized: Dolly the sheep, cloned cattle, mice and human embryos. In his resume of cloning processes he specifically alludes to *Jurassic Park*: where that story is "the stuff of science fiction," the thylacine project is "science fact." Panoramic views of Tasmania's wilderness, described as consisting of craggy mountains, wild rivers, mysterious forests and "valleys which have never known the foot of man" reinforce the isolated, uninhabited island theme. Archer stresses the guilt that humanity should feel for the thylacine's extinction and that his cloning project seeks to "put the thylacine back" into this environment.

The next segment of the DVD is an interview with Robert Lanza who, the narrator says, "has his own Jurassic Park"—a house full of ancient bones and dinosaur eggs. Lanza is Chief Scientific Officer at Advanced Cell Technology ("a biotechnology company focused on applying stem cell technology in the field of regenerative medicine") that in 2001 cloned what the documentary calls an "extinct" species, the gaur (it is, in fact, endangered). Lanza fused the somatic cells of a dead gaur with the enucleated oocytes from domestic cows. The resultant embryos were transferred to surrogate domestic cows (Advanced Cell Technology, 2000, October 8). According to a BBC news item, a total of 692 eggs were used in the experiment, 32 embryo clones were initially implanted in surrogate mothers, but only eight of the adult cows became pregnant. In *End of Extinction*, Noah, the only calf to be born, stumbles jerkily and collapses, drinks milk from a bottle, and gazes at the camera with huge, brown eyes. Noah died 2 days after birth from common dysentery, surely a predictable and preventable disease.

Lanza's justification for cloning attempts, "we do play God when we wreak havoc on the environment ... the least we can do is try to reverse some of that damage, to give these species a fighting chance of surviving in the wild" (BBC News, 2000, October 8) are strikingly similar to Archer's, who says, "I see [the thylacine project] as trying to undo the immoral playing-God act, which was to exterminate the thylacine in the first place. So I do see a rebalancing of nature" (Safe, 2002, June 1). The point of this segment of the DVD is that, while the thylacine project may be ambitious, cloning is possible. At the end of the film, Archer confesses that seeing the thylacine resurrected is his "obsession". He has a "vision" of a population of thylacines restored to the wilderness, just like Reinhold Rau who wanted to see "quaggas, which were once thought to be extinct, come to drink at the waterhole—as they were always meant to do" (Iziko South African Museum, 2007, "The Founder"). Visions such as these are realized in the pivotal scene in the movie *Jurassic Park* when Grant and Sattler first see dinosaurs roaming the landscape. But in contrast to the catastrophic ending of that movie, Archer's comments during the finale of *End of Extinction* are accompanied by visuals that convey *promise*: an animatronic thylacine family gambols in a forest, while the male "roars" on a cliff top. Then the camera pans to the sunset sky and portentous music swells to the final frame.

Bioethics and Extinction Reversal

The story of the two projects demonstrates that the aims and results presented on websites and in newspapers are often strategically selective, vague or sensational, and ignore or mask the problems that could arise for the animals themselves. Ethical issues relating to the re-creation of the thylacine and the retrieval of the quagga are rarely mentioned and, if they are raised in interviews, they are usually given brief or perfunctory answers. Autumn Fiester (2005) identifies a series of ethical issues that arise from reproductive technologies with animals. The most important of these concern the suffering endured during cloning or breeding procedures, the complications often attending the surrogate mother, possible health issues related to the animals produced by these methods, and the suffering they would undergo if used for display or future research purposes, such as investigations into human diseases or pathologies (p. 331).

Little specific information is given about the methods used to breed quaggas, but we do know that Plains Zebras were moved from Etosha National Park in Namibia and from Kwa-Zulu Natal to sites near Cape Town for breeding. Transportation is not easy for zebras: they are first pursued and captured, can be frightened or aggressive, and sometimes unavoidably hurt. They must be persuaded into large crates and often have to be tranquillized (World Association of Zoos and Aquariums, 2008). Sixty–five zebras were originally moved in this way for Reinhold Rau's Quagga Project and then 60 for a separate effort by South African National Parks. In Rau's project, zebras are individually selected as mating pairs (Boroughs, 2000), which would entail more moving of the animals. According to the Project's website, there have been "some losses, due to old age, illness or injury. Some of the less suitable offspring have been sold" (Iziko South African Museum, 2007, "Selective Breeding"). Because the quagga and the Plains Zebra are so similar, there should be few problems of an obstetrical nature and none have been reported. In the case of cloning, the use of domestic species such as cows as surrogates has posed relatively few problems, although the gaur was only accepted by one of many cows that aborted their calves. Possible suffering by mother or calf is ignored by proponents of cloning, but against the background of current practices with animals, such as agricultural and companion breeding and research, this attitude is hardly surprising. The possible suffering of thylacines produced by genetic technology is more difficult to assess. Taking the gaur as an example, there are many tiny embryos or pouch-young that are likely to die before they are placed in the pouch of a surrogate mother. However, Archer ignores any problems, placing the emphasis on long-term results and, while he mentions the Tiger quoll and the Tasmanian devil as possible surrogates, zoologists have noted the difference in size, gestation periods and the nature of uterine walls, which may cause problems for both the mother and the pouch-young (Rickard, 2001, November 17).

Even if the potential good of extinction retrieval—restoring an extinct species to its original habitat—is taken into account, the suffering animals endure during the procedures involved cannot be cancelled out. It has already been noted that the final destination of such animals would almost certainly be the subject of enormous publicity; individuals would be exhibited to the public for at least some of their lives, and it is most unlikely that they would ever be allowed to have an undisturbed existence in a natural environment. It is possible that a herd of quaggas, stringently protected, can be successfully restored to the Karoo. But Don Boroughs (2000) points out that the quagga is just one part of a complex ecosystem that has long been disturbed and that grazing populations have been kept to a minimum in the past to help restore grasses in the Karoo that have been decimated by 300 years of colonization by sheep. This suggests the possibility that grazing by quaggas may also have a deleterious effect on the environment unless strict regulations to protect habitat are put in place. However, Boroughs also points out that the resurrected quagga could "open [the public's] eyes to the value of an entire ecosystem" (np.). He says that Ryder predicts: "you will get [the public's] wonder: you will get their interest, and then from that, their caring. And once they care about something, then they will act to save it" (np.). He seems to be saying that restoring the quagga to its original habitat will persuade people to care about extinct species per se. If this is so, then the Quagga Project, which Rau and others believe may take 10 more years to achieve, will be a worthwhile venture, particularly if the costs, transportation, and handling of animals can be kept to a minimum.

Attempts to clone the thylacine are more problematic. While marsupials present some advantages in terms of cloning because they have a very short gestation period that facilitates culturing, mature animals can be easily traumatized. The thylacine was constantly described as shy and not suited to confinement, although many individuals lived reasonably long lives in zoos in the nineteenth and early twentieth centuries. Research for and initial development of the thylacine project was conducted in a laboratory using glass storage jars, pipettes, and sophisticated molecular biology techniques. It involved long-dead bodies, rather than living animals, so that if active, breathing beings were finally produced they would inevitably be viewed as a highly valuable scientific "achievement" or "result," which may compromise the individual needs of the animal. As Fiester points out, this side-effect of cloning negates the intrinsic value of animals through both objectification and commodification (p. 339). A cloned thylacine or group of thylacines would be highly likely to be disturbed by humans, so artificial habitats, or a secret location would be mandatory. As suggested by Archer (Crawford, 2002, June1), the creation of a thylacine and many other extinct species would be lucrative for the tourism industries and for the governments of the countries in which they live.

While the overall aim of both the quagga and the thylacine projects is the repopulation of the species' original habitats, the publicity that has already surrounded the projects would make it impossible to isolate animals from public attention. The most obvious solution would be a Jurassic-style park: near enough to the stated aim of each project. Is it for this end that anyone concerned about animals wants to see an extinct species retrieved? Both projects have potential commercial benefits for the human communities in which they occur, raising the concerns expressed in the introduction to *Jurassic Park*—that an alliance between business and technoscience is dangerous. A photo gallery on the South African Museum's Quagga Project web pages shows a series of images of the quagga breeding herd. Their healthy bodies, plump buttocks and entrancing stripes scroll by the viewer like animal porn. The South African government, in conjunction with commercial interests, has a page about the Quagga Project on its information website (South Africa.info, 2006, October 3). In Tasmania, the idea of the thylacine is kept alive in images on football jumpers, souvenirs in tourist shops, and advertisements for beer: seductive instruments in keeping longing alive and the economy of Tasmania buoyant. As Archer pointed out, living thylacines would be a boon for tourism to the State.

One of the most concerning features of these projects, then, is how seldom individual animals are mentioned except as stages in a process, or concern for the living creatures involved is expressed. Focusing on retrieving extinct species also diverts attention from discrete populations of animals whose existence is currently threatened. This is because the ultimate goal of both projects is an expression of speciesism. The term can be applied in relation to both Peter Singer's definition, that the projects are promoting the interests of one species over another, with more value placed on the achievements of humans than on the suffering of the animals that are produced (1990, p. 9) and in the sense that they are about the re-creation of a particular "species"—an abstraction constructed by humans who place similar animals in a certain group and then give maintenance of the group priority over the needs of individual animals (Shapiro, 1996, p. 136). Like factory farms and animals used in research, imprecision in many of the details of the projects provided by official websites means that the public is kept ignorant of procedures that may harm the animals involved, or that may result in deaths, or a distressing quality of life. Contrary to discussions relating to human cloning, few objections to cloning animals have been expressed, as indicated by the "overwhelming" support for Archer's project on the online survey.

One of the most persuasive arguments against resurrecting extinct species was a plea for distributive justice raised by Nick Mooney, wildlife manager for the Tasmanian Department of Primary Industries, Water and Environment. Mooney asks "if we have clones, who needs to look after the environment? ... This gives governments and big business an easy way out" (Stubbs, 2002, May 29). Mooney says he is faced with the "day to day realities" of dwindling habitats and suggests that a successful cloning would lead people to think that continuing destruction of species and the environment could be fixed later. He maintains that cloning redirects money away from conservation (Barbeliuk, 1999, September 9). As Australia has one of the highest rates of extinction on the globe and a quarter of the world's mammals and 1,183 species of birds are facing disappearance in the next 30 years (United Nations Environment Programme, 2002), where money goes is of crucial importance. While the cost of cloning was rarely mentioned by the Australian Museum, the AUD$80 million quoted in the *Sydney Morning Herald* and *Daily Telegraph* could save the lives of countless endangered animals (Benson, 2000, May 5; Woodford, 2000, May 5). As Fiester phrases it, "if animals have an intrinsic worth that merits this kind of expenditure, then it is criminal that we would allow so many of the same creatures to die; resources simply need to be redistributed in a way that makes better ethical sense" (p. 335). The situation in Australia and South Africa is similar to that in other parts of the world: species are in danger from pollution, deforestation, burning of habitat and invasion by non-native species. The cost of breeding back the quagga or cloning the thylacine may be better spent in deploying genetic technologies to conserve the species that are left. For example, by using non-invasive reproductive technologies on captive populations; encouraging animal husbandry with semi-captive native species; freezing the eggs of rare or endangered animals; or employing population genetics to identify areas of low genetic diversity in wild populations and then establishing insurance populations.[9] If biological and genetic diversity can be ensured, this will provide the mechanism to respond to environmental change. With the focus of each project so firmly set on resurrecting an extinct species, it seems unlikely that information about the genetic inheritance and biological connections of species or medical breakthroughs that are discovered in the process of cloning would be used to save related animals, as was suggested by Randolph Griffiths. As electronic media becomes more and more the mode of information dissemination and retrieval for the general public as well as researchers, perhaps the most important requirement, as Steven Best suggests in this volume, is that questions about extinction retrieval procedures are raised and addressed on the websites and in other publications of public institutions and government departments that do *not* have a vested interest in the positive outcome of projects.[10] Balanced information, explanation of genetic terms and processes, projections of results, and problems associated with outcomes should be provided so that informed and educated public opinion about individual projects can be canvassed.

Acknowledgments I would like to thank Jules Freeman, Elizabeth Leane and Yvette Watt for their advice and suggestions on previous drafts of this essay, as well as Carol Gigliotti for her patience and support during the editing process.

Notes

1. Genetic technologies are progressing so rapidly that it is difficult to anticipate what will be possible in the future. A paper published in 2006 (Poinar et al.) claims that a high degree of endogenous DNA has now been recovered from the woolly mammoth, and anticipates complete sequencing of the animal's genome. In 2008 it was reported that scientists at Melbourne University had taken skeletal genes from a thylacine specimen and implanted them into a mouse embryo (ABC News, 2008, May 20). Furthermore, PCR (polymerase chain reaction) has recently enabled the amplification of small regions of DNA. Hence, in the future, many of the problems that are associated with fragmented DNA of extinct species may be overcome. For this reason it is vital to examine the issues that are raised by these procedures.
2. Useful consideration of the thylacine cloning project has been made by Turner (2007), who shows how the project reworks evolutionary narratives to satisfy cultural imperatives, and by Fletcher (2008), who focuses on the public reaction to the thylacine cloning project and uses monster theory to analyse how stakeholders endeavoured to "domesticate genetic technologies and bring them into line with cultural norms and conservation paradigms" (1).
3. See Freeman (2007a, b) for discussions about how representations of the thylacine in nineteenth century natural history books, periodicals, and newspapers persistently and actively encouraged eradication of the species, and how photographs published in the early twentieth century failed to adjust these attitudes, which eventually led to the thylacine's extinction.

4. Mike Archer is a vertebrate paleontologist who has been extensively involved in excavations at the Murgon and Riversleigh fossil deposits in South Australia and Queensland, where he discovered the first-known Tertiary skeleton of an extinct thylacinid. Since 1989 he has held a professorship at the University of NSW, where he is currently Dean of Science. According to his career profile, he has published over 300 scientific papers and named or discovered 120 new species. He is a dual citizen of Australia and USA (University of New South Wales, nd.).
5. During a conversation which I had with Owen Griffiths in August 2003, he claimed that the money donated to the thylacine project was not "taking money away" from conservation and commented that cloning programmes were often seen as challenging conversation concerns. The Mauritius giant tortoises were extinct by 1795 and the Madagascar–Seychelles genus *Dipsochelys* that Griffths breeds only survives in a natural state on the atoll of Aldabra. See Chambers (2004) and Gerlach and Canning (2000). The radiated tortoise or sokatra is one of four tortoises also endemic to Madagascar. The IUCN Red List classifies the species as "Vulnerable." See Behler and Iaderosa (1991).
6. In *The Last Dinosaur Book* (1998), W.J.T. Mitchell maintains that the commodity being sold in *Jurassic Park* the movie is the "spectacle of consumption", with the first meal of the film a black worker who is devoured by an unseen raptor through the bars of a cage (p. 221).
7. The Australian Museum and the Griffiths family urged the Tasmanian government to join the research project, but while the vertebrate curator at the Tasmania Museum and Art Gallery, where five thylacine pouch-young are held, initially expressed eagerness to be involved as the "moral keeper of the Tasmanian Tiger", no contact or offers of assistance were forthcoming from Tasmanian government departments (Crawford, 2002, June 1).
8. A newspaper report in March 2000 claimed that the Australian Museum delayed extraction of lung, bone and tissue from the young thylacine so that Discovery Channel could film the procedures. According to the article, a Museum spokesperson stressed that "all footage will be accessible by us. They will not have exclusive rights" (Rose, 2000, March 7).
9. One of many threatened species is another marsupial carnivore, the Tasmanian devil, placed on the State's endangered list in May 2008. The devil population has a facial tumour disease, identified in 1996, that has reduced numbers by more than half. The disease is always fatal, although DNA testing has detected a group of animals in the West of the State that seem to have a resistance to the disease. The value of genetic technologies is obvious, but there is a need for a massive injection of funds to establish insurance populations of genetically diverse animals so that the species does not disappear, as the thylacine did. See Tasmanian Department of Primary Industries and Water (2008) and Tasmanian Government and University of Tasmania (2008).
10. Two Australian newspapers, Hobart's *The Mercury* (2001, March 29) and Adelaide's *Sunday Mail* (2002, June 9), published education specials that used simple, but effective, diagrams to explain the genetic technologies that would be used in cloning the thylacine and outlined other attempts to recreate extinct species.

References

ABC News (2008, May 20). Tasmanian tiger DNA comes alive in mouse [Electroniversion]. Retrieved May 21, 2008 from http://www.abc.net.au/news/stories/2008/05/20/2249778.htm

Ackerman, D. (2007). Galloping ghosts. *Smithsonian*, 38(8) 102–103, 105–107.

Advanced Cell Technology. (2000, October 8). Advanced Cell Technology first to clone endangered species. Retrieved May 7, 2008 from http://www.advancedcell.com/press-release/advanced-cell-technology-first-to-clone-endangered-species

Australian Government (n.d.). The thylacine: A case study. In *Biotechnology Online*. Retrieved May 7, 2008 from http://www.biotechnologyonline.gov.au/enviro/thylacine.cfm

Australian Museum Online (1999, 2002). Rheuben Griffiths Trust launched to help genetically re-create extinct Tasmanian Tiger. In *Australia's Thylacine*. Retrieved April 22, 2008 from http://pandora.nla.gov.au/pan/33031/20030109-0000/www.amonline.net.au/thylacine/15.htm

Australian Museum Online (1999, 2002). The cloning debate: A conversation with Professor Mike Archer, director of the Australian Museum. In *Australia's Thylacine*: Retrieved May 2, 2008 from http://pandora.nla.gov.au/pan/33031/20030109-0000/www.amonline.net.au/thylacine/archer.htm

Australian Museum Online (2002, May 28). News release: Tasmanian tiger cloning breakthrough. In *Australia's Thylacine*. Retrieved April 29, 2008 from http://pandora. nla.gov.au/pan/33031/20030109-0000/www.amonline.net.au/thylacine/newsrelease.htm

Australian Museum Online (2005). Attempting to make a genomic library of an extinct animal. In *Australia's Thylacine*. Retrieved May 8, 2008 from http://www. amonline.net.au/thylacine/summary.htm

Barbeliuk, A. (1999, September 9). A tiger by the tail. *The Mercury*, Hobart, p. 19.

Barbeliuk, A. (1999, September 10). Thylacine cloning hit as mission impossible. *The Mercury*, p. 9.

Barbeliuk, A. (2001, October 30). Expert's fears as vote favors tiger cloning. *The Mercury*, Hobart, p. 5.

BBC News (2000, October 8). Endangered species cloned [Electronic version]. Retrieved May 4, 2008 from http://news.bbc.co.uk/2/hi/science/nature/962159.stm

Behler, J.L. & Iaderosa, J. (1991). A review of the captive breeding program for the radiated tortoise at the New York Zoological Society's Wildlife Survival Center. In K.R. Beaman, F. Caporaso, S. McKeown, and M.D. Graff (Eds.), *Proceedings 1st international symposium turtle & tortoises: Conservation captive husbandry* (pp. 160–162). Van Nuys, CA: California Turtle and Tortoise Club.

Benson, S. (1999, September 8). Tiger cloning project begins. *The Mercury*, Hobart, p. 7.

Benson, S. (2000, May 5). Tasmanian tiger may live again. *The Daily Telegraph*, p. 6.

Benson, S. (2001, March 28). Tickling tiger to life: Thylacine clone bid. *The Mercury*, p. 36.

Best, S. (2009). Genetic science, animal exploitation, and the challenge for democracy. In C. Gigliotti (Ed.). *Leonardo's choice: Genetics technologies and animals*. Dorchedt: Springer.

Boroughs, D. (2000). Stripes and shadows. *Timbila: Magazine of the South African National Parks*, 2 (1), np.

Briggs, L. & Kelber-Kaye, J. (2000). "There is no unauthorised breeding in Jurassic Park": Gender and the uses of genetics. *NSWA Journal*, 12 (3), 92–113.

Brook, S. (2000, May 5). Thylacine clone no longer just a paper tiger. *The Australian*, p. 1.

Chambers, P. (2004). *A sheltered life: The unexpected history of the giant tortoise*. London: John Murray.

Crawford, W. (2002, June 1). Tasmanian pup dream. *The Mercury*, Hobart, p. 25.

Crichton, M. (1991). *Jurassic park*. London: Arrow Books.

Cooper, A., Lalueza-Fox, C., Anderson, S., Rambaut, A., Austin, J. & Ward, R. (2001). Complete mitochondrial genome sequences of two extinct moas clarify ratite evolution. *Nature*, 409, 704–707.

Dayton, L. (2002, May 29). Tiger on comeback trail. *The Australian*, p. 1.

Fiester, A. (2005). Ethical issues in animal cloning. *Perspectives in Biology and Medicine*, 48 (3), 328–343.

Fletcher, A. (2008) Bring 'em back alive: Taming the Tasmanian tiger cloning project, *Technology in Society*, 30 (2), 194–201.

Franklin, S. (2007). Dolly's body: Gender, genetics and the new genetic capital. In L. Kalof & A. Fitzgerald (Eds.), *The animals reader: The essential classic and contemporary writings* (pp. 349–361). Oxford: Berg.

Freeman, C. (2007a). 'In every respect new': European impressions of the thylacine, 1808–1855. *reCollections: Journal of the National Museum of Australia*, 2 (1), 5–24.

Freeman, C. (2007b). Imaging extinction: Disclosure and revision in photographs of the thylacine (Tasmanian tiger). *Society and Animals: Journal of Human-Animal Studies,* 15 (3), pp. 241–256.

Gerlach, J. & Canning, K.L., (2000). Evolution and history of the giant tortoises of the Aldabra Island group. *Phelsumania*. Retrieved April 30, 2008 from http://www. phelsumania.com/public/articles/fauna_dipsochelys_1.html).

Haraway, D. (2003). For the love of a good dog: Webs of action in the world of dog genetics. In A. Goodman (Ed.), Genetic nature/culture: Anthropology and science beyond the two-culture divide (pp. 111–131). Berkeley, CA: University of California Press.

Higuchi, R., Bowman, B., Freiberger, M., Ryder, O.A., & Wilson, A. (1984). DNA sequences from the quagga, an extinct member of the horse family. *Nature*, 312: 282–284.

Higuchi, R., Wrischnik, L.A., Oakes, E., George, M., Tong, B. & Wilson, A. (1987) Mitochondrial DNA of the extinct quagga: Relatedness and extent of postmortem change. *Journal of Molecular Evolution*, 25, 283–287.

Iziko South African Museum (2007). The Quagga Project South Africa. Retrieved April 5- May 7, 2008 from http://www.quaggaproject.org/.

Lee, K. (2001). Can cloning save endangered species? *Current Biology*, 11 (7), R245–246.

McHugh, S. (2002). Bitches from Brazil: Cloning and owning dogs through the Missiplicity Project. In N. Rothfels, (Ed.), *Representing animals*. Bloomington, IN: Indiana University Press.

Max, D.T. (2006, January 1). Can you revive an extinct animal? [Electronic version]. *New York Times Magazine*. Retrieved April 28, 2008 from http://www.nytimes.com/2006/01/01/magazine/01taxidermy.html?_r=2&8hpib&oref=slogin&oref=slogin

Mitchell, R. (2007). Sacrifice, individuation, and the economies of genomics. *Literature and Medicine*, 26 (1), 126–158.

Mitchell, W.J.T. (1998). *The last dinosaur book*. Chicago: University of Chicago Press.

Mitchell, W.J.T. (2003). The work of art in the age of biocybernetic reproduction. *Modernism/Modernity*, 10 (3), 481–500.

Mohr, M. (Producer). (2000, May 4). Bringing the Tasmanian tiger back to life. In *ABC Online: 7.30 Report*. Retrieved May 6, 2008 from http://www.abc.net.au/7.30/stories/s123836.htm

Moore vs The Regents of the University of California (1990) 51 Cal.3d 120; 793 P.2D 479; 271 Cal. Rptr.146.

Museum Victoria (n.d.). Science and discovery: Cloning the thylacine: Fact or fantasy? Retrieved May 8, 2008 from http://museumvictoria.com.au/scidiscovery/dna/cloning.asp

O'Neill, P. (Producer/Director/Writer). (2002). *End of extinction: Cloning of the Tasmanian tiger*. Silver Spring, MD: Discovery Channel/TLC.

Poinar, H., Schwarz, C., Qi, J., Shapiro, B., MacPhee, R.D.E., Buigues, B., Tikhonov, A., Huson, D.H., Tomsho, L.P., Auch, A., Rampp, M., Miller, W. & Schuster S.C. (2006). Metagenomics to paleogenomics: Large-scale sequencing of mammoth DNA. *Science*, 311 (5759), 392–394,

Rickard, P. (Producer). (2001, November 17). Breeding endangered species [Radio transcript]. In *ABC Radio National*: *The Science Show*. Retrieved from http://www.abc.net.au/rn/science/ss/stories/s421153.htm

Rose, D. (2000, March 30). Tiger cloning project delayed by TV deal. *The Mercury*, p. 7.

Safe, M. (2002, June1). Crouching tiger, galloping science. *Australian Magazine*, pp. 10–12.

Schama, S. (1995). *Landscape and memory*. London: Fontana.

Shapiro, K. (1996). The caring sleuth: Portrait of an animal rights activist. In Donovan, J. & Adams, C. (Eds.). *Beyond animal rights: A feminist caring ethic for the treatment of animals*. New York: Continuum.

Singer, P. (1990). *Animal liberation*. New York: Random House.

Skatssoon, J. (Writer). (2005, February 15). Thylacine cloning project dumped. In *ABC Science Online*. Retrieved April 9, 2008 from http://www.abc.net.au/science/news/stories/s1302459.htm

Sleightholme, S. (2006). *International thylacine specimen database* [DVD]. London: Stephen Sleightholme.

South Africa.info (2006, October 3). Bringing back the quagga. Retrieved May 7, 2008 from http://www.southafrica.info/about/animals/quagga.htm

Spielberg, S. (Director) & Crichton, M. (Writer). (1993). *Jurassic park.* United States: Universal Pictures.

Stubbs, B. (2002, May 29). Tiger clone hope burning bright. *The Mercury*, p. 3.

Tasmanian Department of Primary Industries and Water (2008). Native plants and animals: About Tasmania devils. Retrieved May 23, 2008 from http://www.dpiw.tas.gov.au/inter.nsf/WebPages/LBUN-5QF86G?open

Tasmanian Government and University of Tasmania (2008). Save the Tasmanian devil: Tasmanian devil facial tumour disease. Retrieved May 23, 2008 from http://www.tassiedevil.com.au/disease.html#cedric

Telegraph.co.uk (2006, March 24). Obituaries: Reinhold Rau. Retrieved May 7, 2008 from http://www.telegraph.co.uk/news/obituaries/1513790/Reinhold-Rau.html

Turner, S. (2007) Open-ended stories: Extinction narratives in genome time. *Literature and Medicine,* 26 (1), pp. 55–82.

United Nations Environment Programme (2002). GEO 3: Global environmental outlook. Retrieved May 20, 2008 from http://www.unep.org/GEO/geo3/

University of New South Wales (nd.). Faculty of Science: Professor Mike Archer. Retrieved May 23, 2008 from http://profiles.science.unsw.edu.au/marcher/home/

Vuure, C. van (2005). *Retracing the aurochs: History, morphology and ecology of an extinct wild ox*. Sofia-Moscow: Pensoft.

Weidensaul, S. (2002). Raising the dead [Electronic version]. *Audubon*, 5. Retrieved April 25, 2008 from: http://magazine.audubon.org/features0205/thylacine.html

Winstead, R. (2000, October 20). In South Africa the Quagga Project breeds success. *Genome News Network.* Retrieved May 6 from http://www.genomenewsnetwork.org/articles/10_00/Quagga_project.shtml

Wolfe, C. (2003). *Animal rites: American culture, the discourse of species and posthumanist theory*. Chicago: University of Chicago Press.

Woodford, J. (2000, May 5). Get a life, scientists tell extinct tiger. *Sydney Morning Herald*, Late ed. p. 3.

Woodford, J. (2001, November 24–25). Extinction for the time being. *Sydney Morning Herald*, Weekend Edition, Spectrum, 4–5.

World Association of Zoos and Aquariums (2008). Animals: Class mammalia: Hartmann's Mountain Zebra. Retrieved May 8, 2008 from http://www.waza.org/virtualzoo/factsheet.php?id=118-001-001-007a&view=Equids

CPSIA information can be obtained at www.ICGtesting.com
Printed in the USA
LVOW080243211212

312722LV00002B/292/P